Advances in Molecular Biology Methods in Hepatology Research

Advances in Molecular Biology Methods in Hepatology Research

Guest Editors

Dileep G. Nair
Ralf Weiskirchen

MDPI

Basel • Beijing • Wuhan • Barcelona • Belgrade • Novi Sad • Cluj • Manchester

Guest Editors

Dileep G. Nair
Institute of Molecular Pathobiochemistry, Experimental Gene Therapy and Clinical Chemistry
RWTH University Hospital Aachen
Aachen
Germany

Ralf Weiskirchen
Institute of Molecular Pathobiochemistry, Experimental Gene Therapy and Clinical Chemistry
RWTH University Hospital Aachen
Aachen
Germany

Editorial Office

MDPI AG
Grosspeteranlage 5
4052 Basel, Switzerland

This is a reprint of the Special Issue, published open access by the journal *Current Issues in Molecular Biology* (ISSN 1467-3045), freely accessible at: https://www.mdpi.com/journal/cimb/special_issues/4L8BQQE181.

For citation purposes, cite each article independently as indicated on the article page online and as indicated below:

Lastname, A.A.; Lastname, B.B. Article Title. *Journal Name* **Year**, *Volume Number*, Page Range.

ISBN 978-3-7258-5217-8 (Hbk)
ISBN 978-3-7258-5218-5 (PDF)
https://doi.org/10.3390/books978-3-7258-5218-5

Contents

About the Editors

Dileep G. Nair

Dileep G. Nair earned his PhD from the National University of Singapore, followed by 6 years of postdoctoral fellowship in GI disease research at Queen's University, Canada. Prior to taking on roles in startup and pharmaceutical companies, he worked as an Assistant Professor at the University of Technology & Applied Sciences, Oman. Dileep's expertise lies in adhering to the 3R principles while developing sophisticated in vitro models, primarily focusing on drug discovery in diverse therapeutic areas. Some of his major achievements include authoring publications in 16 peer-reviewed journals, mainly on gastrointestinal and hepatology topics that emphasize the utilization of advanced in vitro models characterized by molecular biology tools. He has established complex in vitro models in academic and industry roles to understand disease biology and support drug discovery.

Ralf Weiskirchen

Ralf Weiskirchen is a full professor and the head of the Institute of Molecular Pathobiochemistry, Experimental Gene Therapy and Clinical Chemistry (IFMPEGKC), at RWTH University Aachen. He obtained his PhD from the Institute of Biochemistry, Medical Faculty, University of Cologne, in 1994. With a strong academic background and extensive research experience, he has expertise in various biological disciplines, such as molecular biology, microbiology, biochemistry, pharmacology, and genetics. His primary research focus is on understanding the molecular mechanisms involved in the development of hepatic diseases. Specifically, he investigates the immunobiology of the liver, ranging from its healthy state to chronic inflammation, fibrosis, cirrhosis, and cancer. He places special emphasis on studying cytokine/chemokine activities and identifying new laboratory biomarkers for specific conditions. In recognition of his contributions, he was selected for inclusion in the 2024 Highly Cited Researchers list from Clarivate.

current issues in molecular biology

MDPI

Editorial

Editorial for the Special Issue "Advances in Molecular Biology Methods in Hepatology Research"

Dileep G. Nair and Ralf Weiskirchen *

Institute of Molecular Pathobiochemistry, Experimental Gene Therapy and Clinical Chemistry (IFMPEGKC), RWTH University Hospital Aachen, D-52074 Aachen, Germany; grdileep@gmail.com

* Correspondence: rweiskirchen@ukaachen.de

We are delighted to present this Special Issue of *Current Issues in Molecular Biology*, entitled "Advances in Molecular Biology Methods in Hepatology Research." This collection of ten articles, including five original research papers and five reviews, showcases important technical and conceptual innovations that are poised to shape the future of hepatology. The contributions come from eight countries (Belarus, China, Germany, Italy, Japan, Malaysia, Serbia, and Taiwan), reflecting the truly global and collaborative nature of contemporary hepatology research.

Innovations in molecular biology methods and diagnostics have long underpinned major advances in understanding liver pathophysiology [1,2]. With the increasing complexity and global incidence of liver diseases such as metabolic-dysfunction-associated steatohepatitis (MASH), viral hepatitis, and hepatocellular carcinoma (HCC), it is clear that highly refined and specialized techniques are necessary to dissect the biochemical and cellular underpinnings of these conditions and to identify new molecular markers [3,4]. The research featured in this Special Issue highlights new and refined tools or findings that accelerate our ability to capture and interpret the intricacies of liver function and disease.

The five original research papers break new ground by applying state-of-the-art methodologies to diverse aspects of hepatology. In the first study, patients with EGFR-mutated NSCLC showed a median overall survival of 13 months, whether or not they had liver metastasis, though 9-year survival rates were higher (12.5%) among those with metastases than those without (3.4%) [5]. The second article investigated IL-28B rs8099917 genotypes in HCV patients after liver transplantation, revealing that the IL-28B genotype TT correlated with higher miRNA-122 levels, better outcomes, and a lower rate of acute cellular rejection [6]. In the third study, aspirin reduced lipopolysaccharide-induced liver inflammation and thrombus formation in mice, emphasizing its anti-inflammatory and antiplatelet effects [7]. The fourth article showed that lipid-nanoparticle-delivered siRNA targeting Cyp2e1 alleviated acetaminophen-induced liver injury by modulating PPAR-related pathways and reducing oxidative stress [8]. Lastly, the fifth study identified HBV genotype D as predominant in Belarus, with occasional introductions of other variants, and underscored how local virus circulation fuels the ongoing hepatitis B epidemic (contribution 1).

The five review articles critically evaluate how advancements in molecular biology can effectively be translated into clinical and translational research. They explore emerging techniques in genomics, proteomics, and metabolomics specific to hepatology, providing insights into their potential to lead to breakthroughs in patient stratification, biomarker discovery, and tailored treatment modalities. These reviews highlight the promise of next-generation sequencing platforms, innovative in vitro models, and artificial-intelligence-driven data analysis as essential tools for understanding disease pathways across the spectrum of liver disorders. The first review analyzes how herbal preparations can have hepatoprotective effects by modulating the gut microbiota and improving liver pathology in conditions such as MASLD, MASH, and hepatic steatosis (contribution 2). The second review discusses the challenges of treating NSCLC patients with liver metastases due to the liver's immunotolerant environment, exploring combination therapies, immunotherapies, and surgical resection as potential strategies (contribution 3). In the third review, advanced in vitro systems, particularly organ-on-a-chip platforms, are highlighted as promising tools for predicting drug-induced liver injury and reducing the cost and duration of drug development (contribution 4). The fourth review focuses on lipopolysaccharide (LPS)-mediated liver damage, noting how TLR4-driven inflammation leads to hepatic injury and how agents such as aspirin could mitigate these effects (contribution 5). Finally, the fifth review addresses the role of simple sugars in exacerbating liver disease through inflammation, de novo lipogenesis, oxidative stress, and dysbiosis, pointing to both lifestyle and molecular interventions for prevention and treatment (contribution 6).

By assembling this Special Issue, we aim to highlight the ongoing evolution of molecular biology approaches in addressing the most pressing challenges in hepatology of today. The methods described here have significantly expanded our toolbox, opening up new avenues not only for basic scientists but also for clinicians striving to translate laboratory discoveries into improved patient outcomes. We believe that these advances collectively represent a step forward for the field, propelling hepatology research towards more precise, patient-oriented strategies.

We hope that the diverse perspectives and methodological innovations showcased in this Special Issue will be both stimulating and valuable to our readers, inspiring collaborative efforts to advance the field of molecular biology in hepatology research.

As interest in this field continues to grow, we are pleased to announce a second edition dedicated to "Advances in Molecular Biology Methods in Hepatology Research." We invite clinicians and researchers worldwide to submit their latest findings in the forms of original research, comprehensive reviews, interesting case reports, perspectives, and other relevant manuscripts. We look forward to creating an even broader platform for knowledge exchange, fostering collaborations that will drive the field of hepatology research forward.

Author Contributions: Conceptualization, D.G.N. and R.W.; resources, R.W.; data curation, D.G.N. and R.W.; writing—original draft preparation, D.G.N. and R.W.; writing—review and editing, D.G.N. and R.W. All authors have read and agreed to the published version of the manuscript.

Funding: This research received no external funding.

Institutional Review Board Statement: Not applicable.

Informed Consent Statement: Not applicable.

Data Availability Statement: No new data were created or analyzed in this study. Data sharing is not applicable to this article.

Acknowledgments: We would like to express our gratitude to all the authors for their outstanding contributions to this topic, as well as the supportive team at the Editorial Office of *Current Issues in Molecular Biology* for facilitating a thorough review process and ensuring effective communication with the authors. Additionally, we want to express our thanks for the commendable and efficient efforts of the expert reviewers who evaluated submissions promptly, fairly, and constructively. Finally, our gratitude goes to Arumugam R. Jayakumar (University of Miami Miller School of Medicine) for helping to edit the article where we had conflicts of interest.

Conflicts of Interest: The authors declare no conflicts of interest.

List of Contributions

1. Eremin, V.; Karpenko, F.; Karpov, I.; Semenov, V.; Oiestad, I. Hepatitis B Virus Genotypes and Subgenotypes Circulating in Belarus. *Curr. Issues Mol. Biol.* **2025**, *47*, 415. https://doi.org/10.3390/cimb47060415.
2. Poo, C.L.; Lau, M.S.; Nasir, N.L.M.; Nik Zainuddin, N.A.S.; Rahman, M.R.A.A.; Mustapha Kamal, S.K.; Awang, N.; Muhammad, H. A Scoping Review on Hepatoprotective Mechanism of Herbal Preparations through Gut Microbiota Modulation. *Curr. Issues Mol. Biol.* **2024**, *46*, 11460–11502. https://doi.org/10.3390/cimb46100682.
3. Ćeriman Krstić, V.; Samardžić, N.; Gajić, M.; Savić, M.; Šeha, B.; Roksandić Milenković, M.; Jovanović, D. Treatment Options for Patients with Non-Small Cell Lung Cancer and Liver Metastases. *Curr. Issues Mol. Biol.* **2024**, *46*, 13443–13455. https://doi.org/10.3390/cimb46120802.
4. Nair, D.G.; Weiskirchen, R. Advanced In Vitro Models for Preclinical Drug Safety: Recent Progress and Prospects. *Curr. Issues Mol. Biol.* **2024**, *47*, 7. https://doi.org/10.3390/cimb47010007.
5. Nakadate, K.; Saitoh, H.; Sakaguchi, M.; Miruno, F.; Muramatsu, N.; Ito, N.; Tadokoro, K.; Kawakami, K. Advances in Understanding Lipopolysaccharide-Mediated Hepatitis: Mechanisms and Pathological Features. *Curr. Issues Mol. Biol.* **2025**, *47*, 79. https://doi.org/10.3390/cimb47020079.
6. Parisse, S.; Coltorti, E.; Mischitelli, M.; Ferri, F.; Ginanni Corradini, S. New Insights into the Interplay Between Simple Sugars and Liver Diseases. *Curr. Issues Mol. Biol.* **2025**, *47*, 390. https://doi.org/10.3390/cimb47060390.

References

1. Ma, X.; Huang, T.; Chen, X.; Li, Q.; Liao, M.; Fu, L.; Huang, J.; Yuan, K.; Wang, Z.; Zeng, Y. Molecular mechanisms in liver repair and regeneration: From physiology to therapeutics. *Signal Transduct. Target. Ther.* **2025**, *10*, 63. [CrossRef] [PubMed]
2. Sah, A.K.; Afzal, M.; Elshaikh, R.H.; Abbas, A.M.; Shalabi, M.G.; Prabhakar, P.K.; Babker, A.M.A.; Khalimova, F.T.; Sabrievna, V.A.; Choudhary, R.K. Innovative strategies in the diagnosis and treatment of liver cirrhosis and associated syndromes. *Life* **2025**, *15*, 779. [CrossRef] [PubMed]
3. Gan, C.; Yuan, Y.; Shen, H.; Gao, J.; Kong, X.; Che, Z.; Guo, Y.; Wang, H.; Dong, E.; Xiao, J. Liver diseases: Epidemiology, causes, trends and predictions. *Signal Transduct. Target. Ther.* **2025**, *10*, 33. [CrossRef] [PubMed]
4. Unagolla, J.M.; Das, S.; Flanagan, R.; Oehler, M.; Menon, J.U. Targeting chronic liver diseases: Molecular markers, drug delivery strategies and future perspectives. *Int. J. Pharm.* **2024**, *660*, 124381. [CrossRef] [PubMed]
5. Ćeriman Krstić, V.; Soldatović, I.; Samardžić, N.; Gajić, M.; Kontić, M.; Reljić, A.; Savić, M.; Roksandić Milenković., M.; Jovanović, D. Long-Term Outcomes in Patients with EGFR Positive Lung Adenocarcinoma and Subgroup Analysis Based on Presence of Liver Metastases. *Curr. Issues Mol. Biol.* **2024**, *46*, 13431–13442. [CrossRef] [PubMed]
6. Chiu, K.W.; Lin, Y.C.; Li, W.F.; Huang, K.T.; Hsu, L.W.; Wang, C.C. The Pre-/Post-Transplant Hepatitis C Antibody Associated with the IL-28B RS8099917 TT Genotype and miRNA-122 Expression May Protect Acute Cellular Rejection After LDLT. *Curr. Issues Mol. Biol.* **2024**, *46*, 12772–12783. [CrossRef] [PubMed]
7. Saitoh, H.; Sakaguchi, M.; Miruno, F.; Muramatsu, N.; Ito, N.; Tadokoro, K.; Kawakami, K.; Nakadate, K. Histopathological Analysis of Lipopolysaccharide-Induced Liver Inflammation and Thrombus Formation in Mice: The Protective Effects of Aspirin. *Curr. Issues Mol. Biol.* **2024**, *46*, 14291–14303. [CrossRef] [PubMed]
8. Liu, W.; Huan, L.; Zhang, C.; Yin, R.; Ouyang, Z.; Wei, Y. Amelioration of Acetaminophen-Induced Hepatic Oxidative Stress and Inflammation by RNAi Targeting Cyp2e1 In Vivo. *Curr. Issues Mol. Biol.* **2025**, *47*, 372. [CrossRef]

current issues in molecular biology

MDPI

Article

Toxicity Responses from Tributyltin Chloride on Haarder (*Planiliza haematocheila*) Livers: Oxidative Stress, Energy Metabolism Dysfunction, and Apoptosis

Changsheng Zhao [1,2], Anning Suo [1,2], Dewen Ding [1,2] and Wencheng Song [1,2,*]

1 Institute for the Control of Agrochemicals, Ministry of Agriculture and Rural Affairs, Beijing 100125, China; zhaochangsheng@agri.gov.cn (C.Z.)
2 Key Laboratory of Pesticide Assessment, Ministry of Agriculture and Rural Affairs, Beijing 100125, China
* Correspondence: songwencheng@agri.gov.cn

Abstract: In coastal waters, tributyltin chloride (TBTC), a persistent organic pollutant, is extensively present. It is uncertain, therefore, if exposure to TBTC can harm haarders and how. This study exposed the fish for 60 days in order to investigate the molecular mechanism of haarder following TBTC poisoning. Our findings demonstrated that growth indices dropped, liver tissue was damaged, and the liver's total tin concentration rose following TBTC exposure. Furthermore, we discovered that blood reactive oxygen species rose while total blood cell count decreased. As malondialdehyde levels rose, total antioxidant capacity and antioxidant enzyme activity (superoxide dismutase, catalase, and glutathione peroxidase) were markedly reduced. After being exposed to TBTC, liver cells displayed clear signs of apoptosis. Differentially expressed genes were primarily linked to oxidative stress, energy metabolism, and apoptosis, according to the transcriptome study of livers. Overall, the long-term stress of TBTC resulted in the antioxidant system being harmed, as well as serious malfunction of the energy metabolism and apoptotic response.

Keywords: tributyltin chloride; oxidative stress; energy metabolism disorder; apoptosis

1. Introduction

Since the 1950s, tributyltin (TBT), one of the most hazardous organic pollutants, has been utilized as an antifouling coating for fishing nets and ships as well as an industrial and agricultural pesticide [1]. The continuous release of TBT has led to its highly concentrated presence in coastal waters, especially near ports and docks [2,3]. TBT is thought to be extremely persistent because its half-life in sediments can last for several decades [4]. Since TBT is a frequent plastic stabilizer and catalyst, it has become an additional source of TBT pollution as the issue of plastic pollution has grown more widespread and is now a global concern [5]. Despite a 2008 large-scale ban on TBT as an antifouling paint for ships, its longevity and several extensive sources have resulted in its ubiquitous presence in sediments, water bodies, and even creatures, endangering the aquatic ecosystem and its inhabitants [6,7]. According to international investigations, TBT levels are as high as 111.29 μg Sn/L in water [8] and 10 μg Sn/g in marine sediments [9], respectively. These levels are far higher than the acceptable threshold (1 ng/kg to 50 ng/g organo-tin in seawater) and would undoubtedly result in extremely high ecological concerns [10].

Reactive oxygen species (ROS)-mediated oxidative stress is defined as the disruption of an organism's dynamic balance [11]. Pollutant-induced ROS production and oxidative damage to key oxidative stress molecules are common toxicological mechanisms in teleosts that contribute to their pathogenicity [12]. According to Burgos-Aceves et al. [13], bony fish have developed a comparatively well-developed defense system that allows them to adjust to oxidative stress brought on by contaminants. Fish include a number of antioxidant enzyme systems, including superoxide dismutase (SOD), catalase (CAT), and glutathione peroxidase (GPx), that deal with reactive oxygen-induced damage and reflect the body's stress resistance capabilities [14]. Furthermore, lipid peroxidation is a reaction process exhibited by fish when exposed to contaminants, as indicated by the presence of malondialdehyde (MDA) [15]. According to studies on the damage to the antioxidant defense system in abalone cells or the worsening of lipid peroxidation, TBT induces oxidative stress [16]. Moreover, ATPases (including Na^+-K^+-ATPase and Ca^{2+}-Mg^{2+}-ATPase) are a class of enzymes that regulate energy metabolism. Exogenous stress factors interfere with the metabolic pathways of energy generation in teleost fish, inhibiting the activity of these enzymes [17,18]. Evidence shows that TBT influences mitochondrial Mg ATPase activity in aquatic animals such as Mediterranean mussels [19]. Studies suggest that pollutants in the ocean additionally influence the tissue morphology of various aquatic animals (including bony fish) and induce their apoptosis [20–22]. Numerous studies have found that apoptosis is directly linked to the generation of excessive ROS [23–25]. As a coastal and demersal bony fish, the habitat of haarder (*Planiliza haematocheila*) may overlap with places having high TBT residues [26]. Haarder's habit of feeding on debris in the sediment increases the risk of TBT poisoning [27].

The chronic (60-day) toxic reaction of haarder to TBT (represented by tributyltin chloride, or TBTC) was identified in this investigation. Prior to testing the growth index, antioxidant system response, and blood oxidative stress response indicators, the livers' total tin content was determined. The effects of various TBTC dosages were then investigated by looking at the livers' histological sections and apoptosis. Several sets of functional liver genes were thoroughly examined in order to learn more about the transcriptome-level mechanism of action of TBTC on haarder.

2. Materials and Methods

2.1. Animals and Chemicals

A farm in Zhuhai, Guangdong Province, China, provided *Planiliza haematocheila* (6 months old), with an average length of 13.5 ± 1.12 cm and an average weight of 24.2 ± 4.57 g. A total of 36 haarder individuals were initially reared domestically in 150 L glass tanks for two weeks, receiving the proper commercial feed twice a day to meet the needs of the experiment (at 8:00 a.m. and 16:00 p.m.). The remaining feed and excrement in the tank water were promptly cleaned up, and it was maintained at a temperature of 25.6 ± 0.5 °C and a salinity of 3.2 ± 0.6‰. The Chinese Academy of Sciences' Animal Research and Ethics Committee gave its approval for our work (SCSIO-IACUC-2021-000156), and we followed all ethical and animal welfare guidelines.

2.2. Tributyltin Chloride (TBTC) Exposure Experiments

Treatment groups L, M, and H were established based on the preliminary experiment results of haarder exposed to TBTC [28]. The dose groups were 3.4 (L group), 34.4 (M group), and 344.2 (H group) ng/L, respectively. TBTC was insoluble in water; hence, dimethyl sulfoxide (DMSO) was used as a co-solvent. A V_{DMSO}:V_{water} ratio of $1:10^7$ was used to prepare the experimental water for the control group (C group) and the three treated groups (L, M, and H). For this TBTC acute exposure experiment, 36 fish were chosen at random and split up into a total of 12 tanks (3 replicate tanks per group, with 3 fish in each replication). Every day, a third of the tank water was added to guarantee the stability of each TBTC exposure concentration. At 60 days, three fish were randomly taken out of each tank and put to death on ice using 100 mg/L MS-222. A random sample was taken from each group's surviving fish (n = 9). The fish's livers were sampled for use in subsequent studies when their length and weight were determined.

2.3. Total Tin Content in Haarder

Total tin (Sn) in livers was detected according to modified methods described by the State Standard of the People's Republic of China [29]. A suitable quantity of each homogenized liver sample was digested with 5% HNO_3 overnight in a microwave digestion system, followed by a 30 min heating period at 100 °C in an ultrasonic water bath. The samples were diluted and equally mixed before being analyzed using inductively coupled plasma–mass spectrometry (Agilent 7900, Santa Clara, CA, USA). For total tin, its limit of detection was 1.1 ng/g on a wet weight basis.

2.4. Growth Indices' Measurement

Two morphological indices, hepatosomatic index (HSI) and Fulton's condition factor (FCF), based on the weight and length of haarder respectively, were calculated according to Maddock and Burton [30]. The calculation was performed as follows: HSI = $100W_L/W_T$, where W_L is the liver weight (in g), and W_T is the total weight (in g); and FCF = $100W_T/{L_T}^3$, where L_T is the total length (in cm).

Frozen liver samples for the analysis of ATPase activity (Na^+-K^+-ATPase and Ca^{2+}-Mg^{2+}-ATPase) were homogenized on ice with 1 × phosphate-buffered saline (pH 7.2). At 37 °C, 100 μL of centrifuged supernatant and the appropriate quantity of mixed reagents were subjected to an enzymatic reaction. After standing at room temperature for 5 min, the absorbance was measured using a spectrophotometer (Shimadzu, Kyoto, Japan) at a wavelength of 636 nm.

2.5. Antioxidant Capacity in Livers

Frozen liver samples were used to measure the activities of SOD, CAT, and GPx, the content of MDA, and the total antioxidant capacity (T-AOC). Using appropriate assay kits (Nanjing Jiancheng Bioengineering Institute, Nanjing, China), all of the aforementioned biochemical indicators were measured. One activity unit of SOD was defined as the amount of SOD corresponding to 50% inhibition rate of SOD in each mL of reaction solution [31]. The suspension of homogenized liver samples was mixed into the reaction reagents, and the absorbance was measured at a wavelength of 550 nm. For CAT, its one activity unit was the amount of decomposing one micromole of H_2O_2 per second per milliliter of reaction solution [32]. An amount of 200 μL of the mixed reaction solution was taken out and placed in a 96-well plate, and the absorbance of each well was measured with a microplate reader (Thermo Scientific, Waltham, MA, USA) at 405 nm. One activity unit of GPx was that every 0.1 mg protein reacts at 37 °C for 5 min, and the concentration of glutathione in the reaction system decreased by 1 μmol/L after deducting the non-enzymatic reaction [33]. After the enzymatic reaction and chromogenic reaction, the reaction solution was evenly mixed and stood for 15 min, and its absorbance was measured by a spectrophotometer (Shimadzu, Kyoto, Japan) at the wavelength of 412 nm. The ground liver tissue was diluted into a 10% homogenate, and the supernatant was taken after centrifugation to prepare for the later determination of MDA content. After the supernatant was bathed in boiling water for 40 min, it was centrifuged at 3500 rpm for 10 min, and the absorbance of the supernatant was measured at 532 nm [34]. T-AOC was determined by the ferric reducing ability of plasma method [35]. The reaction solution was reacted at room temperature for 6 min, and the absorbance was measured at the wavelength of 593 nm. The Bradford method was employed to measure the protein in the liver, with bovine serum albumin serving as the reference [36].

2.6. Total Blood Cell Count and Intracellular Reactive Oxygen Species (ROS) Analyses

A 1-mL syringe with an equal volume of pre-cooling anticoagulant solution was used to draw 200 μL of blood from each fish. The total blood cell count value was recorded when the diluted blood sample was placed on a hemocytometer and examined using an optical microscope (Olympus, Tokyo, Japan).

The level of intracellular ROS was usually expressed by the level of respiratory burst (RB) [37]. A probe called 2′,7′-dichloro fluorescein diacetate (DCF-DA) was employed in the detection. An amount of 200 μL of the blood samples (1×10^6 cells/mL) was incubated in triplicate with DCF-DA (20 μM) for 30 min in the dark at room temperature. A FACSCalibur Flow Cytometer (BD Biosciences, San Jose, CA, USA) was used to assess the fluorescence of each sample. RB levels were calculated by averaging the fluorescence of DCF values.

2.7. Histological and Apoptosis Detection

The liver tissues that were taken were fixed for 24 h using a 4% paraformaldehyde solution. After being dried in ethanol, embedded in paraffin, and sectioned into slices, the sections were cleaned, dewaxed, and stained for eight minutes with cold hematoxylin and three minutes with eosin. A light microscope (Leica, Wetzlar, Germany) was used to view each liver section.

To detect liver cell apoptosis, we employed the terminal-deoxynucleotidyl transferase (TdT)-mediated dUTP nick end labeling (TUNEL) method. The fixed liver slices were deparaffinized and rehydrated, and then they were incubated with proteinase K at 25 °C for 20 min. Each section was stained with 4′,6-diamidino-2-phenylindole and incubated with TdT. All sections were then examined under a fluorescence microscope (Zeiss, Oberkochen, Germany) with a 40× lens magnification. Three samples from each treatment or control group were selected for this investigation. Green fluorescein labeling was used to identify the apoptotic nuclei after all liver cells were stained blue. The ImageJ program (1.53k, National Institutes of Health, Bethesda, MD, USA; Java 1.8.0_172) was used to estimate the apoptotic indices [38].

2.8. RNA-Seq and Bioinformatics Analysis

After TBTC exposure, the livers of haarder fish in the H group and the C group were collected for their transcriptome sequencing. The TRIzol Reagent Kit (Invitrogen, Carlsbad, CA, USA) was used to extract total RNA from three liver samples according to the manufacturer's protocol. The end-repaired, poly (A)-added, and ligated pure cDNA fragments were linked to Illumina sequencing adapters (Illumina, San Diego, CA, USA). The library was created by size-selecting the ligation products using agarose gel electrophoresis, PCR amplification, and an Illumina NovaSeq 6000 system. The raw reads were filtered using the fastp program (v0.18.0) [39] to remove adaptor reads and any low-quality reads, leaving only a high-quality clean data set. After that, unigenes were created by de novo assembly using the Trinity short reads assembling tool [40]. The quality of the resulting assembly was evaluated using N50 values, Q20, Q30, sequence length, and Benchmarking Universal Single-Copy Orthologs (BUSCO) (http://busco.ezlab.org/).

Fragments Per Kilobase of transcript per Million mapped reads (FPKM) was used to compute and standardize the unigene expression and abundance values [41]. Based on the expression information of each sample, principal component analysis (PCA) was used to study the distance relationship between samples through the idea of dimension reduction. The annotation of unigenes included their protein functional annotation, pathway annotation, COG/KOG functional annotation, and Gene Ontology (GO) annotation. The analysis of differentially expressed genes (DEGs) between two separate groups was conducted using the DESeq2 package for R (version 4.1.0) [42]. DEGs were defined as genes with a false discovery rate (FDR) < 0.05 and | log2FC | > 1. To acquire GO function categorization annotations for the genes and identify those that are considerably enriched, the GO enrichment analysis was conducted. The most important biochemical and signal transduction pathways were identified by Kyoto Encyclopedia of Genes and Genomes (KEGG) pathway analysis for significance enrichment. Raw sequencing data was deposited with GenBank in a Sequence Read Archive under the BioProject number PRJNA869536.

2.9. *Quantitative Real-Time PCR (qRT-PCR) Detection of Differentially Expressed Genes (DEGs)*

Ten DEGs were selected for the qRT-PCR analysis to verify the accuracy of the transcriptome data. The cDNA samples were extracted from the livers. Their specific primers were designed using Primer Premier software (version 5.00). The elongation factor 1 alpha (*EF1-α*) gene from haarder *P. haematocheila* (GeneBank accession number: AXB71857) was quantified and served as an internal control to normalize the qRT-PCR results. Prior to conducting this qRT-PCR experiment, the specificity and efficiency of the chosen primers were determined using PCR and electrophoresis on a 1.2% agarose gel, respectively. Every reaction was carried out in a 20 μL total volume consisting of 10 μL of 2 × SYBR Premix Ex Taq™ II (TaKaRa, Kyoto, Japan), 2 μL of the cDNA template (50 ng total RNA), 0.4 μL of ROX Reference Dye (50×), 0.4 μL of both sense and anti-sense primers (10 μM), and 6.8 μL of PCR-specific ddH_2O. The PCR program consisted of 1 cycle of 95 °C for 3 min, followed by 38 cycles of 96 °C for 15 s and 59 °C for 30 s. qRT-PCR was performed using the $2^{-\triangle\triangle Ct}$ method in ABI 7500 software (version 2.3, Applied Biosystems, San Jose, CA, USA) [43].

2.10. *Statistical Analysis*

GraphPad Prism v9.0.0 software (San Diego, CA, USA, www.graphpad.com) is used to statistically analyze the data. The mean ± standard error of mean (SEM) is how the findings are displayed. To find significant differences, the sample means are compared using Dunnett's test and one-way analysis of variance (ANOVA). *, **, and *** stand for the thresholds at $p < 0.05$, $p < 0.01$, and $p < 0.001$, respectively.

3. Results

3.1. *Total Tin Contents in Livers*

After 60 days of chronic TBTC exposure, the liver's total tin content was measured. As seen in Figure 1 and Figure S1, the M and H treatment groups' total tin levels were clearly higher than the control group's. The level of total tin was observed to be dose-dependent as the TBTC dose increased in the treatment groups. Specifically, compared to the control group, the M and H groups' liver total tin levels increased by 0.3 and 1.7 times, respectively.

3.2. *Hepatosomatic Index (HSI), Fulton's Condition Factor (FCF), and Histological Index*

As seen in Figure 2, the HSI values rose following a 60-day exposure. The HSI in the treated groups increased 0.8, 1.1, and 0.9 times, respectively, in comparison to the control group ($p < 0.001$).

Compared to the control group, the FCF levels in the M and H groups were lower (Figure 3). In particular, FCF dropped significantly by 9.3% and 13.5%, respectively, in the M ($p < 0.05$) and H ($p < 0.01$) groups.

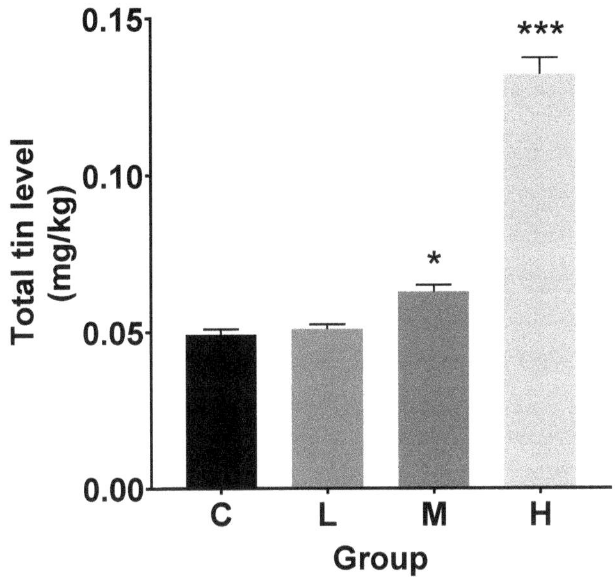

Figure 1. Total tin content in the livers after 60 d of TBTC exposure. The livers were obtained from three individuals in the treatment and control groups (n = 3). A statistically significant difference from the control is shown by an asterisk (* at $p < 0.05$ and *** at $p < 0.001$).

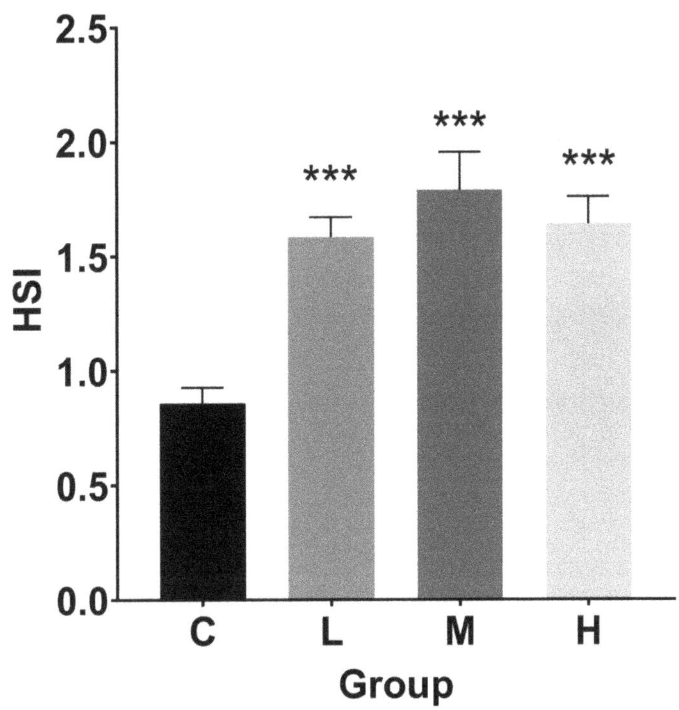

Figure 2. Hepatosomatic index (HSI) for TBTC-treated haarder fish (n = 9). A statistically significant difference from the control is shown by an asterisk (*** at $p < 0.001$).

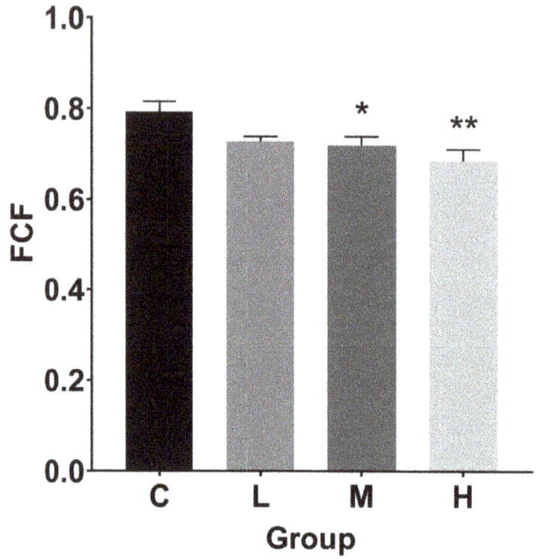

Figure 3. Fulton's condition factor (FCF) for TBTC-treated haarder (n = 9). A statistically significant difference from the control is shown by an asterisk (* at $p < 0.05$ and ** at $p < 0.01$).

Using liver tissues that had been exposed to different TBTC dosages for 60 days, the histopathological alterations were identified (Figure 4a–d). In contrast to the control group's liver, the L group's cell gap widened, and some vacuoles started to form (Figure 4b). The M group's liver cells had more vacuoles (Figure 4c). The nucleus moved to the cell's periphery, and the hepatic vacuolation in the H group increased in Figure 4d.

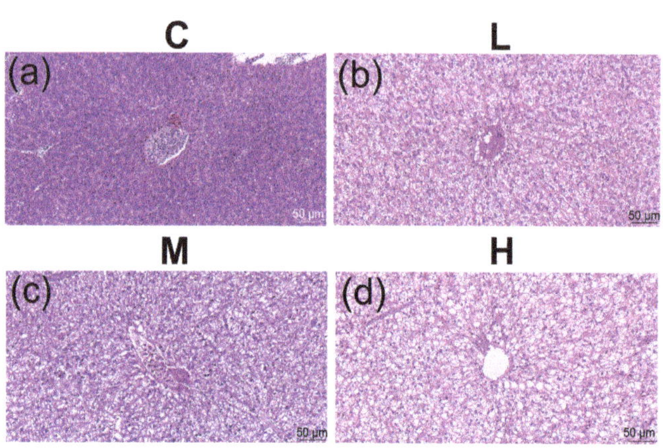

Figure 4. Photomicrographs of hematoxylin-eosin stained livers. (**a**) The control (C) group. (**b**) The L group. (**c**) The M group. (**d**) The H group. The sections were observed under a lens, magnified at 20×.

3.3. *ATPase Activity and Antioxidant Levels in Livers*

In the L, M, and H groups, TBTC significantly decreased Na^{+}-K^{+}-ATPase activity, which decreased by 41.2%, 64.9%, and 71.3%, respectively (Figure 5a). Ca^{2+}-Mg^{2+}-ATPase activity in the treatment group dramatically decreased after 60 days of exposure to TBTC, especially in the M and H groups, which saw declines of 63.9% and 74.2%, respectively (Figure 5b).

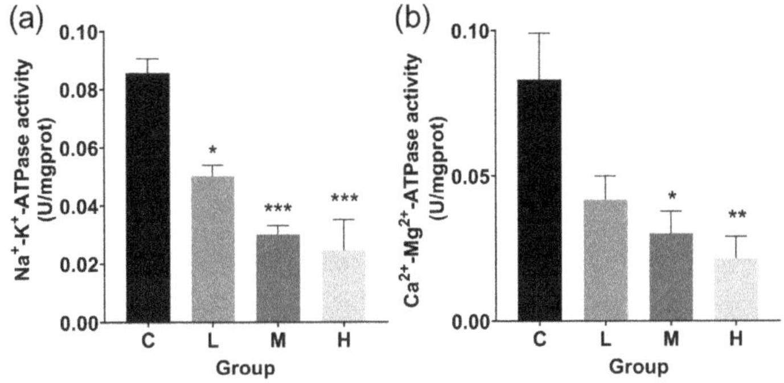

Figure 5. ATPase activity in the TBTC-treated groups of haarder. (**a**) Na^{+}-K^{+}-ATPase activity and (**b**) Ca^{2+}-Mg^{2+}-ATPase activity. An important variation from the control is denoted with an asterisk (* at $p < 0.05$, ** at $p < 0.01$, and *** at $p < 0.001$).

The L group's SOD activity was significantly lower ($p < 0.05$) than that of the control group (Figure 6a). TBTC significantly increased CAT activity with a level of 0.001 in all treated groups (Figure 6b). The GPx activity level dropped sharply at the 0.001 level in every treatment group (Figure 6c).

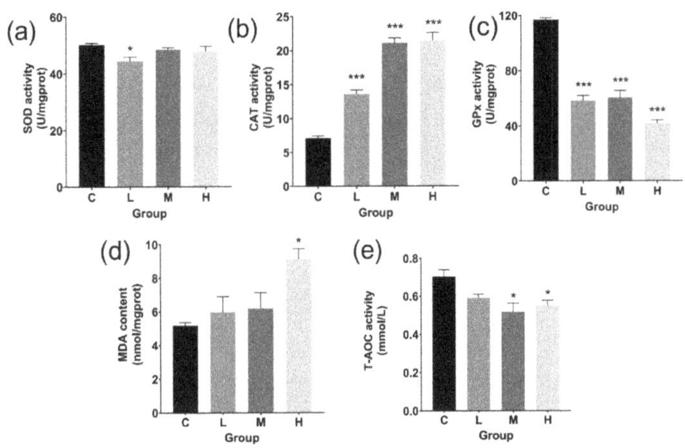

Figure 6. The antioxidant level and lipid peroxidation in the TBTC-treated groups of haarder. (**a**) A bar chart of superoxide dismutase (SOD) activity. (**b**) A bar chart of catalase (CAT) activity. (**c**) A bar chart of glutathione peroxidase (GPx) activity. (**d**) Malondialdehyde (MDA) content in the livers after exposure to TBTC. (**e**) Total antioxidant capacity (T-AOC) level in the liver. All samples were prepared in triplicate per group (n = 3). The asterisks indicated statistically significant differences from the control (* at $p < 0.05$ and *** at $p < 0.001$).

MDA levels increased as the TBTC dosage increased (Figure 6d), especially in the H group, which had a level of 0.05.

Following TBTC exposure, T-AOC activity in the liver dropped by 21.9% in the H group ($p < 0.05$) and 26.6% in the M group ($p < 0.05$), as shown in Figure 6e.

3.4. Total Blood Cell Count and RB Detection

Due to TBTC exposure, the total blood cell count was slightly higher in the L group in Figure 7a. Total blood cell count dropped significantly in the H groups, dropping 47.6% ($p < 0.001$) from the control group.

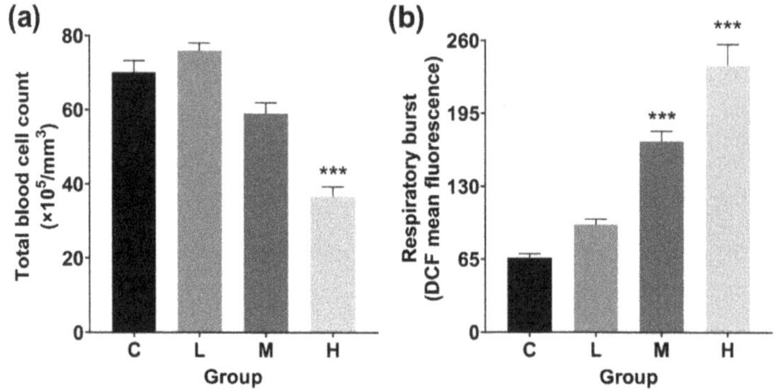

Figure 7. Total blood cell count and intracellular reactive oxygen species (ROS) in the TBTC-treated groups of haarder fish. (**a**) Total blood cell count after 60 d of TBTC exposure. (**b**) Respiratory burst activity in the livers. A statistically significant difference from the control was indicated by the asterisks above the column (*** at $p < 0.001$).

In both the M and H groups, TBTC raised the RB level to a high level ($p < 0.001$). The amount of RB increased by 0.4, 1.6, and 2.6 times in the L, M, and H groups, respectively (Figure 7b).

3.5. Terminal-Deoxynucleotidyl Transferase (TdT)-Mediated dUTP Nick End Labeling (TUNEL) Detection

Liver tissue treated with TBTC for 60 days was examined for apoptosis using the TUNEL assay (Figure 8a–d). The apoptotic indices for the L, M, and H groups (n = 3) were 2.09 ± 0.66, 32.01 ± 3.15, and 50.71 ± 2.27%, respectively (Figure 9). In contrast to the control (0.13 ± 0.003%), the apoptotic indices of the M and H groups significantly increased, by 233.7 ($p < 0.001$) and 393.0 times ($p < 0.001$), respectively.

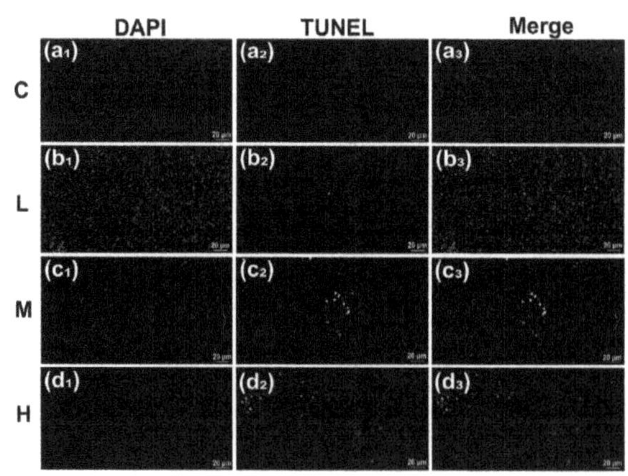

Figure 8. The apoptosis of liver tissue in haarder fish exposed to TBTC. ($\mathbf{a_1}$–$\mathbf{a_3}$) Control liver. TBTC-treated livers in the L ($\mathbf{b_1}$–$\mathbf{b_3}$), M ($\mathbf{c_1}$–$\mathbf{c_3}$), and H ($\mathbf{d_1}$–$\mathbf{d_3}$) groups. The blue fluorescence showed the nuclei of all cells, while the green fluorescence indicated the nuclear regions of apoptotic cells. ($\mathbf{a_1}$,$\mathbf{b_1}$,$\mathbf{c_1}$,$\mathbf{d_1}$) showed the DAPI staining results of the C group, L group, M group and H group, respectively; ($\mathbf{a_2}$,$\mathbf{b_2}$,$\mathbf{c_2}$,$\mathbf{d_2}$) showed the staining results of the apoptotic liver cells in the C group, L group, M group and H group, respectively; ($\mathbf{a_3}$,$\mathbf{b_3}$,$\mathbf{c_3}$,$\mathbf{d_3}$) showed the merged nuclear regions of the cells stained with DAPI and the apoptotic cells in the C group, L group, M group and H group, respectively.

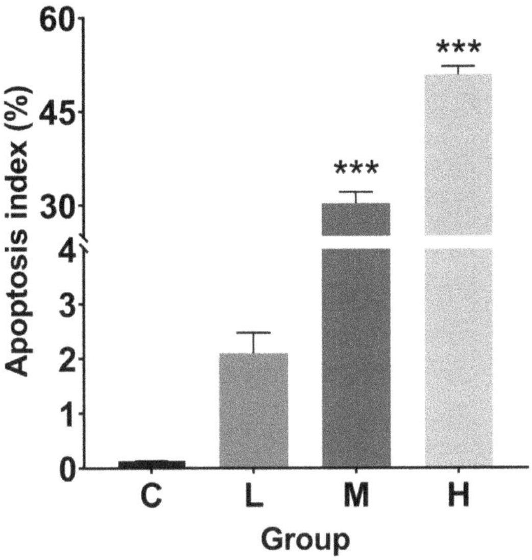

Figure 9. The apoptosis index of the liver tissue. The sections were observed under a lens, magnified at 40× for apoptosis livers. A statistically significant difference from the control was indicated by the asterisks above the column (*** at $p < 0.001$).

3.6. Transcriptional Levels in the Livers of TBTC-Treated Haarder

An average of 6.3 Gb clean bases per sample, or 544,443,632 clean reads, were gathered. Each sample's Q20 and Q30 values were above 98.20% and 94.68%, respectively, indicating good transcriptome sequencing data quality (Table S1). PCA analysis of the transcriptome data showed good repeatability within each group and a significant difference between the TBTC treatment group and the C group (Figure 10). Group H had 5442 up-regulated DEGs and 5340 down-regulated DEGs in pairwise comparisons between the TBTC treatment groups and the control group ($|\log2FC| > 1$, FDR < 0.05, Figure 11). Three categories—biological process (BP), cellular component (CC), and molecular function (MF)—are associated with the top 20 differential enrichment GO keywords in the H group (Figure 12). Among these, the BP category accounted for 55% of the DEGs, including "Response to stimulus" (GO: 0050896), "Signaling" (GO: 0023052), "Biological regulation" (GO: 0065007), "Immune system process" (GO: 0002376), "Regulation of biological process" (GO: 0050789), "Positive regulation of biological process" (GO: 0048518), "Developmental process" (GO: 0032502)", etc. The category of CC included "Extracellular region" (GO: 0005576), "Extracellular region part" (GO: 0044421), "Membrane part" (GO: 0044425), etc. "Molecular function regulator" (GO: 0098772), "Molecular transducer activity" (GO: 0060089), and "Binding" (GO: 0005488) were three notable terms, belonging to the MF category. The DEGs related to oxidative stress, energy metabolism, and apoptosis were enriched in four classes (organismal systems, human diseases, environmental information processing, and cellular processes). The top-20 differential enrichment KEGG terms mainly included "Leukocyte transendothelial migration" (ko04670), "Proteoglycans in cancer" (ko05205), "Chemokine signaling pathway" (ko04062), and "JAK-STAT signaling pathway" (ko04630), etc.

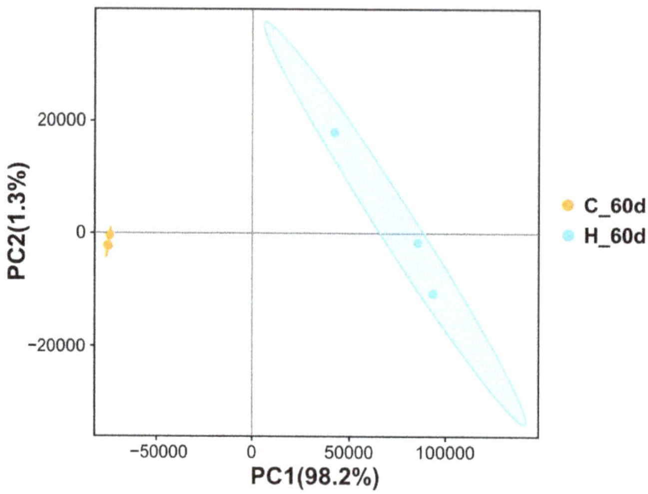

Figure 10. Principal component analysis (PCA) of haarder liver between the 60 d treatment of TBTC (H group, blue) and the control (C group, orange).

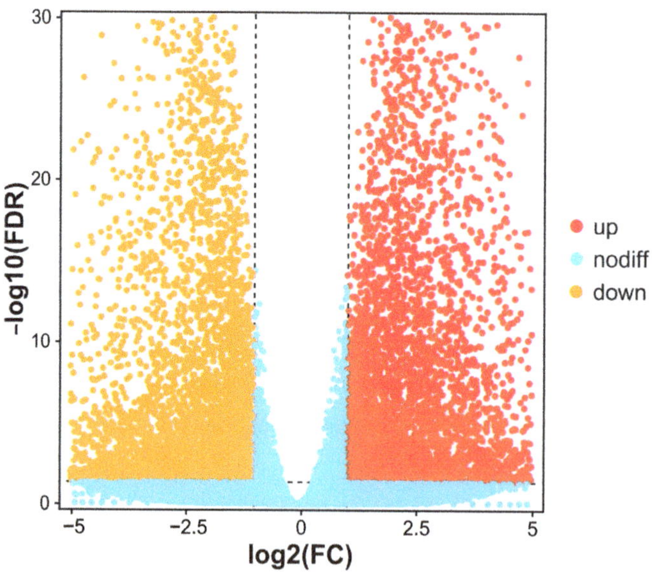

Figure 11. The up- and down-regulated genes in the H group depicted in a volcano plot of the differentially expressed genes (DEGs). The mean value of the log2FC is used as the ordinate (n = 3). Red dots represent up-regulated genes, while orange dots represent down-regulated genes. FDR, false discovery rate; FC, fold change.

Seven apoptosis-related DEGs were identified from the transcriptome data of the H group using the KEGG database (Figure 13). Of these, genes that were up-regulated were caspase 8 (*casp8*), BCL2-associated X, apoptosis regulator (*bax*), and apoptotic protease-activating factor 1 (*apaf1*) (Figure 14). In addition, TBTC in the H group also induced DEGs expression related to oxidative stress. Four up-regulated genes, including peroxiredoxin 2 (*prdx2*), peroxiredoxin 6 (*prdx6*), matrix metalloproteinase-9 (*mmp9*), and heat shock protein 70 (*hsp70*), and one down-regulated genes, such as heat shock protein 90 (*hsp90*), was identified (Figure 14). Besides, six genes associated with energy metabolism were also significantly affected by TBTC after 60 d of exposure, including three up-regulated genes, such as sodium/potassium-dependent ATPase subunit beta-233 (*atnb233*), cystic fibrosis transmembrane conductance regulator (*cftr*), and protein kinase AMP-activated non-catalytic subunit gamma 2 (*prkag2*), and three down-regulated genes, such as acyl-CoA synthettase short chain family member 1 (*acss1*), ATP-binding cassette subfamily D member 2 (*abcd2*), and SWI/SNF-related matrix-associated actin-dependent regulator of chromatin subfamily B member 1 (*smarcb1*) (Figure 14). The expression trends of these selected genes obtained from the qRT-PCR experiment in two representative differential enrichment pathways in the H group are consistent with their corresponding transcriptomic results (Figure S2). The related information of primers was listed in Table S2.

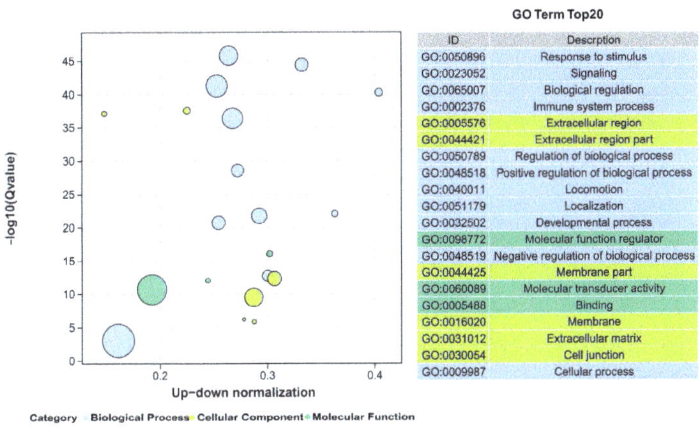

Figure 12. Gene Ontology (GO) enrichment analysis for haarder liver in the H group. The enrichment of the differential bubble plot was created using the significance value −log10 (Qvalue), and the top 20 GO terms are also listed on the right.

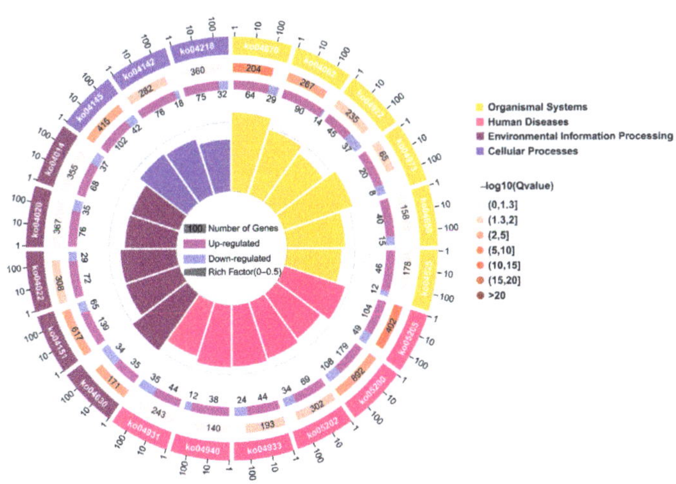

Figure 13. KEGG enrichment analysis for haarder liver in the H group. The circular plot of KEGG between the H group and the control group was plotted with three groups (energy metabolism dysfunction, oxidative stress, and apoptosis).

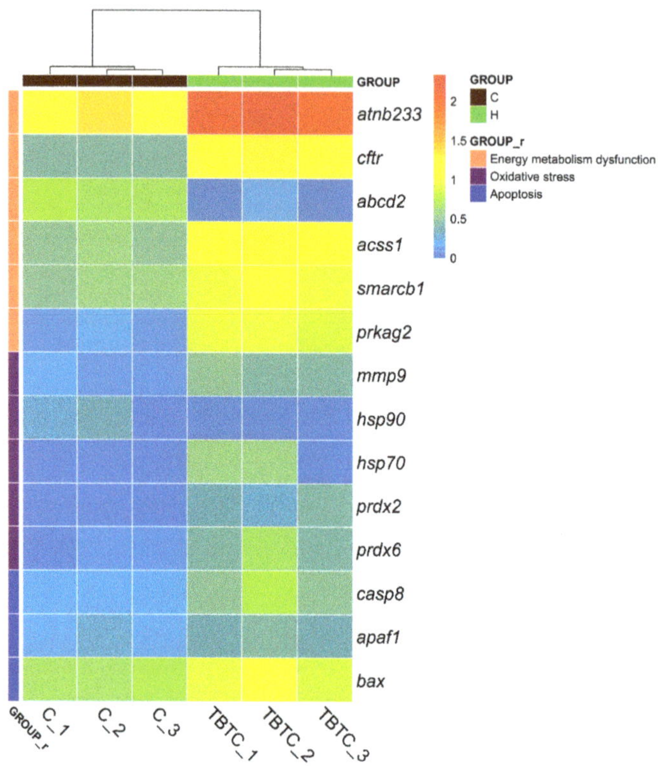

Figure 14. The heat map of DEGs between the H group and the control group.

4. Discussion

4.1. The Effect of TBTC on Haarder Growth and Energy Metabolism

The process by which a chemical substance enters an organism through all exposure pathways, just as it does in the environment, is known as bioaccumulation [44]. As one of the most toxic persistent organic pollutants, the bioaccumulation of TBTC in aquatic animals may cause great ecological risks [45]. TBTC has been proven to bioaccumulate in many aquatic animals, like gastropods [46], zebrafish [47], seahorses [48], and the accumulation rate is very high. In our study, TBTC in all treatment groups had a high accumulation in the livers, and its trend became more and more obvious with the increase of concentration. This may be due to the fact that the liver, as the most important organ of fat accumulation, will lead to a large number of fat-soluble TBTC combined with liver cells [49]. This phenomenon will cause TBTC to accumulate rapidly in haarder's body and not be discharged easily, which will have a series of effects on the whole fish body.

Many pollutants can slow down the growth of fish [50]. Two widely used growth indices, HSI and FCF, are used to assess fish growth when adverse circumstances are present [51]. In zebrafish, HSI was found to rise in female individuals, while FCF decreased when exposed to 10 and 50 ng Sn/L TBT [52]. Our study showed a similar trend in haarder after exposure to TBTC for 60 d, and showed a dose-dependent manner. The above phenomenon may be attributed to the higher bioaccumulation of TBTC in the haarder body. In addition, ATPase related to energy indirectly affected the growth and development of fish [53]. ATPase activity was found to be inhibited in TBT exposed common carp and Japanese medaka [54,55]. In this study, the enzyme activities of Na^{+}-K^{+}-ATPase and Ca^{2+}-Mg^{2+}-ATPase decreased in the treatment groups, which indicated that TBTC could inhibit their activities at the μg/L level. This indicates that the chronic exposure to TBTC may have led to haarder's growth and development based on energy metabolism.

ATNB233 participates in osmoregulation as a subunit of sodium/potassium-dependent ATPase [56]. Additionally, PRKAG2 is the $\gamma 2$ regulatory subunit of 5′ AMP-activated protein kinase, which plays a significant role in the regulation of cellular ATP metabolism [57]. The expressions of *atnb233* and *prkag2* were both up-regulated after TBTC exposure in haarder might be due to their counterbalances of ATP depletion. Many organs' salt, fluid, and pH balance are controlled by the anion channel CFTR, which is also implicated in hypo-osmoregulation in teleosts that are acclimated to seawater. In the gills of Japanese medaka, cortisol elevated the mRNA expression of *cftr* [58]. The TBTC's role in the hypotonic regulation of haarder may account for its increased expression in livers treated with TBTC. During the remediation of pollutants, certain ATP-dependent molecules may sustain harm [59,60]. ACSS1 is essential for the metabolism of acetate, which produces energy [61]. The SWI/SNF ATP-dependent chromatin remodelling complex includes the *smarcb1* gene as a core subunit. Each subunit contains a single ATPase, which is involved in both activating and repressing gene transcription [62]. Following TBTC exposure in haarders, the transcriptional levels of *smarcb1* and *acss1* were down-regulated along with *abcd2*, suggesting that their structures were altered to decrease expression levels.

4.2. Antioxidant Response of Haarder to TBTC

The antioxidant enzyme system serves as a barrier against the entry of dangerous compounds and is employed as a biomarker to indicate the level of toxicity of those substances [63]. CAT and GPx catalyze the reduction of H_2O_2, while SOD and other common antioxidant enzymes cleanse cells by turning toxic superoxide anion into H_2O_2 [64–66]. According to our findings, the livers' activity of three enzymes (SOD, CAT, and GPx) declined as the dosage of TBTC increased, suggesting that TBTC contributed to the inhibition of these three enzymes' activity. Similarly, the activities of SOD and GPx in *Perna viridis*'s gills and hepatopancreas dramatically decreased following a 24 h exposure to TBTC [67]. Furthermore, 10 and 100 ng/L of TBT decreased SOD and CAT activities in the livers of zebrafish, which once again proved the inhibitory effect of TBTC on antioxidant enzyme activities of fish and shellfish [68]. A high level of cytotoxicity and inhibition of protective enzymes is shown by the rise in lipid peroxides like MDA in fish [69]. MDA content significantly increased in the intestine of juvenile common carp exposed to TBT at a level of 7.5 μg/L for 60 days [70]. For TBTC exposed haarder, MDA content increased at 60 d, indicating that TBTC might induce ROS-mediated lipid peroxidation. T-AOC is a comprehensive indicator for measuring the functional status of the body's antioxidant system in response to external stimulations [71]. In previous studies, TBT could weaken T-AOC in liver tissues of juvenile grass carp [72]. Our findings indicated that T-AOC was suppressed in every group treated with TBTC, indicating that the haarder's antioxidant system had some degree of damage following the TBTC attack. We hypothesized that the haarder could first start antioxidant and detoxifying responses to counteract the negative effects of TBTC, in addition to the histological damage produced in the haarder liver. However, other toxicologically significant processes, like metabolic disorders and tissue damage, might be brought on by TBTC when antioxidation and detoxification systems fail.

Following brief focal cerebral ischemia, oxidative stress triggers *mmp9* to mediate the blood-brain barrier's collapse [73]. *Mmp9* expression was up-regulated significantly in the 10 and 100 ng/L of TBT groups in zebrafish [47]. Our result of *mmp9* expression levels in TBTC-exposed haarder livers was found to be elevated, which implied that TBTC initiated *mmp9*-mediated oxidative stress in haarder livers. *Prdx2* is an endogenous peroxidase and has been found to reduce the oxidative burden in cells [74]. As members involved in redox processes, *prdx2* and *cat* are highly expressed compared to normal biliary epithelium cell line [75]. This phenomenon was consistent with the increases of CAT activity and *prdx2* mRNA level detected in haarder exposed to TBTC. *Prdx6*, an antioxidant, was deemed to attenuate oxidative stress on human Trabecular Meshwork cells [76]. *Prdx6* expression in haarder was up-regulated, which suggested that TBTC might reduce the oxidative stress of the liver by activating PRDX6. As an important member of heat shock proteins, the inhibition of HSP90 induced oxidative stress. The transcription levels of *hsp90* in haarder decreased after TBTC stress, which might explain that TBTC inhibited *hsp90*, which disturbed H_2O_2 balance and led to oxidative stress. As another member of heat shock proteins, the induction of HSP70 appeared to be a physiological response to oxidative stress of exercise [77]. The increased mRNA level of *hsp70* in haarder after TBTC exposure might explain the excessive accumulation of *hsp70* caused by oxidative stress.

4.3. Apoptosis Reaction of Haarder to TBTC

Apoptosis is recognized as an important toxicological feature regulated by many factors. CASP8 is an aspartate-specific cysteine protease that acts as an initiator caspase in extrinsic cell death pathways induced by tumor necrosis factor family members. The elevated expression of *casp8* in TBTC-treated haarder might indicate that the signal triggered by TBTC stimulated the expression of *casp8*, and then transmitted apoptosis signals to mitochondria [78]. *Apaf1*, as a pro-apoptotic factor, is activated by cytochrome c released from the mitochondria into the cytoplasm under the induction of the apoptosis signal. The increased mRNA level of *apaf1* in haarder after TBTC exposure confirmed that TBTC induced *apaf1* activation and promoted apoptosis in haarder. *Bax*, a member of the Bcl-2 protein family, accelerated apoptosis by adverse factors [79]. The increased mRNA level of *bax* in haarder after TBTC exposure, indicating apoptosis was exacerbated in the liver. The TUNEL results of haarder's liver also confirmed that TBTC induced liver apoptosis.

5. Conclusions

The harmful effects of long-term TBTC exposure on haarders were demonstrated by this investigation. The total tin concentration in the livers increased significantly after 60 days of TBTC treatment, suggesting that TBTC had a high level of bioaccumulation in the livers and may also be the cause of a number of later bodily reactions. The histological alterations of the livers confirmed that TBTC generated severe oxidative stress and significantly impacted ATPase activity in the haarder body. Related genes at the transcriptome level demonstrated that TBTC severely harmed the haarder antioxidant enzyme system. Additionally, TBTC caused apoptosis in the liver and set off a chain reaction of functional gene responses. The distribution ranges of haarder and TBTC may overlap in some water areas, and the harmful toxicological effects of TBTC on haarder may lead to an imbalance in ecological stability [80,81]. The potential ecological hazards of TBTC in aquatic environments to fish growth and reproduction were referenced in this work.

Supplementary Materials: The following supporting information can be downloaded at: https://www.mdpi.com/article/10.3390/cimb47070526/s1.

Author Contributions: Conceptualization, C.Z. and D.D.; methodology, A.S.; software, C.Z.; validation, A.S. and W.S.; formal analysis, C.Z.; investigation, C.Z.; resources, A.S.; data curation, C.Z.; writing—original draft preparation, C.Z.; writing—review and editing, W.S. and A.S.; visualization, C.Z.; supervision, D.D. All authors have read and agreed to the published version of the manuscript.

Funding: This research received no external funding.

Institutional Review Board Statement: The animal study protocol was approved by the Ethics Committee of the Chinese Academy of Sciences (protocol code SCSIO-IACUC-2021-000156 and date of approval 15 April 2021).

Informed Consent Statement: Not applicable.

Data Availability Statement: The data presented in this study are available on request from the corresponding author due to the protection policy of the research institution.

Acknowledgments: We express our gratitude to Guangdong Longsheng Fish Farm for providing the excellent fish fry.

Conflicts of Interest: The authors declare no conflicts of interest.

Abbreviations

TBT: tributyltin; TBTC, tributyltin chloride; ROS, reactive oxygen species; GPx, glutathione peroxidase; CAT, catalase; SOD, superoxide dismutase; MDA, malondialdehyde; HSI, hepatosomatic index; FCF, Fulton's condition factor; T-AOC, total antioxidant capacity; DMSO, dimethyl sulfoxide; RB, respiratory burst; DCF-DA, 2′,7′-dichloro fluorescein diacetate; TUNEL, terminal-deoxynucleotidyl transferase (TdT)-mediated dUTP nick end labeling; qRT-PCR, quantitative real-time PCR; BUSCO, Benchmarking Universal Single-Copy Orthologs; FPKM, Fragments Per Kilobase of transcript per Million mapped reads; PCA, principal component analysis; GO, Gene Ontology; KEGG, Kyoto Encyclopedia of Genes and Genomes; DEGs, differentially expressed genes; FDR, false discovery rate; BP, biological process; CC, cellular component; MF, molecular function; casp8, caspase 8; bax, BCL2 associated X, apoptosis regulator; apaf1, apoptotic protease-activating factor 1; prdx2, peroxiredoxin 2; prdx6, peroxiredoxin 6; mmp9, matrix metalloproteinase-9; hsp70, heat shock protein 70; hsp90, heat shock protein 90; atnb233, sodium/potassium-dependent ATPase subunit beta-233; cftr, cystic fibrosis transmembrane conductance regulator; prkag2, protein kinase AMP-activated non-catalytic subunit gamma 2; acss1, acyl-CoA synthettase short chain family member 1; abcd2, ATP binding cassette subfamily D member 2; smarcb1, SWI/SNF-related matrix-associated actin-dependent regulator of chromatin subfamily B member 1.

References

1. Hoch, M. Organotin compounds in the environment—An overview. *Appl. Geochem.* **2001**, *16*, 719–743. [CrossRef]
2. Burton, E.D.; Phillips, I.R.; Hawker, D.W. Sorption and desorption behavior of tributyltin with natural sediments. *Environ. Sci. Technol.* **2004**, *38*, 6694. [CrossRef] [PubMed]
3. Luan, T.G.; Jin, J.; Chan, S.M.N.; Wong, Y.S.; Tam, N.F.Y. Biosorption and biodegradation of tributyltin (TBT) by alginate immobilized *Chlorella vulgaris* beads in several treatment cycles. *Process Biochem.* **2006**, *41*, 1560–1565. [CrossRef]
4. Gadd, G.M. Microbial interactions with tributyltin compounds: Detoxification, accumulation, and environmental fate. *Sci. Total Environ.* **2000**, *258*, 119–127. [CrossRef]
5. Wang, X.H.; Fang, C.; Hong, H.S.; Wang, W.X. Gender differences in TBT accumulation and transformation in *Thais clavigera* after aqueous and dietary exposure. *Aquat. Toxicol.* **2010**, *99*, 413–422. [CrossRef]
6. Anastasiou, T.I.; Chatzinikolaou, E.; Mandalakis, M.; Arvanitidis, C. Imposex and organotin compounds in ports of the Mediterranean and the Atlantic: Is the story over? *Sci. Total Environ.* **2016**, *569–570*, 1315–1329. [CrossRef]
7. Batista, R.M.; Castro, I.B.; Fillmann, G. Imposex and butyltin contamination still evident in Chile after TBT global ban. *Sci. Total Environ.* **2016**, *566–567*, 446–453. [CrossRef]
8. Okoro, H.K.; Fatoki, O.S.; Adekola, F.A.; Ximba, B.J.; Snyman, R.G. Spatio-temporal variation of organotin compounds in seawater and sediments from Cape Town harbour, South Africa using gas chromatography with flame photometric detector (GC-FPD). *Arab. J. Chem.* **2016**, *9*, 95–104. [CrossRef]
9. Briant, N.; Bancon-Montigny, C.; Elbaz-Poulichet, F.; Freydier, R.; Delpoux, S.; Cossa, D. Trace elements in the sediments of a large *Mediterranean marina* (Port Camargue, France): Levels and contamination history. *Mar. Pollut. Bull.* **2013**, *73*, 78–85. [CrossRef]
10. Fung, W.H.; Can, S.Z.; Tse, K.C.W.; Tsoi, Y.T.; Lee, W.W.; Arı, B.; Tunç, M.; Merrick, J.; de Sena, R.C.; Nadeau, K.; et al. CCQM-K155: Elements and tributyltin in seawater. *Metrologia* **2025**, *62*, 08007. [CrossRef]
11. Morris, J.J.; Rose, A.L.; Lu, Z. Reactive oxygen species in the world ocean and their impacts on marine ecosystems. *Redox biol.* **2022**, *52*, 102285. [CrossRef] [PubMed]

12. Livingstone, D.R.; Mitchelmore, C.L.; O'Hara, S.C.M.; Lemaire, P.; Sturve, J.; Förlin, L. Increased potential for NAD(P)H-dependent reactive oxygen species production of hepatic subcellular fractions of fish species with in vivo exposure to contaminants. *Mar. Environ. Res.* **2000**, *50*, 57–60. [CrossRef] [PubMed]
13. Burgos-Aceves, M.A.; Cohen, A.; Smith, Y.; Faggio, C. Estrogen regulation of gene expression in the teleost fish immune system. *Fish Shellfish. Immun.* **2016**, *58*, 42–49. [CrossRef] [PubMed]
14. Lushchak, V.I. Environmentally induced oxidative stress in aquatic animals. *Aquat. Toxicol.* **2011**, *101*, 13–30. [CrossRef]
15. Tsikas, D. Assessment of lipid peroxidation by measuring malondialdehyde (MDA) and relatives in biological samples: Analytical and biological challenges. *Anal. Biochem.* **2017**, *524*, 13–30. [CrossRef]
16. Zhou, J.; Zhu, X.; Cai, Z. Tributyltin toxicity in abalone (*Haliotis diversicolor supertexta*) assessed by antioxidant enzyme activity, metabolic response, and histopathology. *J. Hazard. Mater.* **2010**, *183*, 428–433. [CrossRef]
17. Almeida, E.A.; Bainy, A.C.D.; Dafre, A.L.; Gomes, O.F.; Medeiros, M.H.G.; Di Mascio, P. Oxidative stress in digestive gland and gill of the brown mussel (*Perna perna*) exposed to air and re-submersed. *J. Exp. Mar. Biol. Ecol.* **2005**, *318*, 21–30. [CrossRef]
18. Sun, H.J.; Zhao, W.J.; Teng, X.Q.; Shu, S.P.; Li, S.W.; Hong, H.C.; Guan, D.X. Antioxidant responses and pathological changes in the gill of zebrafish (*Danio rerio*) after chronic exposure to arsenite at its reference dose. *Ecotoxicol. Environ. Saf.* **2020**, *200*, 110743. [CrossRef]
19. Ventrella, V.; Nesci, S.; Trombetti, F.; Bandiera, P.; Pirini, M.; Borgatti, A.R.; Pagliarani, A. Tributyltin inhibits the oligomycin-sensitive Mg-ATPase activity in *Mytilus galloprovincialis* digestive gland mitochondria. *Comp. Biochem. Phys. C* **2011**, *153*, 75–81. [CrossRef]
20. Suzuki, N.; Honda, M.; Sato, M.; Yoshitake, S.; Kawabe, K.; Tabuchi, Y.; Omote, T.; Sekiguchi, T.; Furusawa, Y.; Toriba, A.; et al. Hydroxylated benzo[c]phenanthrene metabolites cause osteoblast apoptosis and skeletal abnormalities in fish. *Ecotoxicol. Environ. Saf.* **2022**, *234*, 113401. [CrossRef]
21. Wang, J.X.; Gao, X.Z.; Liu, F.X.; Fang, W.; Dong, J.Q.; Zhao, P.P. Difenoconazole causes cardiotoxicity in common carp (*Cyprinus carpio*): Involvement of oxidative stress, inflammation, apoptosis and autophagy. *Chemosphere* **2022**, *306*, 135562. [CrossRef] [PubMed]
22. Zhang, H.R.; Zhao, F.Q.; Gai, X.X.; Cai, J.Z.; Zhang, X.T.; Chen, X.M.; Zhu, Y.; Zhang, Z.W. Astilbin attenuates apoptosis induced by cadmium through oxidative stress in carp (*Cyprinus carpio* L.) head kidney lymphocyte. *Fish Shellfish. Immun.* **2022**, *125*, 230–237. [CrossRef] [PubMed]
23. Cui, Y.L.; Song, M.; Xiao, B.N.; Liu, M.L.; Liu, P.L.; Han, Y.F.; Shao, B.; Li, Y.F. ROS-mediated mitophagy and apoptosis are involved in aluminum-induced femoral impairment in mice. *Chem. Biol. Interac.* **2021**, *349*, 109663. [CrossRef] [PubMed]
24. Ding, C.; Shi, X.; Guan, Y.L.; Li, X.J. Deoxynivalenol induces carp neutrophil apoptosis and necroptosis via CYP450s/ROS/PI3K/AKT pathway. *Aquaculture* **2021**, *545*, 737182. [CrossRef]
25. Jia, Z.L.; Cen, J.; Wang, J.B.; Zhang, F.; Xia, Q.; Wang, X.; Chen, X.Q.; Wang, R.C.; Hsiao, C.D.; Liu, K.C.; et al. Mechanism of isoniazid-induced hepatotoxicity in zebrafish larvae: Activation of ROS-mediated ERS, apoptosis and the Nrf2 pathway. *Chemosphere* **2019**, *227*, 541–550. [CrossRef]
26. Meng, W.; Gao, T.X.; Zheng, B. Genetic analysis of four populations of redlip mullet (*Chelon haematocheilus*) collected in China seas. *J. Ocean. Univ. China* **2007**, *6*, 72–75. [CrossRef]
27. Kang, B.; Xian, W.W.; Liu, R.Y. Transporting nutrients from sediments: Nitrogen and carbon as indicators to measure scavenging ability of a detritivorous fish, *Liza haematocheila*. *Environ. Biol. Fish* **2005**, *74*, 283–290. [CrossRef]
28. Xu, D.P.; Fang, D.A.; Zhao, C.S.; Jiang, S.L.; Hu, H.Y. Effect of tributyltin chloride (TBT-Cl) exposure on expression of HSP90β1 in the river pufferfish (*Takifugu obscurus*): Evidences for its immunologic function involving in exploring process. *Gene* **2018**, *666*, 9–17. [CrossRef]
29. *GB 5009.268-2016*; Food Safety National Standard—Determination of Multiple Elements in Food. National Health and Family Planning Commission of the People's Republic of China and the State Food and Drug Administration: Beijing, China, 2016.
30. Maddock, D.M.; Burton, M.P.M. Gross and histological observations of ovarian development and related condition changes in American plaice. *J. Fish Biol.* **1999**, *53*, 928–944. [CrossRef]
31. Mao, G.X.; Zheng, L.D.; Cao, Y.B.; Chen, Z.M.; Lv, Y.D.; Wang, Y.Z.; Hu, X.L.; Wang, G.F.; Yan, J. Antiaging effect of pine pollen in human diploid fibroblasts and in a mouse model induced by D-galactose. *Oxid. Med. Cell. Longev.* **2012**, *2012*, 750963. [CrossRef]
32. Zhang, J.Q.; Shen, M.; Zhu, C.C.; Yu, F.X.; Liu, Z.Q.; Ally, N.; Sun, S.C.; Li, K.; Liu, H.L. 3-Nitropropionic acid induces ovarian oxidative stress and impairs follicle in mouse. *PLoS ONE* **2014**, *9*, e86589. [CrossRef] [PubMed]
33. Deng, Y.; Wang, W.; Yu, P.F.; Xi, Z.J.; Xu, L.J.; Li, X.L.; He, N.Y. Comparison of taurine, GABA, Glu, and Asp as scavengers of malondialdehyde in vitro and in vivo. *Nanoscale Res. Lett.* **2013**, *8*, 190. [CrossRef] [PubMed]
34. Wang, H.M.; Liu, X.Y.; Yang, P.; Wu, R.Z.; Wang, S.Y.; He, S.N.; Zhou, Q.H. Potassium application promote cotton acclimation to soil waterlogging stress by regulating endogenous protective enzymes activities and hormones contents. *Plant Physiol. Bioch.* **2022**, *185*, 336–343. [CrossRef] [PubMed]

35. Liu, F.; Li, X.L.; Lin, T.; He, D.W.; Wei, G.H.; Liu, J.H.; Li, L.S. The cyclophosphamide metabolite, acrolein, induces cytoskeletal changes and oxidative stress in Sertoli cells. *Mol. Biol. Rep.* **2012**, *39*, 493–500. [CrossRef]
36. Bradford, M.M. A rapid and sensitive method for the quantification of microgram quantities of protein utilizing the principle of protein-dye binding. *Anal. Biochem.* **1976**, *72*, 248–254. [CrossRef]
37. Xian, J.A.; Wang, A.L.; Tian, J.X.; Huang, J.W.; Ye, C.X.; Wang, W.N.; Sun, R.Y. Morphologic, physiological and immunological changes of haemocytes from *Litopenaeus vannamei* treated by lipopolysaccharide. *Aquaculture* **2009**, *298*, 139–145. [CrossRef]
38. Stein, M.; Lin, H.; Jeyamohan, C.; Dvorzhinski, D.; Gounder, M.; Bray, K.; Eddy, S.; Goodin, S.; White, E.; DiPaola, R.S. Targeting tumor metabolism with 2-deoxyglucose in patients with castrate-resistant prostate cancer and advanced malignancies. *Prostate* **2010**, *70*, 1388–1394. [CrossRef]
39. Chen, S.F.; Zhou, Y.Q.; Chen, Y.R.; Gu, J. fastp: An ultra-fast all-in-one FASTQ preprocessor. *Bioinformatics* **2018**, *34*, 884. [CrossRef]
40. Grabherr, M.G.; Haas, B.J.; Yassour, M.; Levin, J.Z.; Thompson, D.A.; Amit, I.; Adiconis, X.; Fan, L.; Raychowdhury, R.; Zeng, Q.; et al. Full-length transcriptome assembly from RNA-Seq data without a reference genome. *Nat. Biotechnol.* **2011**, *29*, 644–652. [CrossRef]
41. Ran, M.X.; Huang, H.X.; Hu, B.; Hu, S.Q.; Hu, J.W.; Li, L.; He, H.; Liu, H.H.; Wang, J.W. Comparative analysis of testicular histology and lncRNA-mRNA expression patterns between Landes geese (*Anser anser*) and Sichuan White geese (*Anser cygnoides*). *Front. Genet.* **2021**, *12*, 627384. [CrossRef]
42. Love, M.I.; Huber, W.; Anders, S. Moderated estimation of fold change and dispersion for RNA-seq data with DESeq2. *Genome Biol.* **2014**, *15*, 550. [CrossRef] [PubMed]
43. Schmittgen, T.D.; Livak, K.J. Analyzing real-time PCR data by the comparative C_T method. *Nat. Protoc.* **2008**, *3*, 1101–1108. [CrossRef] [PubMed]
44. Arnot, J.A.; Gobas, F.A.P.C. A review of bioconcentration factor (BCF) and bioaccumulation factor (BAF) assessments for organic chemicals in aquatic organisms. *Environ. Rev.* **2006**, *14*, 257–297. [CrossRef]
45. van den Dungen, M.W.; Murk, A.J.; Kok, D.E.; Steegenga, W.T. Persistent organic pollutants alter DNA methylation during human adipocyte differentiation. *Toxicol. Vitr.* **2017**, *40*, 79–87. [CrossRef]
46. Berto, D.; Giani, M.; Boscolo, R.; Covelli, S.; Giovanardi, O.; Massironi, M.; Grassia, L. Organotins (TBT and DBT) in water, sediments, and gastropods of the southern Venice lagoon (Italy). *Mar. Pollut. Bull.* **2007**, *55*, 425–435. [CrossRef]
47. Zhang, J.L.; Sun, P.; Kong, T.; Yang, F.; Guan, W.C. Tributyltin promoted hepatic steatosis in zebrafish (*Danio rerio*) and the molecular pathogenesis involved. *Aquat. Toxicol.* **2016**, *170*, 208–215. [CrossRef]
48. Tang, L.; Zhang, Y.H.; Wang, X.; Zhang, C.C.; Qin, G.; Lin, Q. Effects of chronic exposure to environmental levels of tributyltin on the lined seahorse (*Hippocampus erectus*) liver: Analysis of bioaccumulation, antioxidant defense, and immune gene expression. *Sci. Total Environ.* **2021**, *801*, 149646. [CrossRef]
49. Maisano, M.; Cappello, T.; Oliva, S.; Natalotto, A.; Giannetto, A.; Parrino, V.; Battaglia, P.; Romeo, T.; Salvo, A.; Spanò, N.; et al. PCB and OCP accumulation and evidence of hepatic alteration in the Atlantic bluefin tuna, *T. thynnus*, from the Mediterranean Sea. *Mar. Environ. Res.* **2016**, *121*, 40–48. [CrossRef]
50. Forrester, G.E.; Fredericks, B.I.; Gerdeman, D.; Evans, B.; Steele, M.A.; Zayed, K.; Schweitzer, L.E.; Suffet, I.H.; Vance, R.R.; Ambrose, R.F. Growth of estuarine fish is associated with the combined concentration of sediment contaminants and shows no adaptation or acclimation to past conditions. *Mar. Environ. Res.* **2003**, *56*, 423–442. [CrossRef]
51. Hauser-Davis, R.A.; Bordon, I.C.; Kannan, K.; Moreira, I.; Quinete, N. Perfluoroalkyl substances associations with morphometric health indices in three fish species from differentially contaminated water bodies in Southeastern Brazil. *Environ. Technol. Innov.* **2021**, *21*, 101198. [CrossRef]
52. Lyssimachou, A.; Santos, J.G.; André, A.; Soares, J.; Lima, D.; Guimarães, L.; Almeida, C.M.R.; Teixeira, C.; Castro, L.F.C.; Santos, M.M. The mammalian "obesogen" tributyltin targets hepatic triglyceride accumulation and the transcriptional regulation of lipid metabolism in the liver and brain of zebrafish. *PLoS ONE* **2015**, *10*, e0143911. [CrossRef] [PubMed]
53. Musharraf, M.; Khan, M.A. Dietary potassium requirement of fingerling Indian major carp, *Labeo rohita* (Hamilton), based on growth parameters, gill Na^+-K^+ ATPase activity, tissue mineralization and antioxidant activities. *Aquacult. Nutr.* **2019**, *25*, 271–280. [CrossRef]
54. Li, Z.H.; Li, P.; Shi, Z.C. Chronic effects of tributyltin on multiple biomarkers responses in juvenile common carp, *Cyprinus carpio*. *Environ. Toxicol.* **2016**, *31*, 937–944. [CrossRef] [PubMed]
55. Shi, Y.H.; Chen, C.; Li, M.; Liu, L.; Dong, K.J.; Chen, K.; Qiu, X.C. Oral exposure to tributyltin induced behavioral abnormality and oxidative stress in the eyes and brains of juvenile Japanese medaka (*Oryzias latipes*). *Antioxidants* **2021**, *10*, 1647. [CrossRef]
56. Major, K.M.; DeCourten, B.M.; Li, J.; Britton, M.; Settles, M.L.; Mehinto, A.C.; Connon, R.E.; Brander, S.M. Early life exposure to environmentally relevant levels of endocrine disruptors drive multigenerational and transgenerational epigenetic changes in a fish model. *Front. Mar. Sci.* **2020**, *7*, 471. [CrossRef]

57. Gollob, M.H.; Green, M.S.; Tang, A.S.-L.; Gollob, T.; Karibe, A.; Hassan, A.-S.; Ahmad, F.; Lozado, R.; Shah, G.; Fananapazir, L.; et al. Identification of a gene responsible for familial Wolff–Parkinson–White syndrome. *N. Engl. J. Med.* **2001**, *344*, 1823–1831. [CrossRef]
58. Bossus, M.C.; Bollinger, R.J.; Reed, P.J.; Tipsmark, C.K. Prolactin and cortisol regulate branchial claudin expression in Japanese medaka. *Gen. Comp. Endocr.* **2017**, *240*, 77–83. [CrossRef]
59. Bhagat, J.; Zang, L.Q.; Nishimura, N.; Shimada, Y. Application of omics approaches for assessing microplastic and nanoplastic toxicity in fish and seafood species. *TrAC Trend Anal. Chem.* **2022**, *154*, 116674. [CrossRef]
60. de Souza, S.S.; da Silva Castro, J.; Campos, D.F.; Pereira, R.S.; Bataglion, G.A.; da Silva, G.S.; de Almeida-Val, V.M. Temporal exposure to malathion: Biochemical changes in the Amazonian fish tambaqui, *Colossoma macropomum*. *Aquat. Toxicol.* **2021**, *241*, 105997. [CrossRef]
61. Calhoun, S.; Duan, L.; Maki, C.G. Acetyl-CoA synthetases ACSS1 and ACSS2 are 4-hydroxytamoxifen responsive factors that promote survival in tamoxifen treated and estrogen deprived cells. *Transl. Oncol.* **2022**, *19*, 101386. [CrossRef]
62. Kalimuthu, S.N.; Chetty, R. Gene of the month: SMARCB1. *J. Clin. Pathol.* **2016**, *69*, 484–489. [CrossRef] [PubMed]
63. Emmanouil, C.; Green, R.M.; Willey, F.R.; Chipman, J.K. Oxidative damage in gill of *Mytilus edulis* from Merseyside, UK, and reversibility after depuration. *Environ. Pollut.* **2008**, *151*, 663–668. [CrossRef] [PubMed]
64. Glorieux, C.; Calderon, P.B. Catalase, a remarkable enzyme: Targeting the oldest antioxidant enzyme to find a new cancer treatment approach. *Biol. Chem.* **2017**, *398*, 1095–1108. [CrossRef] [PubMed]
65. Margis, R.; Dunand, C.; Teixeira, F.K.; Margis-Pinheiro, M. Glutathione peroxidase family—An evolutionary overview. *FEBS J.* **2008**, *275*, 3959–3970. [CrossRef]
66. McRdj, M.; Fridovich, I. Superoxide dismutase: An enzyme function for erychrocuprein. *J. Biol. Chem.* **1969**, *244*, 6049.
67. Chen, H.G.; Jia, X.P.; Cai, W.G.; Lin, Q.; Ma, S.W. Antioxidant responses and bioaccumulation in green-lipped mussels (*Perna Viridis*) under acute tributyltin chloride exposure. *Bull. Environ. Contam. Tox.* **2011**, *87*, 506. [CrossRef]
68. Zhang, C.N.; Zhang, J.L.; Ren, H.T.; Zhou, B.H.; Wu, Q.J.; Sun, P. Effect of tributyltin on antioxidant ability and immune responses of zebrafish (*Danio rerio*). *Ecotoxicol. Environ. Saf.* **2017**, *138*, 1–8. [CrossRef]
69. Rundle, A.; Robertson, A.B.; Blay, A.M.; Butler, K.M.A.; Callaghan, N.I.; Dieni, C.A.; MacCormack, T.J. Cerium oxide nanoparticles exhibit minimal cardiac and cytotoxicity in the freshwater fish *Catostomus commersonii*. *Comp. Biochem. Physiol. Part C Toxicol. Pharmacol.* **2016**, *181–182*, 19–26. [CrossRef]
70. Li, Z.H.; Li, P.; Shi, Z.C. Molecular responses in digestive tract of juvenile common carp after chronic exposure to sublethal tributyltin. *Ecotoxicol. Environ. Saf.* **2014**, *109*, 10–14. [CrossRef]
71. Wang, Y.X.; Wu, P.P.; Shen, H.Y. Effect of ampicillin wastewater on the T-AOC and LDH activities of zebrafish. In Proceedings of the IOP Conference Series: Earth and Environmental Science, Bristol, UK, 7–9 June 2019; Volume 295, p. 012018.
72. Mu, W.N.; Li, Z.H.; Zhong, L.Q.; Wu, Y.H. Effect of tributyltin, cadmium, and their combination on physiological responses in juvenile grass carp. *J. Aquat. Anim. Health* **2016**, *28*, 181–186. [CrossRef]
73. Kamada, H.; Yu, F.; Nito, C.; Chan, P.H. Influence of hyperglycemia on oxidative stress and matrix metalloproteinase-9 activation after focal cerebral ischemia/reperfusion in rats. *Stroke* **2007**, *38*, 1044–1049. [CrossRef] [PubMed]
74. Li, H.; Yang, H.; Wang, D.; Zhang, L.; Ma, T. Peroxiredoxin2 (prdx2) reduces oxidative stress and apoptosis of myocardial cells induced by acute myocardial infarction by inhibiting the TLR4/nuclear factor kappa B (NF-κB) signaling pathway. *Med. Sci. Monit.* **2020**, *26*, e926281. [CrossRef] [PubMed]
75. Cavalloni, G.; Peraldo-Neia, C.; Massa, A.; Bergamini, C.; Trentini, A.; De Rosa, G.; Daniele, L.; Ciccosanti, F.; Cervellati, C.; Leone, F.; et al. Proteomic analysis identifies deregulated metabolic and oxidative-associated proteins in Italian intrahepatic cholangiocarcinoma patients. *BMC Cancer* **2021**, *21*, 865. [CrossRef] [PubMed]
76. Fatma, N.; Kubo, E.; Toris, C.B.; Stamer, W.D.; Camras, C.B.; Singh, D.P. PRDX6 attenuates oxidative stress- and TGFβ-induced abnormalities of human trabecular meshwork cells. *Free. Radic. Res.* **2009**, *43*, 783–795. [CrossRef]
77. Salo, D.C.; Donovan, C.M.; Davies, K.J.A. HSP70 and other possible heat shock or oxidative stress proteins are induced in skeletal muscle, heart, and liver during exercise. Free Radic. *Biol. Med.* **1991**, *11*, 239–246. [CrossRef]
78. MacPherson, G.; Healey, C.S.; Teare, M.D.; Balasubramanian, S.P.; Reed, M.W.; Pharoah, P.D.P.; Ponder, B.A.J.; Meuth, M.; Bhattacharyya, N.P.; Cox, A. Association of a common variant of the CASP8 gene with reduced risk of breast cancer. *J. Natl. Cancer Inst.* **2004**, *96*, 1866–1869. [CrossRef]
79. Wolter, K.G.; Hsu, Y.-T.; Smith, C.L.; Nechushtan, A.; Xi, X.-G.; Youle, R.J. Movement of bax from the cytosol to mitochondria during apoptosis. *J. Cell Biol.* **1997**, *139*, 1281–1292. [CrossRef]

80. Gao, J.M.; Wu, L.; Chen, Y.P.; Zhou, B.; Guo, J.S.; Zhang, K.; Ouyang, W.J. Spatiotemporal distribution and risk assessment of organotins in the surface water of the Three Gorges Reservoir Region, China. *Chemosphere* **2017**, *171*, 405–414. [CrossRef]
81. Hai Le, L.T.; Takahashi, S.; Saeki, K.; Nakatani, N.; Tanabe, S.; Miyazaki, N.; Fujise, Y. High percentage of butyltin residues in total tin in the livers of cetaceans from Japanese coastal waters. *Environ. Sci. Technol.* **1999**, *33*, 1781–1786. [CrossRef]

current issues in molecular biology

MDPI

Article

Hepatitis B Virus Genotypes and Subgenotypes Circulating in Belarus

Vladimir Eremin [1,*], Fedor Karpenko [1], Igor Karpov [2], Valery Semenov [3] and Ina Oiestad [3]

1 Republican Scientific and Practical Center for Transfusiology and Medical Biotechnology, 220053 Minsk, Belarus; fedor-doc@tut.by
2 Department of Infectious Diseases, Belarusian State Medical University, 220116 Minsk, Belarus; igorkarpov57@outlooc.com
3 Department of Infectious Diseases, Vitebsk State Medical University, 210009 Vitebsk, Belarus; vmsemenov@mail.ru (V.S.); inaoiestad@gmail.com (I.O.)
* Correspondence: eremin.vf@gmail.com

Abstract: Approximately 800–900 new cases of chronic forms of hepatitis B are reported in Belarus annually. Compulsory vaccination, introduced to the country in the mid-90s, produces a certain positive effect on reducing the number of HBV cases. However, around 75,000 patients with chronic hepatitis B are estimated to live in the country at the moment. The main goal of this research was to determine the genotypes/subgenotypes of the hepatitis B virus and establish their role in maintaining the epidemic process in the country. Serological (CMIA, ELISA), molecular biological (PCR, sequencing), and bioinformatic (phylogenetic analysis) methods have been used to obtain results. As studies have shown, in 722 (81.7%) samples, genotype D has been determined; in 156 (17.7%), genotype A; in 3 (0.3%), genotype C; and only in 1 sample (0.1%), genotype B. In two cases (0.2%), a recombinant form of the hepatitis B virus has been detected. The epidemic process of hepatitis B in the country is supported mainly by the circulation of "local" variants of the virus. At the same time, there are occasional introductions of new genetic variants from outside of the country.

Keywords: HBV; genotypes/subgenotypes; sequencing; phylogenetic analysis

1. Introduction

Hepatitis B virus (HBV) infection remains a serious health concern despite the HBV vaccine being introduced in the late 1990s. Approximately two billion cases of the disease were reported worldwide from the time of HBV discovery to the beginning of the 21st century. More than 400 million of them subsequently became chronic HBV carriers [1]. Chronic HBV infection is considered a major cause of many liver-related medical complications such as hepatocellular carcinoma, liver cirrhosis, and liver failure [2–4]. Hepatitis B virus, like HIV, has three main transmission mechanisms: parenteral (through blood and its products and/or co-injection of drugs), vertical (from infected mother to child), and sexual transmission. HBV belongs to the Hepadnaviridae family, with a genome of approximately 3.2 kb in length. Due to the virus's complexity, the HBV genome contains an overlapping region that encodes four different genes (S, C, P, and X). The S gene encodes exclusively the HBV surface envelope protein in its long open reading frame [5]. The presence of start codons allows the gene splicing into small, medium and large polypeptides, which are typically used either as a single target gene or in combination with other HBV genes to detect HBV DNA [5]. HBV core antigen is commonly encoded by the C gene, whereas the X gene encodes protein X [6]. The P gene encodes a polymerase protein, an integral component of the reverse transcription process during HBV replication. The HBV replication cycle lacks proofreading properties, resulting in progeny with a high genomic variability [7].

Due to significant genetic diversity, HBV is categorized into ten genotypes (A–J) with 7.5% intergroup variation [8]. In addition to E and G, all genotypes are classified into 25 subgenotypes (A1–A3, B1–B5, C1–C6, D1–D6, and F1–F4) with 4% amino acid variability [9,10]. HBV genotypes are distributed differently depending on the geography: HBV-B, HBV-C, and HBV-E are most common in Oceania and East Asia, while HBV-E is most prevalent in Central and West Africa. HBV-F and HBV-H are found only in Alaska and Latin America. In contrast, HBV-D is a global pandemic. In Australia, Europe, Indonesia, North Africa, and Western Asia, HBV-D1 is the most common virus, while HBV-D2 is found in Albania, Japan, Malaysia, Northeastern Europe, Russia, and the UK [11–16]. A systematic genotype and subgenotype re-ranking of hepatitis B virus under a novel classification standard [16].

The progression and natural course of the disease vary among different HBV genotypes. These factors can complicate HBV infection treatment since the efficacy of known therapeutics turns out to be ineffective against certain genotypes and new genotypic variants [8,9]. Thus, there is a global exigency for constant updates in genotypic information and surveys of HBV-infected patients [15].

In Belarus, selective hepatitis B vaccination for epidemic indications has been organized since 1996 and included in the national preventive vaccination calendar since 2000. Vaccination of newborns, children aged 13 years, and certain risk groups (health care workers and persons in domestic contact with infected persons) has helped to reduce the incidence of hepatitis B infection. Vaccination against viral hepatitis B has made it possible to reduce the incidence of acute hepatitis B almost six-fold over the past 10 years (from 5.9 to 1.02 per 100,000 population) and to consider Belarus as a country with a low incidence of acute viral hepatitis B (less than 2% of the population). This means that vaccination coverage against hepatitis B has been over 97 percent for several decades. This is a good result. In 2022, a WHO-supported study of the child population in Belarus showed how many of those vaccinated had HBsAg, the surface antigen of viral hepatitis B—it was detected in one child only. The most affected age groups are teenagers and adults 15 to 40 years old. The incidence rate in this specific age group is significantly higher than in other age groups of the population of the Republic. Approximately 800–900 new cases of chronic viral hepatitis B are detected annually in the country, despite a significant decrease in the occurrence of acute viral hepatitis B. According to the epidemiological essays, 75,000 patients with chronic hepatitis B currently live in Belarus.

2. Materials and Methods

Samples. In this study, 884 blood serum/plasma samples were tested for HBV markers with CMIA/ELISA and PCR methods. The samples were obtained from patients and blood donors with acute and chronic HBV from different regions of the country. The age of the patients ranged from 4 years to 78 years. Furthermore, 428 samples were obtained from males and 374 from females. In 82 cases, gender and age were not specified.

The results of the epidemiologic investigation showed that out of 884 patient samples, 367 (41.5%) were infected through sexual contact, 211 (23.9%) indicated medical manipulation, 49 (5.5%) were children born to infected mothers, 36 (4.1%) were injecting drug users, 21 (2.4%) indicated household contacts, 18 (2.0%) indicated non-medical manipulation (tattoos, etc.), and 182 (20.60%) patients were unable to specify any route of infection. Serum/plasma samples were collected and analyzed during 2013–2024.

Serological tests. The commercial kit chemiluminescent microparticle immunoassay (CMIA) Architect HBsAg Qualitative II Reagent Kit (Abbott, Lake County, IL, USA) was applied to detect HBsAg. Positive results were confirmed using the Architect HBsAg Qualitative II Confirmatory Reagent Kit (Abbott, USA), as well as ELISA "Vectogep B-HBs antigen", manufactured by "Vector-Best", Novosibirsk, Russia.

PCR. The polymerase chain reaction was used to determine the DNA of the hepatitis B virus using test systems produced by Vector-Best: "RealBest HBV DNA quantitative" and "RealBest HIV RNA quantitative" in accordance with the manufacturer's instructions. Viral load in the samples ranged from 1.2×10^2 to 3.4×10^6 IU/mL HBV DNA.

DNA isolation. Viral RNA/DNA from the blood serum/plasma samples were isolated using the "Kit of reagents for the isolation of NK" (manufactured by "Vector-Best", Russia) and the "Kit of reagents for the isolation of RNA/DNA from clinical material" "RIBO-sorb", "RNA-prep", Moscow, Russia. The manufacturers' instructions were strictly followed.

Nested PCR was performed using previously described primer pairs p1/pR5 (p1: 5-CCTGCTGGTGGCTCCAGTTC-3 at nucleotide position 55—6, and primer pR5: 5-GGT TGC GTC AGC AAA CAC TTG-3 at position 1197—1178) and p4/pR2 (p4 5-CTC ACA ATA CCG CAG AGT CTA GAC T-3 at nucleotide position 230—254, and pR2: 5-AAA GCC CAA AAG ACC CAC AAT-3 at position 1017—997) [17] according to the following protocol in a volume of 25 µL: 2.5 µL 10× buffer + $MgCl_2$; 0.25 µL 25 mM dNTPs; 0.5 µL 10 µM p1; 0.5 µL 10 µM pR5; 0.8 µL 10 µM Taq polymerase; and 18.45 µL bdH_2O. The first round of PCR was carried out according to the following protocol: one denaturation cycle at 95 °C for 2 min and 30 cycles of amplification were performed with denaturation at 95 °C for 30 s, annealing at 53 °C for 30 s, and extension at 68 °C for 1 min, then one cycle at 68 °C for 5 min and storage at 10 °C. In the second round of PCR, the annealing temperature was 50 °C; the rest of the parameters were not changed. As a result, we obtained a fragment of about 900 base pairs.

Amplified DNA fragments were analyzed on a 2% agarose gel (Condalab, Madrid, Spain). Electrophoresis was carried out at 10 w/cm of gel in TRIS-acetate buffer, pH 8.0. DNA was visualized using the Vitran Photo gel documentation system (Biocom Company LLC, Stavropol, Russia). The fragment size was determined according to the molecular weight of a marker of 100–1000 bp (Fermentas, Vilnius, Lithuania).

The resulting DNA fragments were purified using the NimaGen ExS-Pure kit, Nijmegen, The Netherlands.

PCR sequencing was carried out using the second-round primer pair p4/pR2 in the volume of 10 µL according to the following protocol: 5× seqbuf-2 µL; BigDye Terminator v. 3.1 (Applied Biosystems)—1 µL; p4/pR4—2 µL; bdH_2O—3 bdH_2O; and DNA—2 µL.

Electrophoresis of HBV DNA fragments obtained and purified after PCR sequencing was performed on an AB 3500 genetic analyzer, USA.

Phylogenetic analysis of the obtained HBV DNA fragments was conducted using the computer programs Sequencing Analysis v.6, BioEdit, SeqScape v3, MEGAX, and Genious v.8.1. Phylogenetic trees were built using the ML (maximum likelihood) algorithm in the PHYML V.3.0 program. The SH-aLRT test was performed to calculate the statistical significance of the clusters. Clusters with a support node ≥ 0.9 were considered reliable.

Mutations in the HBV genome in the S and P regions were determined using the following programs: https://www.geno2pheno.org, http://www.hiv-grade.de/hbv_grade/deployed/grade.pl?program=hbvalg, and https://hivdb.stanford.edu/HBV/HBVseq, accessed on 22 October 2024.

3. Results

Our studies showed the following: 722 (81.7%) out of all 884 sequenced and analyzed samples had genotype D; 156 (17.7%)—Genotype A; 3 (0.3%)—Genotype C; and only 1 (0.1%)—Genotype B. For two cases (0.2%), a recombinant form of the hepatitis B virus was detected. Genotype D was found to be represented by subgenotypes D1 (145/20.1%, ±3.26), D2 (371/51.4%, ±5.3), D3 (201/27, 8%, ±3.5), and D4 (5/0.7%, ±0.3); A-A2 (156/17.7%, ±2.9); C1-1 (0.11%) and C2-2 (0.23%); B-B4 (0.11%); and recombinant forms C2/D (0.11%) and A2/C2 (0.11%) were detected (Figure 1).

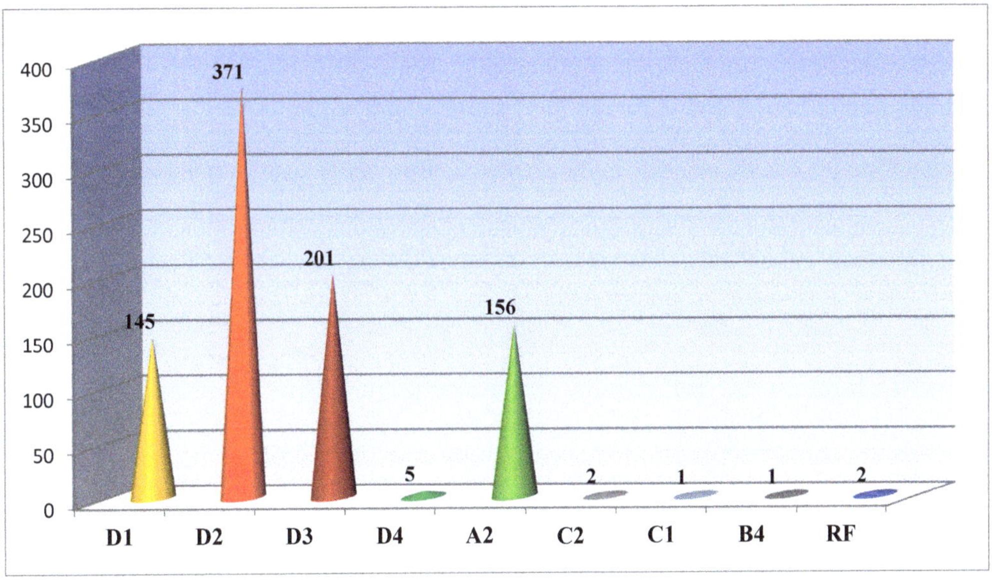

Figure 1. HBV subgenotypes identified in Belarus.

In 12 cases, viruses with mutations to reverse transcriptase inhibitors Lamivudine, Zeffix®, Telbivudine, Tyzeka®, and Sebivo® and partial resistance to Entecavir and Baraclude® were identified. Moreover, in four cases, resistance mutations were identified in subgenotypes A2 and D2; in three cases—in the genome of subgenotype D1; and in one case—on subgenotype D3. Mutations were most often recorded at positions 180 M and 204 V—in eight cases, 204 I—in three cases, 80 I and 173 L—two times each, and 80 V and 181 T in one case (Table 1).

Table 1. Resistance of mutations identified in different HBV subgenotypes.

N	Subgenotypes	Mutations
1	A2	80 V−1; **180 M**-3; **181 T**-1; **204 V**-3;
2	D1	173 L-1; **180 M**-3; **204 V**-3;
3	D2	80 I-2; 173 L-1; **180 M**-2; **204 V**-2; **204 I**-2;
4	D3	**204 I**-1;

Determining mutations is very important for infectious disease doctors, since substitutions at positions 180 M, 181 T, 204 V, and 204 I lead to resistance to Lamivudine (Zeffix), Adefovir (Hepsera), and Telbivudine (Tyzeka, Sebivo) and require a change in treatment regimen.

Eleven clusters were identified while analyzing samples of subgenotype D1, Figure 2. Five large clusters (1–5) containing 14 or more HBV DNA sequences, as well as small clusters 9–11, contained samples of HBV detected in citizens permanently residing in Belarus. Clusters 2, 4, and 6–8 contained HBV DNA sequences detected in patients from Ukraine (cluster 4), Pakistan (cluster 6), Turkmenistan and Afghanistan (clusters 2 and 7), and Azerbaijan (cluster 8).

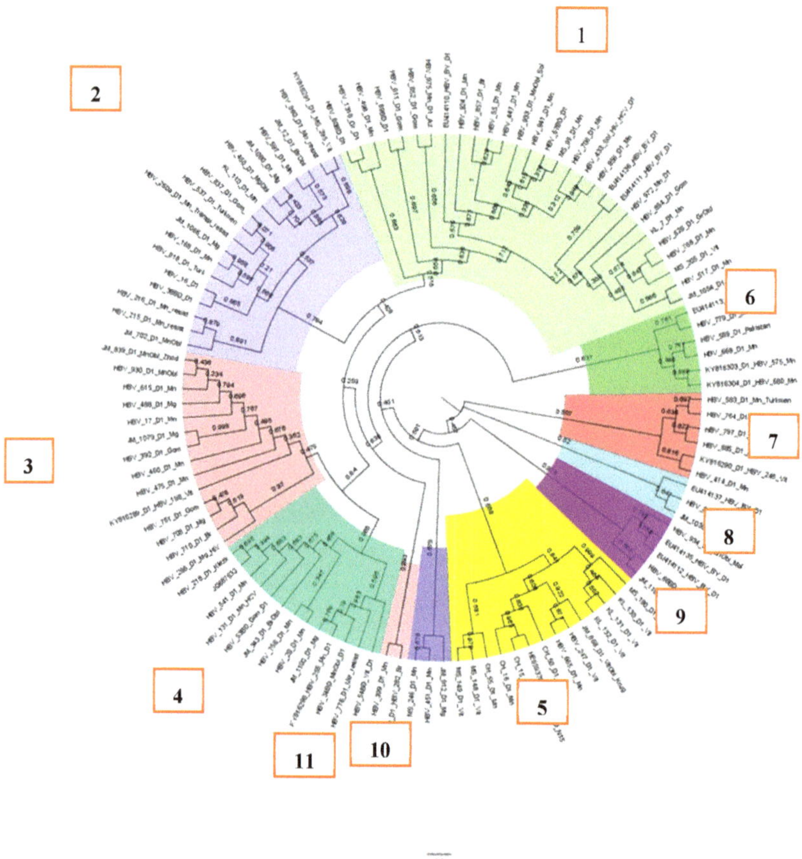

Figure 2. Phylogenetic analysis of HBV subgenotype D1.

Overall, the samples, as can be seen in Figure 2, were not related to each other, as the support node was less than 0.9. At the same time, in cluster 5, samples MS_148_D1_Vit and MS_149_D1_Vit had a support node of 0.976, indicating a possibly single source of origin of the virus in the patients. Indeed, the samples were from a mother and her vertically infected child.

The D2 subgenotype consisted of nine clusters, and it should be noted that more than 80% of all HBV DNA sequences were concentrated in clusters 1–3, which contained 35 or more sequences (Figure 3).

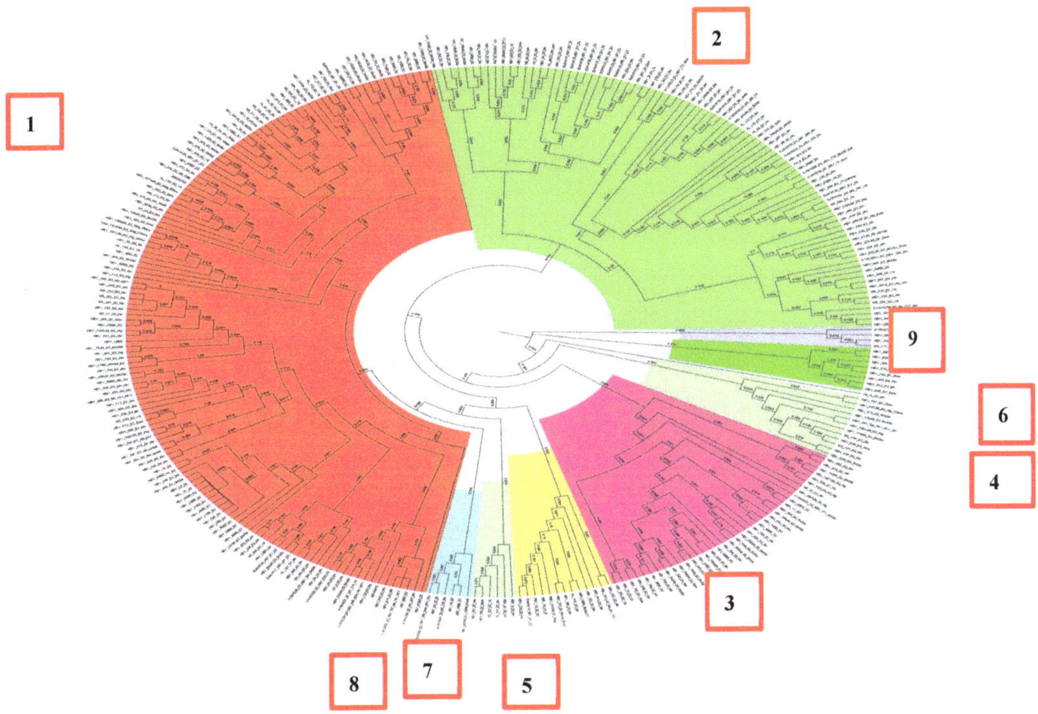

Figure 3. Phylogenetic analysis of HBV subgenotype D2.

In this study, 20% of the remaining samples were concentrated in the remaining six clusters. It is important to note that the D2 subgenotype was exclusively formed by samples obtained from patients permanently residing in Belarus. Most of the samples were unrelated and had a support node less than 0.9.

HBV DNA sequences of subgenotype D3 formed eight clusters and consisted mainly of samples obtained from patients living permanently in Belarus (Figure 4). The first four clusters contained almost 90% of all samples. The seventh cluster was formed by samples obtained from patients from one family (a mother and two children) who came to Belarus for treatment from the Kyrgyz Republic.

HBV DNA sequences of subgenotype A2 formed at least six clusters, with the first three accounting for almost 90% of all samples (Figure 5). All analyzed sequences were obtained from patients permanently residing in Belarus.

All five HBV DNA sequences of subgenotype D4 were obtained from epidemiologically unrelated patients from Belarus. Samples of subgenotypes C1 and C2 were identified in adult patients and a child from China. The subtype B4 sample was revealed in a student from Vietnam who came to study in our country.

Finally, both samples with recombinant forms of HBV were identified in residents of Belarus.

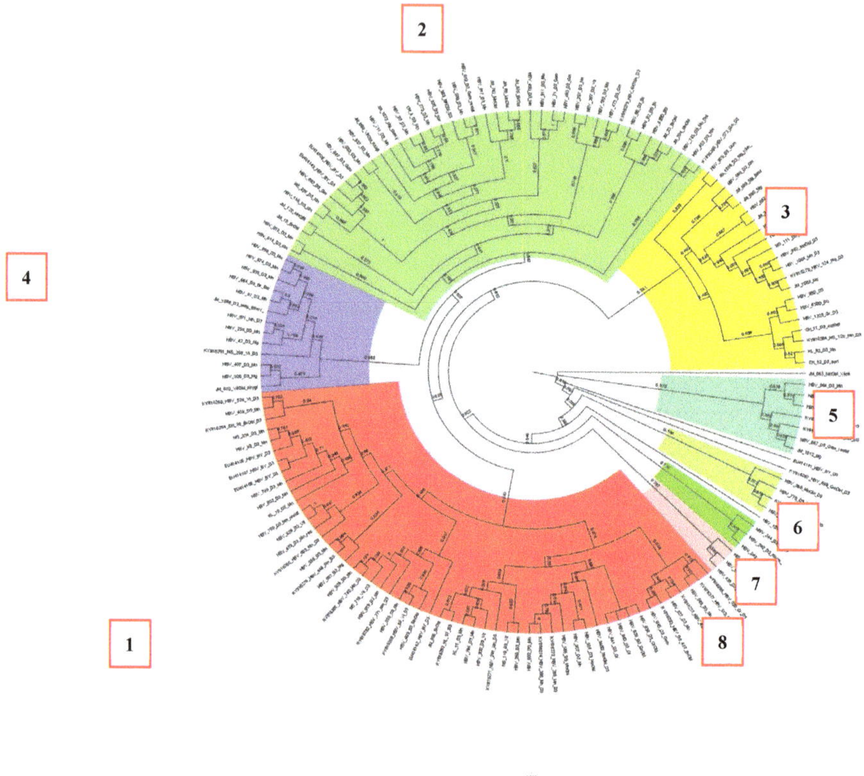

Figure 4. Phylogenetic analysis of HBV subgenotype D3.

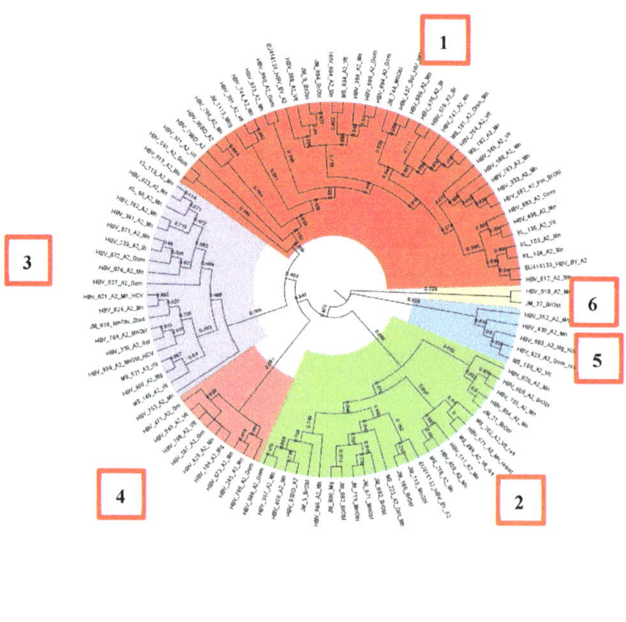

Figure 5. Phylogenetic analysis of HBV subgenotype A2.

4. Discussion

Approximately 800–900 new cases of chronic forms of hepatitis B are reported in Belarus annually. Compulsory vaccination, introduced to the country in the mid-90s, produces a certain positive effect on reducing the number of HBV cases. However, around 75,000 patients with chronic hepatitis B are estimated to live in the country at the moment.

Our previous study showed that subgenotype D2 (55.8%) was dominant in Belarus, followed by D3 (18.3%), D1 (11.6%), and A2 (11.6%) [18]. The conducted studies show that the structure of genotypes/subgenotypes in the country has actually remained unchanged. Genotype D still dominates in the country and accounts for 81.7% of all analyzed cases. Genotype D was found to be represented by subgenotypes D1 (145/20.1%), D2 (371/51.4%), D3 (201/27.8%), and D4 (5/0.7%); A-A2; C-C1 and C2; B-B4; and recombinant forms C2/D and A2/C2 were detected. Possible causes of the infection in patients were determined, i.e., they occurred due to the circulation of "domestic" variants of the virus and/or the introduction of the virus from outside. The findings obtained show that the epidemic transmission of viral hepatitis B in the country is maintained mainly due to the circulation of local variants of the virus. Subgenotypes C1 and C2, as well as B4, were identified in patients from China and Vietnam who came to Belarus to study and work. Two recombinant forms, C2/D and A2/D, were identified in the residents of our country.

The large number of clusters identified in groups of different subgenotypes may be due to repeated introductions of the virus into the country. This occurred both in pre-vaccine times and at present, as evidenced by samples from Asian countries.

Comparing the distribution of different HBV genotypes in neighboring countries and even in different regions of the same country, we may conclude that they vary. For example, genotype D significantly dominates in the European part of the Russian Federation, while, for example, in Yakutia, genotype A occupies one of the leading positions (36.4% A and 58.6% D) [19]. In Asia, particularly in its eastern part, genotypes B and C dominate, and the C2 subgenotype is epidemic in China [20,21]. In Western Europe, in particular in Portugal, HBV genotype A (HBV/A) was the most prevalent genotype (41.5%), followed by D [HBV/D; (33.8%)] and E [HBV/E; (24.6%)]. Subgenotypes A1 (HBV/A1) and D4 (HBV/D4) were almost equally prevalent, with 23.1% and 22.3%, respectively, followed by A2 (HBV/A2) with 16.2% and D3 (HBV/D3) with 11.5% [22].

On the African continent, genotype E was most represented and significantly outnumbered all other genotypes. By region, genotype A had the highest cumulative prevalence in eastern and southern Africa, E in western Africa, and D in northern Africa ($p < 0.0001$). Regarding the new genotypes B and C on the African continent, genotype B was significantly higher in southern Africa than C ($p < 0.001$). In contrast, genotype C was significantly higher in eastern Africa than in western Africa ($p < 0.0001$). A1 and D/E were the most diverse subgenotypes and mixtures of genotypes, respectively [23].

According to the European Centre for Disease Prevention and Control, 34 (2.6 per 100,000 population) cases of viral hepatitis B were registered in Estonia in 2022, of which 2 were acute (0.2 per 100,000 population) and 32 chronic (2.4 per 100,000 population). In neighboring Latvia during the same period, 191 (10.2 per 100,000 population) cases of viral hepatitis B were registered, of which 17 (0.9 per 100,000 population) were acute cases and 174 (9.3 per 100,000 population) were chronic viral (1.0 per 100,000 population) cases of viral hepatitis B, of which 7 (0.2 per 100,000 population) were acute and 20 (0.7 per 100,000 population) were chronic viral hepatitis B cases [24].

The Republic of Belarus is geographically located in the center of Europe. The residents of the country are actively traveling, and many people from other countries relocate to Belarus for work and study. This fact can easily explain the introduction of the new variants of HBV to the Republic.

The results of this study contribute valuable information on the molecular genetic characteristics of HBV prevailing in the Republic of Belarus. Possible routes of the introduction of the new variants of HBV have been identified. The main mutations in the P region of the virus genome leading to resistance to antiviral drugs have been detected. Identification of the mutations at positions 181 T, 204 V, and 204 I, which determine resistance not only to Lamivudine (Zefffix) but also to Adefovir (Hepsera) and Telbivudine (Tyzeka and Sebivo), made it possible to use Tenofovir DF for treatment of the patients. The compensatory mutations 80 V, 173 L, and 80 I have been found in the complex with the mutations leading to virus resistance to antiviral drugs.

5. Conclusions

Summarizing the results obtained, we can once again note that the epidemic process of viral hepatitis B in Belarus is maintained mainly due to the circulation of local variants of the virus. Even in spite of vaccine prophylaxis, about 1000 new cases of chronic viral hepatitis B are detected annually in the country. The detection of new cases, for example, by blood donor testing, may indirectly confirm this. At the same time, as our results show, new variants of hepatitis B virus are entering the country from outside the country with workers and students, for example, from Southeast Asian countries, as well as with citizens of our country who worked outside Belarus. Currently, these subgenotypes of hepatitis B virus are not involved in the epidemiologic process and are detected in individual patients, but their active inclusion in the epidemiologic process cannot be excluded in the future.

Author Contributions: Conceptualization, V.E.; Methodology, V.E. and I.O.; Software, F.K. and V.S.; Validation, I.K. and I.O.; Formal analysis, I.K.; Investigation, V.E.; Resources, F.K and V.S. All authors have read and agreed to the published version of the manuscript.

Funding: This research received no external funding.

Institutional Review Board Statement: Not applicable.

Informed Consent Statement: Not applicable.

Data Availability Statement: Data are contained within the article.

Acknowledgments: We would like to show our gratitude to the Abbott Transfusion Medicine Medical affairs team for providing financial support for this publication.

Conflicts of Interest: The authors declare no conflict of interest.

References

1. Thun, M.J.; DeLancey, J.O.; Center, M.M.; Jemal, A.; Ward, E.M. The global burden of cancer: Priorities for prevention. *Carcinogenesis* **2010**, *31*, 100–110. [CrossRef]
2. Philips, C.A.; Ahamed, R.; Abduljaleel, J.K.; Rajesh, S.; Augustine, P. Critical Updates on Chronic Hepatitis B Virus Infection in 2021. *Cureus* **2021**, *13*, e19152. [CrossRef]
3. Kao, J.-H.; Chen, P.-J.; Lai, M.-Y.; Chen, D.-S. Hepatitis B genotypes correlate with clinical outcomes in patients with chronic hepatitis B. *Gastroenterology* **2000**, *118*, 554–559. [CrossRef]
4. Aghakhani, A.; Hamkar, R.; Zamani, N.; Eslamifar, A.; Banifazl, M.; Saadat, A.; Sofian, M.; Adibi, L.; Irani, N.; Mehryar, M. Hepatitis B virus genotype in Iranian patients with hepatocellular carcinoma. *Int. J. Infect. Dis.* **2009**, *13*, 685–689. [CrossRef]
5. Pollicino, T.; Cacciola, I.; Saffioti, F.; Raimondo, G. Hepatitis B virus PreS/S gene variants: Pathobiology and clinical implications. *J. Hepatol.* **2014**, *61*, 408–417. [CrossRef]
6. Ie, S.I.; Thedja, M.D.; Roni, M.; Muljono, D.H. Prediction of conformational changes by single mutation in the hepatitis B virus surface antigen (HBsAg) identified in HBsAg-negative blood donors. *Virol. J.* **2010**, *7*, 326. [CrossRef]
7. Beck, J.; Nassal, M. Hepatitis B virus replication. *World J. Gastroenterol.* **2007**, *13*, 48–64. [CrossRef]
8. Lin, C.-L.; Kao, J.-H. The clinical implications of hepatitis B virus genotype: Recent advances. *J. Gastroenterol. Hepatol.* **2011**, *26* (Suppl. S1), 123–130. [CrossRef]

9. Zhu, S.S.; Zhang, H.F.; Wang, H.B.; Dong, Y.; Chen, D.; Jia, W.; Gan, Y.; Chen, J. Relation of viral genotypes to clinical features in children with chronic hepatitis B. *Chin. J. Exp. Clin. Virol.* **2008**, *22*, 192–194.
10. Shamseer, L.; Moher, D.; Clarke, M.; Ghersi, D.; Liberati, A.; Petticrew, M.; Shekelle, P.; Stewart, L.A. Preferred reporting items for systematic review and meta-analysis protocols (PRISMA-P) 2015: Elaboration and explanation. *BMJ* **2015**, *349*, g7647. [CrossRef]
11. George, B.J.; Aban, I.B. An application of meta-analysis based on DerSimonian and Laird method. *J. Nucl. Cardiol.* **2016**, *23*, 690–692. [CrossRef]
12. Fletcher, J. What is heterogeneity and is it important? *BMJ* **2007**, *334*, 94–96. [CrossRef]
13. Higgins, J.P.T.; Thompson, S.G.; Deeks, J.J.; Altman, D.G. Measuring inconsistency in meta-analyses. *BMJ* **2003**, *327*, 557–560. [CrossRef]
14. Wallace, B.C.; Dahabreh, I.J.; Trikalinos, T.A.; Lau, J.; Trow, P.; Schmid, C.H. Closing the Gap between Methodologists and End-Users: R as a Computational Back-End. *J. Stat. Softw.* **2012**, *49*, 2–15. [CrossRef]
15. Munn, Z.; MClinSc, S.M.; Lisy, K.; Riitano, D.; Tufanaru, C. Methodological guidance for systematic reviews of observational epidemiological studies reporting prevalence and cumulative incidence data. *Int. J. Evid. Based Healthc.* **2015**, *13*, 147–153. [CrossRef]
16. Yin, Y.; He, K.; Wu, B.; Xu, M.; Du, L.; Liu, W.; Liao, P.; Liu, Y.; He, M. A systematic genotype and subgenotype re-ranking of hepatitis B virus under a novel classification standard. *Heliyon* **2019**, *5*, e02556. [CrossRef]
17. Serfaty, L.; Thabut, D.; Zoulim, F.; Andreani, T.; Eres, O.C.; Carbonell, N.; Loria, A.; Poupon, R. Sequential treatment with lamivudine and interferon monotherapies in patients with chronic hepatitis B not responding to interferon alone: Results of a pilot study. *Hepatology* **2001**, *34*, 573–577. [CrossRef]
18. Olinger, C.M.; Lazouskaya, N.V.; Eremin, V.F.; Muller, C.P. Multiple genotypes of hepatitis B and C viruses in Belarus: Similarities with Russia and European influences. *Clin. Microbiol. Infect.* **2008**, *14*, 575–581. [CrossRef]
19. Karlsen, A.A.; Kyuregyan, K.K.; Isaeva, O.V.; Kichatova, V.S.; Asadi Mobarkhan, F.A.; Bezuglova, L.V.; Netesova, I.G.; Manuylov, V.A.; Pochtovyi, A.A.; Gushchin, V.A.; et al. Different evolutionary dynamics of hepatitis B virus genotypes A and D, and hepatitis D virus genotypes 1 and 2 in an endemic area of Yakutia, Russia. *BMC Infect. Dis.* **2022**, *22*, 452. [CrossRef]
20. Bello, K.E.; Mat Jusoh, T.N.A.; Irekeola, A.A.; Abu, N.; Mohd Amin, N.A.Z.; Mustaffa, N.; Shueb, R.H. A Recent Prevalence of Hepatitis B Virus (HBV) Genotypes and Subtypes in Asia: A Systematic Review and Meta-Analysis. *Healthcare* **2023**, *11*, 1011. [CrossRef]
21. Sun, B.; Andrades Valtueña, A.; Kocher, A.; Gao, S.; Li, C.; Qiu, Y.; Zhang, Q.; Ma, J.; Chen, S.; Xiao, X.; et al. Origin and dispersal history of Hepatitis B virus in Eastern Eurasia. *Nat. Commun.* **2024**, *15*, 2951. [CrossRef]
22. Marcelino, R.; Ezeonwumelu, I.J.; Janeiro, A.; Mimoso, P.; Matos, S.; Briz, V.; Pimentel, V.; Pingarilho, M.; Tato Marinho, R.; Maria Marcelino, J.; et al. Phylogeography of hepatitis B virus: The role of Portugal in the early dissemination of HBV worldwide. *PLoS ONE* **2022**, *17*, e0276618. [CrossRef]
23. Kafeero, H.M.; Ndagire, D.; Ocama, P.; Kato, C.D.; Wampande, E.; Walusansa, A.; Kajumbula, H.; Kateete, D.; Ssenku, J.E.; Sendagire, H. Mapping hepatitis B virus genotypes on the African continent from 1997 to 2021: A systematic review with meta-analysis. *Sci. Rep.* **2023**, *13*, 5723. [CrossRef]
24. Hepatitis, B. Annual Epidemiological Report for 2022. Available online: https://www.ecdc.europa.eu/sites/default/files/documents/AER%20HEPB%202022_0.pdf (accessed on 22 October 2024).

current issues in
molecular biology

MDPI

Article

Amelioration of Acetaminophen-Induced Hepatic Oxidative Stress and Inflammation by RNAi Targeting *Cyp2e1* In Vivo

Wenwen Liu, Liwen Huan, Cai Zhang, Runting Yin, Zhen Ouyang and Yuan Wei *

School of Pharmacy, Jiangsu University, Zhenjiang 212013, China; wwen0239@163.com (W.L.); hlw102077@163.com (L.H.); caizhang@ujs.edu.cn (C.Z.); yinrunting@126.com (R.Y.); zhenouyang@ujs.edu.cn (Z.O.)
* Correspondence: ywei@ujs.edu.cn

Abstract: The overdose of acetaminophen (APAP) has become the leading cause of acute liver failure in the United States and some Western countries. As a principal member of the cytochrome P450 enzymes (CYPs), CYP2E1 is a vital enzyme in regard to the production of toxic APAP metabolites and in the development of APAP-induced liver injury (AILI). In this study, we investigated the therapeutic effects and mechanisms of lipid nanoparticle (LNP)-based delivery of small interfering RNA targeting *Cyp2e1* (si-*Cyp2e1* LNPs) on AILI in mice. C57BL/6J male mice were injected with 300 mg/kg APAP to establish an AILI model, and si-*Cyp2e1* LNPs were administered via the tail vein. The results showed that the levels of serum alanine aminotransferase and aspartate aminotransferase were lower than those in APAP mice after treatment with si-*Cyp2e1* LNPs immediately. Moreover, si-*Cyp2e1* LNPs significantly inhibited liver necrosis and oxidative stress in APAP mice. RNA sequencing revealed that si-*Cyp2e1* LNPs exerted regulatory effects on pathways and genes related to peroxisome proliferator-activated receptor (PPAR). Consistent with this finding, we also proved that si-*Cyp2e1* LNPs markedly regulated the expressions of genes involved in the PPAR signaling pathway (CYP4A, PPARα, FABP 1, and CD36) in APAP mice, as well as inflammatory factors (*Il-6*, *Il-1β*, and *Tnf-α*). These findings suggested that si-*Cyp2e1* LNPs may alleviate APAP-induced oxidative stress and inflammation by regulating lipid metabolism via PPAR-related pathways.

Keywords: acetaminophen; hepatotoxicity; *Cyp2e1*; RNAi; lipid nanoparticle; oxidative stress; inflammation; lipid metabolism

1. Introduction

Drug-induced liver injury (DILI), characterized by hepatic damage caused by drugs or their metabolites, is one of the most common and severe adverse reactions. Many drugs, including anticancer, anti-tuberculosis, and non-steroidal anti-inflammatory drugs (NSAIDs), can cause DILI to varying degrees [1,2]. Acute liver injury induced by excess acetaminophen (APAP; Supplementary Figure S1A), which is the most commonly used medication for pain and fever, is the leading cause of acute liver failure in the United States and the United Kingdom [3,4].

APAP-induced liver injury (AILI) is initiated by the conversion of N-acetyl-p-benzoquinone imine (NAPQI), which is primarily metabolized by cytochrome P450 enzymes (CYPs) [5]. High doses of APAP and NAPQI deplete glutathione (GSH). Excessive NAPQI can combine with mitochondrial proteins to form APAP protein adducts (APAP-ADs), thus triggering electron leakage, leading to the formation of mitochondrial superoxide, initiating oxidative stress and activating redox-sensitive mitogen-activated protein kinase (MAPK), which ultimately leads to the phosphorylation of c-jun N-terminal kinase (JNK) [6]. Moreover, peroxynitrite, generated by the rapid reaction of superoxide radicals with nitric oxide, further amplifies mitochondrial dysfunction and hepatocyte necrosis [7]. *N*-acetyl-*L*-cysteine (NAC; Supplementary Figure S1B) is the only clinically approved antidote for APAP overdose, which works by restoring GSH levels in the early phase [8]. However, the mortality rate of the overall cases of APAP overdose has changed slightly over the past 20 years, despite the presence of NAC [9]. Therefore, novel alternatives to NAC or supplements for treating APAP overdose need to be developed.

CYP2E1 has been confirmed as a vital enzyme responsive to NAPQI generation and to the development of AILI [10]. Enhanced sensitivity to APAP hepatotoxicity and oxidative stress is observed after CYP2E1 induction [11,12]. Oxidative stress and liver injury are markedly ameliorated after the inhibition or knockout of CYP2E1 in many hepatic diseases [13–15]. Moreover, 4-methylpyrazole, an inhibitor of CYP2E1, is already undergoing clinical trials as a supplement to NAC for the treatment of APAP overdose [9], demonstrating the potential of drugs targeting CYP2E1 inhibition.

RNA interference (RNAi) is a new clinical therapy used to treat diseases by silencing specific target genes and blocking protein expression in vivo. However, it is difficult for small interfering RNA (siRNA) to enter cells because of their large molecular size, strong hydrophilicity, and anionic charge. Therefore, different drug carriers are required for delivery [16]. Lipid nanoparticle (LNP) delivery systems are characterized by high safety, low immunogenicity, and strong targeting abilities. Our previous research showed that an LNP-delivered siRNA targeting *Cyp2e1* (si-*Cyp2e1* LNPs) could effectively inhibit the development of chronic and subacute alcoholic liver disease in mouse models [17,18]. However, the effect of si-*Cyp2e1* LNPs in acute liver injury induced by APAP overdose in mice has not yet been investigated.

In this study, the therapeutic potential of si-*Cyp2e1* LNPs in a mouse model of AILI was demonstrated, and the underlying mechanisms were explored through transcriptomics.

2. Materials and Methods

2.1. Chemicals

APAP and NAC were purchased from Yuanye Biotechnology (Shanghai, China). (6Z, 9Z, 28Z, 31Z)-heptatriacont-6,9,28,31-tetraene-19-yl 4-(dimethylamino) butanoate (Dlin-MC3-DMA) and cholesterol were purchased from AVT Pharmaceutical Tech Co., Ltd. (Shanghai, China). Distearoyl phosphatidylcholine (DSPC) was purchased from Macklin Biochemical Technology (Shanghai, China). 1,2-dimyristoyl-sn-glycero-3-phosphoethanolamine-*N*-[methoxy(polyethylene glycol)-2000] (C14-PEG2000) was obtained from Sigma Aldrich (St. Louis, MO, USA).

2.2. Preparation and Characterization of LNPs

As previously described, an siRNA targeting mouse *Cyp2e1* was designed (Supplementary Table S1) [17]. Single-stranded RNAs were synthesized and then annealed into double-stranded siRNAs by GenScript (Nanjing, China). LNPs were prepared as previously described [19]. Briefly, siRNAs were dissolved in 10 mmol/L sodium acetate buffer, and Dlin-MC3-DMA, cholesterol, DSPC, and C14-PEG2000 were dissolved and mixed in ethanol at a molar ratio of 50/38.5/10/1.5, respectively. Then, the lipid and siRNAs were mixed at a volume and velocity ratio of 1/3 through a microfluidic device (FluidicLab, Shanghai, China). Lastly, the mixture was dialyzed against Dulbecco's phosphate-buffered saline (DPBS; pH 7.4) for 2 h using a 20,000 molecular weight cut-off (MWCO) dialysis bag at 25 °C. The LNPs were characterized using a NanoBrook 90Plus PALS system (Brookhaven, Holtsville, NY, USA), and the encapsulation efficiency was measured according to established procedures [20]. Morphology was observed by transmission electron microscopy (TEM) (Hitachi, Tokyo, Japan).

2.3. Animal Experiments

Healthy 6–8-week-old male C57BL/6J mice (18–22 g) were sourced from and housed in the Laboratory Animal Research Center of Jiangsu University (license nos. SCXK (Su) 2023-0017 and SYXK (Su) 2023-0081). Experimental mice were housed in a standard environment, with a room temperature of 20–24 °C, relative humidity of 50–60%, and 12 h light/dark cycles. All animal experiments were approved by the Institutional Animal Care and Use Committee of Jiangsu University (approval number: UJS-IACUC-2023122802).

In this study, si-*Cyp2e1* was generated by *Cyp2e1* knockdown using RNAi. The mice were injected with 0.5 mg/kg si-*Cyp2e1* LNPs or si-Control LNPs dissolved in DPBS via the tail vein to confirm their silencing effect [18]. The mice injected with the same dose of DPBS served as a negative control group. Each group consisted of four mice, and each mouse received a single injection. The mice were euthanized via intraperitoneal administration of sodium pentobarbital 24 h after tail vein injection [17]; then, the livers were harvested. Subsequently, *Cyp2e1* expression in the liver was assessed.

The AILI mouse model used in this study was modified from a previous report [21]. The experimental mice were randomly divided into five different groups, including the control group (Ctrl), APAP group (APAP), APAP + si-*Cyp2e1* LNPs group (si-*Cyp2e1* LNPs), APAP + si-Control LNPs group (si-Control LNPs), and APAP + NAC group (NAC) ($n = 4$). All the experimental animals were fasted for 12 h before drug administration. In addition to the Ctrl group, the mice in the other four groups received intraperitoneal injections of 300 mg/kg APAP dissolved in hot saline. The mice in the Ctrl group were administered the same volume of normal saline. Subsequently, the mice in the si-*Cyp2e1* and si-Control treatment groups were immediately injected with 0.5 mg/kg LNPs through the tail vein. The mice in the positive drug treatment group received immediate intragastric administration of NAC at a dose of 200 mg/kg. A total of 24 h after model establishment, all the mice were anesthetized by intraperitoneal injection of pentobarbital sodium and subsequently euthanized after blood collection. Complete fresh liver tissues were collected, one lobe of liver tissue was soaked in 4% paraformaldehyde fixing solution, and the remaining liver tissues were stored in a refrigerator at −80 °C for further use.

To investigate the effect of delayed treatment, all the mice were divided into five groups: Ctrl, APAP, and three treatment groups ($n = 4$). The administration of 300 mg/kg APAP and normal saline was consistent with the experiment above. In contrast to the previous, 0.5 mg/kg LNPs or 200 mg/kg NAC was administered once, 2 h after APAP overdose. After 24 h of modeling, blood and fresh liver samples were collected, and one lobe of the liver was soaked in paraformaldehyde.

A previous study demonstrated that a 2 h delayed treatment with 100 mg/kg NAC every 12 h for three consecutive days delayed the recovery of AILI mice in the late stage [22]. To determine whether delayed treatment with si-*Cyp2e1* LNPs affects liver recovery of AILI mice, the modeling period was extended to 72 h. All the mice were divided into four groups: Ctrl, APAP, si-*Cyp2e1* LNPs delayed treatment, and NAC delayed treatment ($n = 4$). The administration of 300 mg/kg APAP and saline was consistent with the above experiments, and the mice were treated 2 h after APAP overdose. The si-*Cyp2e1* LNPs-treated mice received a single tail vein injection of 0.5 mg/kg LNPs, which was previously shown to have a significant gene silencing effect for 72 h [23]. The NAC treatment group received 100 mg/kg NAC by oral gavage every 12 h. A total of 72 h after APAP injection, the livers were collected for subsequent pathological analysis.

2.4. Serum and Liver Analysis

The liver index was calculated as the ratio of liver weight to body weight. Serum was isolated by centrifuging whole blood, and the homogenate was obtained by mixing and grinding the tissue with a buffer solution. Levels of serum alanine aminotransferase (ALT), aspartate aminotransferase (AST), hepatic glutathione (GSH), malondialdehyde (MDA), and superoxide dismutase (SOD) were determined according to the instruction of kits purchased from Nanjing Jiancheng Bioengineering Institute (Nanjing, China). The levels of oxidative stress indicators in the liver were normalized to the corresponding tissue protein content, which was measured using the BCA Protein Assay Kit (GlpBio, Montclair, CA, USA).

In order to determine reactive oxygen species (ROS) levels, fresh liver tissue was homogenized on ice for 30 s after adding the buffer. The samples were centrifuged once ($100\times g$, 4 °C, 10 min), and the supernatant was removed. According to the instructions of the ROS activity detection kit (BestBio Technology, Shanghai, China), the supernatant was mixed with the O06 probe and incubated at 37 °C for 30 min, and the fluorescence intensity of the product was detected using a fluorescent enzyme spectrometer (excitation wavelength 488 nm, emission wavelength 526 nm). The corresponding tissue protein concentration was determined, and the ratio of fluorescence intensity to tissue protein content was used as the level of reactive oxygen species in the tissue for data analysis.

2.5. Liver Pathology

The liver tissue was fixed in 4% paraformaldehyde, embedded in paraffin, and sectioned (4 μm). After dewaxing and dehydration, the slices were stained with hematoxylin and eosin (H&E; Servicebio, Wuhan, China) for 4 min and 15 s, respectively. Finally, the sections were sealed with neutral gum after dehydration. Representative images were captured using a pathological slide scanner (3DHISTECH, Budapest, Hungary).

2.6. RNA Sequencing and Enrichment Analysis

Total RNA was extracted using TRIzol (Thermo Fisher Scientific, Waltham, MA, USA), and the concentration was determined. RNA enrichment and sequencing library construction were performed using mRNA capture and fast RNA library kits (Tiangen Biotech, Beijing, China). Qualified libraries were sequenced on a NovaSeq6000 platform (Illumina, San Diego, CA, USA). The raw data were filtered to obtain high-quality clean reads, which were mapped to the reference genome using HISAT2 [24]. Gene expression quantification was performed using StringTie [25]. Limma was used to screen differentially expressed genes (DEGs) in the APAP and Ctrl groups, as well as in the si-*Cyp2e1* LNPs-treated and APAP groups [26]. The DEGs were screened based on the criteria $p < 0.05$ and $\log_2 | \text{Fold Change} | > 0.585$. Finally, enrichment analysis of DEGs was performed using Gene Ontology (GO), Kyoto Encyclopedia of Genes and Genomes (KEGG), and gene set enrichment analysis (GSEA). The enrichplot package in R v4.4.2 was used for visualization, with a Q value < 0.05 as the screening criterion.

2.7. Quantitative Real-Time Polymerase Chain Reaction (qPCR)

RNA reverse transcription and qPCR were performed according to a previous study [17]. Each program ran 45 cycles and was carried out three times. Gene expression levels were calculated according to the $2^{-\Delta\Delta Ct}$ method, and glyceraldehyde-3-phosphate dehydrogenase (*Gapdh*) was used as an internal reference. The PCR primers (Supplementary Table S2) were synthesized by Sangon Biotech (Shanghai, China).

2.8. Western Blotting

The mouse livers were lysed by RIPA buffer to obtain total proteins. The proteins were quantified using a BCA Protein Assay Kit (GK10009, GlpBio). The extracted proteins and loading buffer (5×) were mixed and dissolved in a ratio of 4:1. The proteins (20 μg) were resolved by SDS-PAGE and transferred to a PVDF membrane, which was closed for 1.5 h using 5% skim milk powder at room temperature. Next, the membrane was separately probed with the primary antibodies CYP2E1 (67263-1-Ig, Proteintech, Rosemont, Il, USA, 1:1000), CYP4A (sc-374537, Santa Cruz, Dallas, TX, USA, 1:1000), CD36 (sc-7309, Santa Cruz, 1:1000), L-FABP (sc-374537, Santa Cruz, 1:1000), and GAPDH (10494-1-AP, Proteintech, 1:1000) overnight at 4 °C, and the non-specifically bound proteins were washed off. Subsequently, the bands were incubated with the corresponding secondary antibodies (Beyotime Biotechnology, Shanghai, China, 1:3000) for 1 h. After washing off the unbound proteins, the membrane was visualized using an enhanced chemiluminescence kit (GK10008, GlpBio) and a ChemiDoc XRS imaging system (Bio-Rad, Hercules, CA, USA). Gray values were analyzed using ImageJ v1.51, and GAPDH was used as an internal reference to calculate the relative expression of the target proteins.

2.9. Statistical Analysis

The data are presented as means ± standard deviations (SDs) and were analyzed using one-way analysis of variance (ANOVA) in GraphPad Prism 8.3.0. A p-value < 0.05 was considered to indicate a statistically significant difference. All experiments were performed in triplicate.

3. Results

3.1. Characterization and Validation of si-Cyp2e1 LNPs In Vivo

The characterization results are presented in Table 1. The TEM images revealed that the si-*Cyp2e1* LNPs exhibited a spherical morphology with a particle size of approximately 80 nm, which can be used to target the liver (Figure 1A).

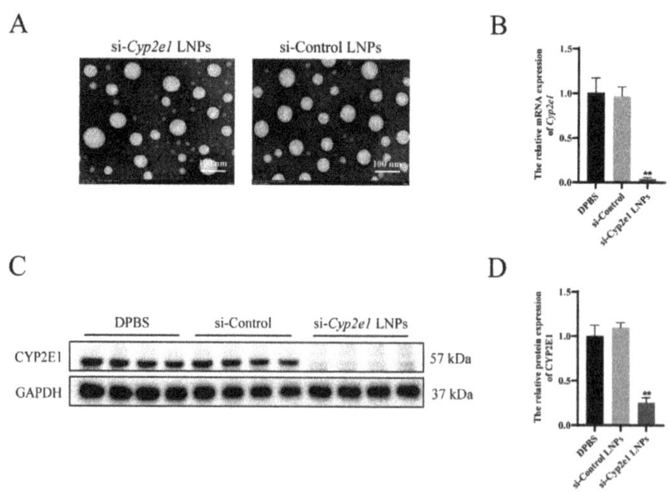

Figure 1. Characterization of si-*Cyp2e1* LNPs and knockdown efficiency in vivo (n = 4). (**A**) Representative image under transmission electron microscopy (scale = 100 nm). (**B**) *Cyp2e1* mRNA expression in the liver 24 h after a single injection of si-*Cyp2e1* LNPs. (**C**,**D**) CYP2E1 protein expression and quantification. ** p < 0.01 vs. the DPBS group.

Table 1. Characterization of lipid nanoparticles (LNPs).

Name	Size (nm)	Polydispersity Index	Zeta Potential (mV)	Encapsulation Rate (%)
si-*Cyp2e1* LNPs	78.04 ± 0.29	0.143 ± 0.031	−0.45 ± 0.1	89.05 ± 0.38
si-Control LNPs	72.47 ± 2.00	0.144 ± 0.013	−0.37 ± 0.1	85.86 ± 1.07

The hepatic expression levels of *Cyp2e1* were quantitatively assessed 24 h after a single tail vein injection. qPCR analysis showed that *Cyp2e1* mRNA expression was reduced by over 90% in the mice injected with si-*Cyp2e1* LNPs compared to that in the mice receiving DPBS (Figure 1B). Western blotting also indicated that the CYP2E1 protein level decreased by approximately 70% 24 h after tail vein injection of si-*Cyp2e1* LNPs (Figure 1C,D).

3.2. Immediate Treatment with si-Cyp2e1 LNPs Could Alleviate AILI in Mice

All of the mice survived throughout the whole experiment. The liver appearance of the mice was observed after treatment with different drugs immediately after APAP administration; NAC served as a positive control. The surface texture of the liver in the APAP and si-Control LNPs groups was slightly rough with an obvious granular appearance (Figure 2A). However, the liver surfaces of the mice treated with si-*Cyp2e1* LNPs or NAC were smooth, without a granular appearance, and showed no evident difference from the Ctrl group (Figure 2A). When compared to the Ctrl group, the mice injected with APAP showed a decrease in body weight at 24 h (Figure 2B). The results of mouse body weight and liver index analysis indicated that the livers of the APAP mice showed significant hepatomegaly compared to those of the Ctrl mice, and the livers returned to normal size after treatment with si-*Cyp2e1* LNPs or NAC immediately (Figure 2B,C).

The serum levels of ALT and AST were used as sensitive indicators of liver injury [27,28]. When compared to the Ctrl group, these indices sharply increased in the APAP group, indicating that the DILI mouse model was successfully established. However, si-*Cyp2e1* LNPs significantly reversed this increase in the APAP mice (Figure 2D,E). Consistent with these findings, H&E staining revealed disordered hepatic lobules in the APAP mice, with numerous necrotic hepatocytes around the portal area and inflammatory infiltration. In contrast, edema and inflammation were considerably alleviated in the si-*Cyp2e1* LNPs and NAC groups, and the hepatic lobules were clearly organized and closely arranged (Figure 2F). In summary, the progression of AILI was slowed by si-*Cyp2e1* LNPs.

To elucidate the effects of si-*Cyp2e1* LNPs on the oxidative status, we evaluated the levels of ROS, GSH, SOD, and MDA in the liver. Excessive ROS accumulation triggers oxidative stress and mitochondrial dysfunction [29]. Elevated hepatic MDA levels are indicative of increased lipid peroxidation, potentially culminating in hepatocellular injury and diminished antioxidant defenses [30]. Hepatic ROS and MDA levels significantly increased 24 h after APAP injection, whereas si-*Cyp2e1* LNPs significantly inhibited this increase and oxidation effect (Figure 2G,H). In contrast to ROS and MDA, the levels of SOD and GSH reflect the antioxidant capacity of the body. The results showed that the liver SOD content was significantly reduced after APAP treatment compared to that in the Ctrl group, while si-*Cyp2e1* LNPs restored the antioxidant capacity of the APAP mice (Figure 2I). The GSH levels in the APAP mice recovered to levels higher than those in the Ctrl group 24 h after intraperitoneal injection of APAP. Interestingly, the hepatic GSH levels in the si-*Cyp2e1* LNPs and NAC groups were higher than those in the APAP group (Figure 2J).

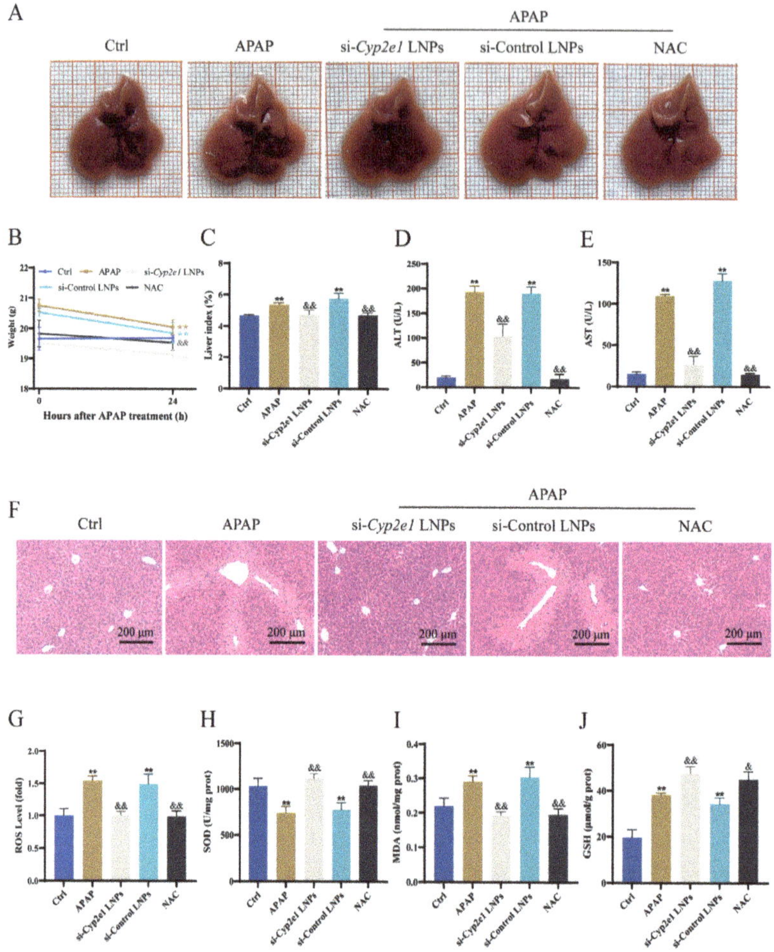

Figure 2. Effects of immediate treatment with si-*Cyp2e1* LNPs on APAP-injected mice (n = 4). (**A**) Appearance morphology of mouse livers. (**B**) Changes in the body weight of the mice. (**C–E**) Levels of liver index, serum ALT, and AST. (**F**) H&E staining of mouse livers (scale bars = 200 μm). (**G–J**) Levels of ROS, SOD, MDA, and GSH in mouse livers 24 h after a single dose of 300 mg/kg APAP. ** $p < 0.01$ vs. the Ctrl group; & $p < 0.05$ and && $p < 0.01$ vs. the APAP group.

3.3. *Transcriptomics Revealed the Possible Mechanism by Which si-Cyp2e1 LNPs Attenuate AILI in Mice*

The above results indicated that timely treatment with si-*Cyp2e1* LNPs could alleviate AILI in mice. Next, the mouse liver samples were sequenced to further explore the potential mechanism.

The DEGs between the groups were compared and are presented (Supplementary Figure S2A,B). In contrast to the Ctrl group, there were 1053 upregulated genes and 1217 downregulated genes in the APAP group. A total of 1242 DEGs with 583 downregulated and 659 upregulated genes were present between the si-*Cyp2e1* LNPs and APAP groups (Supplementary Figure S2C and Figure 3A,B). The GO and KEGG analyses showed that APAP exerted regulatory effects on several physiological processes and signaling pathways. These included reactive oxygen species, protein processing in the endoplasmic reticulum (ER), misfolded protein binding, oxidoreductase activity, and the p53 and peroxisome proliferator-activated receptor (PPAR) signaling pathways. In the meantime, si-*Cyp2e1* LNPs affected cytochrome P450, fatty acid degradation and metabolism, unsaturated fatty acid biosynthesis, glutathione metabolism, cholesterol metabolism, and the PPAR signaling pathway (Figure 3C–F). From the results of GSEA, we found that the production of cytokines and tumor necrosis factor (TNF), glutathione metabolism, the ERK1 and ERK2 cascade, the MAPK cascade, and the regulation of the apoptotic signaling pathway in the APAP model group were notably upregulated compared to those in the Ctrl group (Supplementary Figure S3A). After si-*Cyp2e1* LNPs treatment, fatty acid degradation, unsaturated fatty acid biosynthesis, and the PPAR signaling pathway were significantly inhibited compared with those in the APAP group (Supplementary Figure S3B and Figure 3G). The heatmap further revealed a significant downregulation of genes associated with the PPAR signaling pathway (Figure 3H).

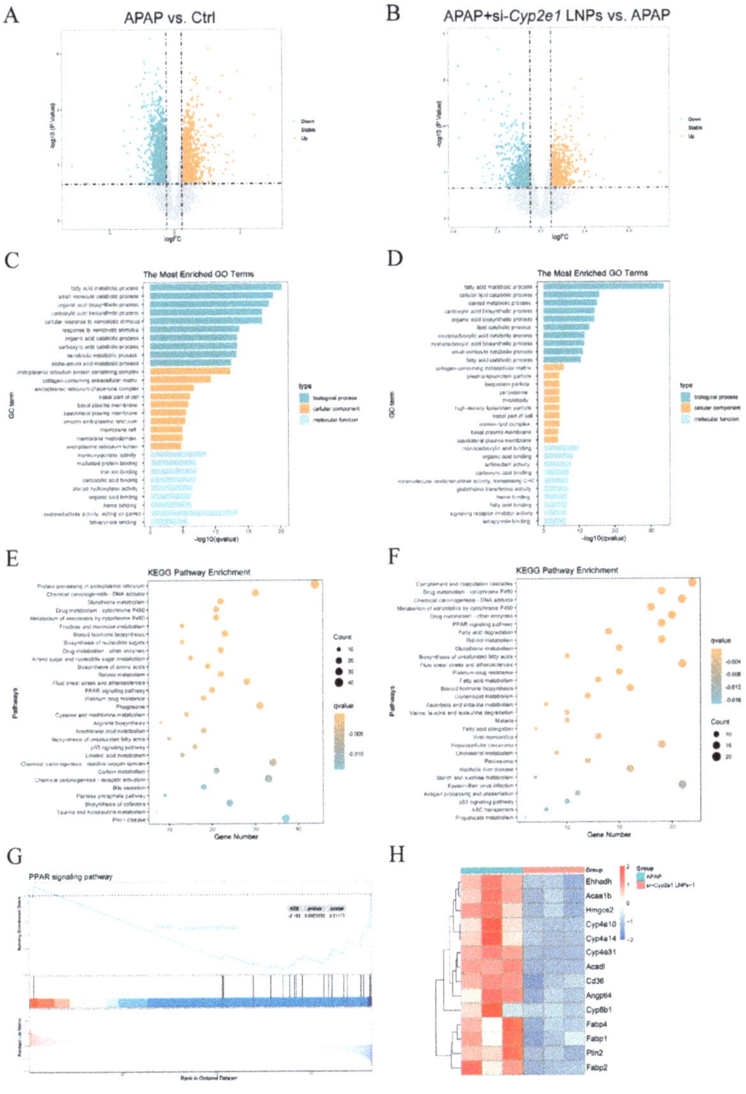

Figure 3. Transcriptomic analysis of si-*Cyp2e1* LNPs on AILI mice (n = 3). (**A**,**B**) Volcano map of the APAP group vs. the Ctrl group and the si-*Cyp2e1* LNPs treatment group vs. the APAP group. (**C**,**D**) GO

analysis chart. (**E**,**F**) KEGG analysis. (**G**) GSEA of the PPAR signaling pathway between the si-*Cyp2e1* LNPs treatment and APAP groups. (**H**) Heatmap of the PPAR signaling pathways related gene expression.

3.4. si-*Cyp2e1* LNPs Could Regulate the PPAR Signaling Pathway

To further validate the findings of RNA sequencing, we verified the mRNA and protein levels separately. Similar to the previous results, the qPCR analysis indicated that the mRNA expression levels of *Cyp4a10*, *Cyp4a14*, fatty acid binding protein 1 (*Fabp1*), and cluster of differentiation 36 (*Cd36*) were significantly decreased in the si-*Cyp2e1* LNPs treatment group compared to those in the APAP group (Figure 4A–D). Notably, si-*Cyp2e1* LNPs upregulated the mRNA of *Pparα* (Figure 4E). Furthermore, inflammatory cytokines were significantly inhibited in the si-*Cyp2e1* LNPs-treated mice compared with the APAP mice, including interleukin (*Il)-6*, *Il-1β*, and *Tnf-α* (Figure 4F–H). Western blotting further demonstrated that si-*Cyp2e1* LNPs markedly downregulated CYP4A, CD36, and FABP1 proteins, while the PPARα levels were significantly increased compared with the APAP group (Figure 4I–L). These results suggest that si-*Cyp2e1* LNPs could regulate the genes involved in the PPAR signaling pathway and inflammation.

3.5. Delayed Treatment with si-*Cyp2e1* LNPs for 2 h Did Not Ameliorate AILI in Time

We further investigated whether delayed treatment with si-*Cyp2e1* LNPs protected the mice from APAP hepatotoxicity. We injected si-*Cyp2e1* LNPs into the tail veins of the mice 2 h after APAP overdose. The therapeutic effect after 24 h of modeling is shown in Supplementary Figure S4. As shown in the mouse liver appearance and liver index, a 2 h delay in the si-*Cyp2e1* LNPs treatment did not alter the granular appearance or hepatomegaly caused by APAP (Supplementary Figure S4A–C). In addition, the serum ALT and AST levels did not exhibit a statistically significant difference between the APAP and si-*Cyp2e1* LNPs delayed treatment groups (Supplementary Figure S4D,E). Consistent with these findings, H&E staining indicated that delayed treatment with si-*Cyp2e1* LNPs did not improve local liver necrosis (Supplementary Figure S4F).

3.6. si-*Cyp2e1* LNPs Might Accelerate Liver Recovery Compared with NAC

In both cellular and mouse models, previous reports have shown that delayed and continuous treatment with NAC may inhibit hepatocyte regeneration, affecting the recovery of liver function in the late stage of AILI [9,31]. To investigate the effect of delayed treatment with a single dose of 0.5 mg/kg si-*Cyp2e1* LNPs on liver recovery in the AILI mice, pathological analysis of the liver was performed 72 h after modeling. As shown in Supplementary Figure S5, significant inflammatory infiltration was still observed in the liver tissue of the APAP group when compared to that in the Ctrl group, as indicated by the black arrow. Continuous administration of NAC did not improve this inflammation, and areas of hepatocyte edema were observed, as indicated by the red arrow. Notably, si-*Cyp2e1* LNPs significantly ameliorated the hepatic pathological changes induced by APAP, and there were no significant differences between the si-*Cyp2e1* LNPs delayed treatment group and the Ctrl group, implying that si-*Cyp2e1* LNPs may accelerate liver recovery in the late phase of AILI in mice.

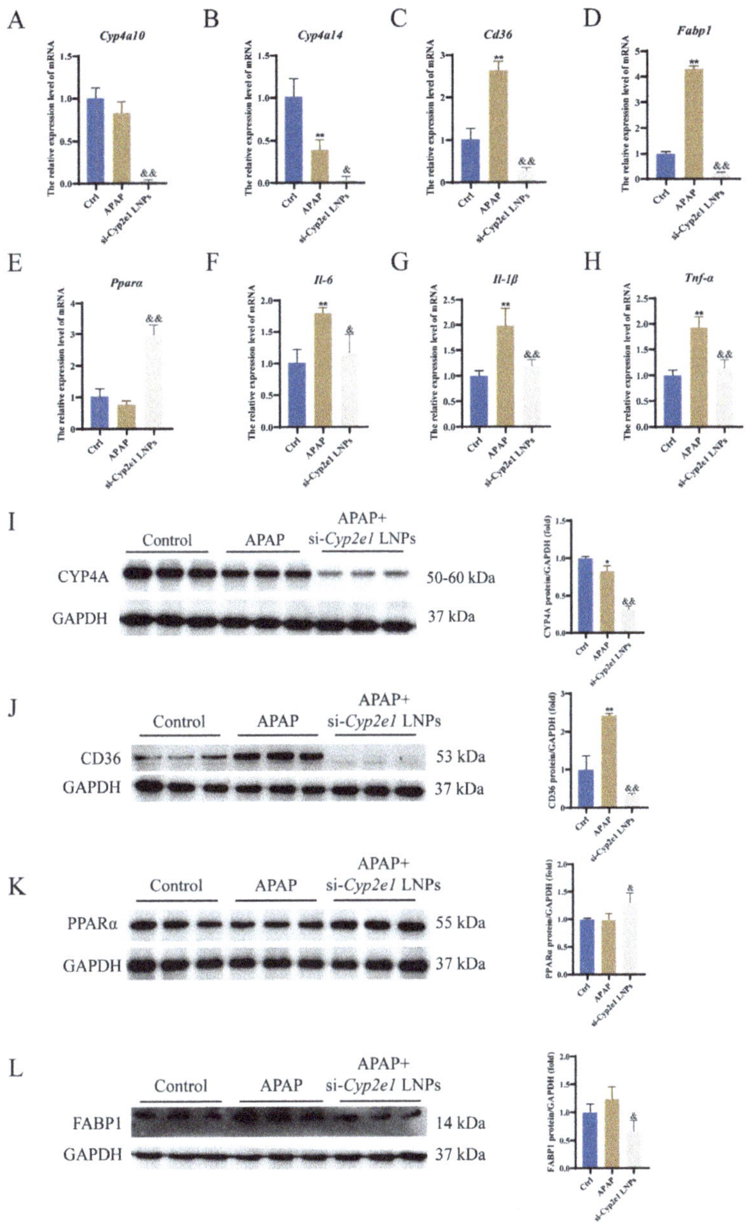

Figure 4. Effects of si-*Cyp2e1* LNPs on the genes involved in the PPAR signaling pathway and inflammation in the livers of APAP mice ($n = 3$). (**A–H**) mRNA expression of *Cyp4a10*, *Cyp4a14*, *Pparα*, *Cd36*, *Fabp1*, *Il-6*, *Il-1β*, and *Tnf-α*. (**I–L**) Protein levels of CYP4A, PPARα, CD36, and FABP1. * $p < 0.05$, ** $p < 0.01$ vs. the Ctrl group; & $p < 0.05$ and && $p < 0.01$ vs. the APAP group.

4. Discussion

Previous studies have demonstrated that excessive APAP is metabolized by CYPs into bioactive metabolites, leading to GSH depletion and ROS formation, triggering oxidative stress, mitochondrial permeability transition, and ATP loss, eventually resulting in hepatocyte necrosis [10]. During the development of AILI, CYP2E1 is a key enzyme in the generation of the hepatotoxic metabolite NAPQI, and the hepatotoxicity of APAP is enhanced after CYP2E1 induction [32]. In the present study, siRNA-LNPs targeting *Cyp2e1* were used to investigate their therapeutic effects against AILI in mice. The measured particle size of si-*Cyp2e1* LNPs was approximately 80 nm, demonstrating effective hepatic targeting [33]. In subsequent pharmacodynamic experiments, the levels of liver injury and oxidative stress-related indicators significantly changed 24 h after APAP administration. Fortunately, all indices, except GSH, tended to be normal after immediate treatment with si-*Cyp2e1* LNPs, and the area of liver tissue necrosis was markedly reduced. According to previous reports, GSH is depleted 2 h after APAP overdose, gradually recovers from 6 h, and returns to above-normal levels by 24 h [34,35]. This phenomenon of body self-replenishment of GSH was reflected in our experimental results and was more significant in the mice treated with si-*Cyp2e1* LNPs. These results implied that targeted inhibition of *Cyp2e1* could effectively alleviate oxidative stress and AILI in mice.

RNA sequencing and enrichment analysis showed that APAP could induce significant changes in protein processing in the ER, misfolded protein binding, ROS, production of cytokine and TNF, ERK and MAPK cascades, apoptosis, and the p53 signaling pathway, indicating that APAP could induce ER stress, oxidative stress, the inflammatory response, and hepatocyte apoptosis. In the development of APAP hepatotoxicity, initial ROS formation leads to the activation of MAPK and the phosphorylation of JNK, followed by p-JNK translocation to the mitochondria, which results in further amplification of mitochondrial ROS release [36]. Subsequently, the p53 signaling pathway is activated in response to cellular stress and injury [6]. These events are consistent with our findings. We found that the PPAR signaling pathway might be an important pathway for si-*Cyp2e1* LNPs to alleviate AILI in mice, as verified by subsequent research.

Similar to CYP2E1, the CYP4A family comprises microsomal oxidases involved in fatty acid oxidation, both of which can reduce molecular oxygen to produce prooxidants, leading to oxidative stress if antioxidants are ineffective. Mouse CYP4A10 and CYP4A14 are important components of CYP4A, which not only catalyze the ω-hydroxylation of fatty acids [37] but also activate the expression of inflammatory factors through the ROS/NF-κB pathway [38]. Studies have shown that CYP2E1 can be induced by long-term alcohol consumption or a high-fat diet, while the knockout or continuous inhibition of *Cyp2e1* leads to a compensatory increase in CYP4A [39,40]. Our previous research showed that the simultaneous inhibition of *Cyp2e1*, *Cyp4a10*, and *Cyp4a14* significantly inhibited oxidative stress in mice with alcoholic liver injury [41]. Interestingly, the present study indicated that short-term targeted inhibition of CYP2E1 in vivo resulted in significant suppression of CYP4A expression in mice exposed to APAP. Under such a condition, the alleviation of APAP-induced oxidative stress by si-*Cyp2e1* LNPs may be attributed to the effective inhibition of CYP2E1 and CYP4A. These observations provided new insights into the interactions between CYPs under varying circumstances in vivo.

In the present study, we found that si-*Cyp2e1* LNPs not only downregulated the expressions of CD36, FABP1, and inflammatory factors but also increased the level of PPARα in the liver. CD36 is a scavenger receptor that mediates lipid intake, the adhesion of immune recognition cells, inflammation, and apoptosis [42], and it is a key factor involved in fat degeneration and liver damage [43]. Studies have proved that CYP2E1 inhibitors would decrease CD36 expression in vivo and in vitro [44] and that CD36 deficiency attenuates AILI in mice by inhibiting JNK activation and HMGB1-mediated inflammatory responses [45]. PPARs are critical for fatty acid catabolism, and PPARα is a key molecule that controls lipid synthesis and fatty acid oxidation pathways [46]. PPARα is activated in *Cyp2e1* knockout mice, which is attributed to the accumulation of its endogenous substrates, identified as direct PPARα agonists [47,48]. A PPARα activator attenuated APAP hepatotoxicity in wild-type mice, whereas this protective effect was absent in *Ppar*$\alpha^{-/-}$ mice [49,50]. FABP1 is downstream of the PPAR signaling pathway, and the silencing of FABP1 in the liver ameliorated hepatic oxidative stress and inflammation in a mouse model of nonalcoholic fatty liver disease [51]. Lipid peroxidation is often considered an important mechanism by which ROS regulate hepatocyte death induced by APAP overdose [52]. ERK, JNK, and p38 form adducts with the products of lipid peroxidation to activate the MAPK signaling pathway [53]. We found that si-*Cyp2e1* LNPs regulated lipid metabolism and inhibited lipid peroxidation by activating PPARα while inhibiting unsaturated fatty acid biosynthesis and fatty acid degradation. These results suggest that si-*Cyp2e1* LNPs may protect the mouse liver from inflammation and injury triggered by APAP through the regulation of PPAR-related pathways. However, the effects of si-*Cyp2e1* LNPs on other pathways, such as the JNK, MAPK, and p38 signaling pathways, and their interactions remain to be studied.

Next, we explored the effect of a 2 h delayed treatment with si-*Cyp2e1* LNPs. Unfortunately, the results of delayed treatment did not meet our expectations, probably due to the rapid metabolism of APAP in vivo. As previously mentioned, GSH levels recovered 6 h after APAP overdose [34], whereas our recent study showed that si-*Cyp2e1* LNPs required 6 h after injection to significantly inhibit CYP2E1 expression [23], implying that delayed treatment for 2 h might not inhibit the formation of NAPQI and APAP adducts. Therefore, delayed si-*Cyp2e1* LNPs treatment did not have the same therapeutic effect as immediate administration during the early phase of AILI. Interestingly, we observed that si-*Cyp2e1* LNPs could accelerate hepatic recovery in the late stages of AILI. However, this conclusion is preliminary, and changes in the biomarkers of liver regeneration remain to be further determined.

As the only antidote for APAP overdose in clinical practice, NAC still has many drawbacks, such as delayed liver recovery after prolonged treatment and adverse gastrointestinal reactions [22]. Our present study has demonstrated that LNP-delivered siRNA targeting *Cyp2e1* could alleviate acute liver injury induced by APAP overdose, providing a new therapeutic strategy for rescuing APAP poisoning and broadening the application of RNAi in acute liver disease. In the future, RNAi therapy could be considered for the treatment of other drug metabolism-related diseases.

5. Conclusions

In summary, our study demonstrated that RNAi targeting *Cyp2e1* in vivo could not only alleviate APAP-induced oxidative stress and inflammation in mice by regulating lipid metabolism through PPAR-related pathways, but it could also contribute to liver recovery in the later stages of the injury, illustrating the therapeutic potential of RNAi therapy in acute liver disease and hepatic metabolism-related diseases in the future.

Supplementary Materials: The following supporting information can be downloaded at: https://www.mdpi.com/article/10.3390/cimb47050372/s1.

Author Contributions: Conceptualization, Y.W.; data curation, W.L. and L.H.; formal analysis, L.H. and C.Z.; funding acquisition, Y.W.; investigation, W.L., L.H., C.Z., R.Y. and Z.O.; methodology, W.L., R.Y. and Z.O.; supervision, Y.W.; validation, R.Y. and Z.O.; visualization, C.Z.; writing—original draft, W.L.; writing—review and editing, Y.W. All authors have read and agreed to the published version of the manuscript.

Funding: This research was funded by Key project at central government level: The ability establishment of sustainable use for valuable Chinese medicine resources (2060302), and The National Key R&D Program of China: Chinese-Australian 'Belt and Road' Joint Laboratory on Traditional Chinese Medicine for the Prevention and Treatment of Severe Infectious Diseases (2020YFE0205100).

Institutional Review Board Statement: The animal study protocol was approved by the Institutional Animal Care and Use Committee of Jiangsu University (UJS-IACUC-2023122802; approved on 28 December 2023).

Informed Consent Statement: Not applicable.

Data Availability Statement: The raw sequencing data were deposited in the NCBI under the BioProject number PRJNA1211279. The data will be made available on request.

Acknowledgments: The authors would like to acknowledge Figdraw (www.figdraw.com) for providing a platform for drawing the graphical abstract and Hongyu Lu (Subei People's Hospital) for his technical support.

Conflicts of Interest: The authors declare no conflicts of interest.

Abbreviations

APAP	acetaminophen
AILI	APAP-induced liver injury
NAPQI	*N*-acetyl-*p*-benzoquinone imine
CYPs	cytochrome P450 enzymes
RNAi	RNA interference
siRNA	small interfering RNA
LNPs	lipid nanoparticles
ALT	alanine aminotransferase
AST	aspartate aminotransferase
ROS	reactive oxygen species
MDA	malondialdehyde
SOD	superoxide dismutase
GSH	glutathione
H&E	hematoxylin and eosin
DEGs	differentially expressed genes
GO	Gene ontology
KEGG	Kyoto encyclopedia of genes and genomes
GSEA	gene set enrichment analysis
qPCR	quantitative real-time polymerase chain reaction
GAPDH	glyceraldehyde-3-phosphate dehydrogenase
PPAR	peroxisome proliferator-activated receptor
CD36	cluster of differentiation 36
FABP1	fatty acid binding protein 1
IL	interleukin
TNF	tumor necrosis factor

References

1. Fu, L.; Qian, Y.; Shang, Z.; Sun, X.; Kong, X.; Gao, Y. Antibiotics enhancing drug-induced liver injury assessed for causality using Roussel Uclaf Causality Assessment Method: Emerging role of gut microbiota dysbiosis. *Front. Med.* **2022**, *9*, 972518. [CrossRef]
2. LaForge, J.M.; Urso, K.; Day, J.M.; Bourgeois, C.W.; Ross, M.M.; Ahmadzadeh, S.; Shekoohi, S.; Cornett, E.M.; Kaye, A.M.; Kaye, A.D. Non-steroidal Anti-inflammatory Drugs: Clinical Implications, Renal Impairment Risks, and AKI. *Adv. Ther.* **2023**, *40*, 2082–2096. [CrossRef]
3. Ostapowicz, G.; Fontana, R.J.; Schiødt, F.V.; Larson, A.; Davern, T.J.; Han, S.H.; McCashland, T.M.; Shakil, A.O.; Hay, J.E.; Hynan, L.; et al. Results of a prospective study of acute liver failure at 17 tertiary care centers in the United States. *Ann. Intern. Med.* **2002**, *137*, 947–954. [CrossRef]
4. Bunchorntavakul, C.; Reddy, K.R. Acetaminophen (APAP or N-Acetyl-p-Aminophenol) and Acute Liver Failure. *Clin. Liver Dis.* **2018**, *22*, 325–346. [CrossRef]
5. Lee, S.S.; Buters, J.T.; Pineau, T.; Fernandez-Salguero, P.; Gonzalez, F.J. Role of CYP2E1 in the hepatotoxicity of acetaminophen. *J. Biol. Chem.* **1996**, *271*, 12063–12067. [CrossRef]

6. Yan, M.; Huo, Y.; Yin, S.; Hu, H. Mechanisms of acetaminophen-induced liver injury and its implications for therapeutic interventions. *Redox Biol.* **2018**, *17*, 274–283. [CrossRef]
7. Saito, C.; Lemasters, J.J.; Jaeschke, H. c-Jun N-terminal kinase modulates oxidant stress and peroxynitrite formation independent of inducible nitric oxide synthase in acetaminophen hepatotoxicity. *Toxicol. Appl. Pharmacol.* **2010**, *246*, 8–17. [CrossRef]
8. Heard, K.J. Acetylcysteine for acetaminophen poisoning. *N. Engl. J. Med.* **2008**, *359*, 285–292. [CrossRef]
9. Akakpo, J.Y.; Ramachandran, A.; Curry, S.C.; Rumack, B.H.; Jaeschke, H. Comparing N-acetylcysteine and 4-methylpyrazole as antidotes for acetaminophen overdose. *Arch. Toxicol.* **2022**, *96*, 453–465. [CrossRef]
10. Hinson, J.A.; Roberts, D.W.; James, L.P. Mechanisms of acetaminophen-induced liver necrosis. *Handb. Exp. Pharmacol.* **2010**, *196*, 369–405. [CrossRef]
11. van Rongen, A.; Välitalo, P.A.J.; Peeters, M.Y.M.; Boerma, D.; Huisman, F.W.; van Ramshorst, B.; van Dongen, E.P.A.; van den Anker, J.N.; Knibbe, C.A.J. Morbidly Obese Patients Exhibit Increased CYP2E1-Mediated Oxidation of Acetaminophen. *Clin. Pharmacokinet.* **2016**, *55*, 833–847. [CrossRef]
12. Wang, X.; Lu, Y.; Cederbaum, A.I. Induction of cytochrome P450 2E1 increases hepatotoxicity caused by Fas agonistic Jo2 antibody in mice. *Hepatology* **2005**, *42*, 400–410. [CrossRef]
13. Lu, Y.; Wu, D.; Wang, X.; Ward, S.C.; Cederbaum, A.I. Chronic alcohol-induced liver injury and oxidant stress are decreased in cytochrome P4502E1 knockout mice and restored in humanized cytochrome P4502E1 knock-in mice. *Free Radic. Biol. Med.* **2010**, *49*, 1406–1416. [CrossRef]
14. Schattenberg, J.M.; Czaja, M.J. Regulation of the effects of CYP2E1-induced oxidative stress by JNK signaling. *Redox Biol.* **2014**, *3*, 7–15. [CrossRef]
15. Wang, C.; Li, H.; Meng, Q.; Du, Y.; Xiao, F.; Zhang, Q.; Yu, J.; Li, K.; Chen, S.; Huang, Z.; et al. ATF4 deficiency protects hepatocytes from oxidative stress via inhibiting CYP2E1 expression. *J. Cell. Mol. Med.* **2014**, *18*, 80–90. [CrossRef]
16. Kanasty, R.; Dorkin, J.R.; Vegas, A.; Anderson, D. Delivery materials for siRNA therapeutics. *Nat. Mater.* **2013**, *12*, 967–977. [CrossRef]
17. Wang, Y.; Chen, Q.; Wu, S.; Sun, X.; Yin, R.; Ouyang, Z.; Yin, H.; Wei, Y. Amelioration of ethanol-induced oxidative stress and alcoholic liver disease by in vivo RNAi targeting Cyp2e1. *Acta Pharm. Sin. B* **2023**, *13*, 3906–3918. [CrossRef]
18. Wu, S.; Chen, Q.; Wang, Y.; Yin, H.; Wei, Y. Lipid nanoparticle delivery of siRNA targeting Cyp2e1 gene attenuates subacute alcoholic liver injury in mice. *J. Zhejiang Univ. (Med. Sci.)* **2023**, *52*, 306–317. [CrossRef]
19. Bogorad, R.L.; Yin, H.; Zeigerer, A.; Nonaka, H.; Ruda, V.M.; Zerial, M.; Anderson, D.G.; Koteliansky, V. Nanoparticle-formulated siRNA targeting integrins inhibits hepatocellular carcinoma progression in mice. *Nat. Commun.* **2014**, *5*, 3869. [CrossRef]
20. Zeigerer, A.; Gilleron, J.; Bogorad, R.L.; Marsico, G.; Nonaka, H.; Seifert, S.; Epstein-Barash, H.; Kuchimanchi, S.; Peng, C.G.; Ruda, V.M.; et al. Rab5 is necessary for the biogenesis of the endolysosomal system in vivo. *Nature* **2012**, *485*, 465–470. [CrossRef]
21. McGill, M.R.; Jaeschke, H. Animal models of drug-induced liver injury. *Biochim. Biophys. Acta Mol. Basis Dis.* **2019**, *1865*, 1031–1039. [CrossRef]
22. Yang, R.; Miki, K.; He, X.; Killeen, M.E.; Fink, M.P. Prolonged treatment with N-acetylcystine delays liver recovery from acetaminophen hepatotoxicity. *Crit. Care* **2009**, *13*, R55. [CrossRef]
23. Huan, L.; Zhang, Y.; Liu, W.; Lu, H.; Zhang, C.; Yin, R.; Ouyang, Z.; Wei, Y. In vivo inhibition of CYP2E1 fails to reduce pulegone-induced liver injury in mice. *Xenobiotica* **2025**, 1–10. [CrossRef]
24. Kim, D.; Langmead, B.; Salzberg, S.L. HISAT: A fast spliced aligner with low memory requirements. *Nat. Methods* **2015**, *12*, 357–360. [CrossRef]
25. Trapnell, C.; Williams, B.A.; Pertea, G.; Mortazavi, A.; Kwan, G.; van Baren, M.J.; Salzberg, S.L.; Wold, B.J.; Pachter, L. Transcript assembly and quantification by RNA-Seq reveals unannotated transcripts and isoform switching during cell differentiation. *Nat. Biotechnol.* **2010**, *28*, 511–515. [CrossRef]
26. Diboun, I.; Wernisch, L.; Orengo, C.A.; Koltzenburg, M. Microarray analysis after RNA amplification can detect pronounced differences in gene expression using limma. *BMC Genom.* **2006**, *7*, 252. [CrossRef]
27. Zimmerman, H.J.; West, M. Serum Enzyme Levels in the Diagnosis of Hepatic Disease. *Am. J. Gastroenterol.* **1963**, *40*, 387–404.
28. Wei, Y.; Li, M.; Feng, Z.; Zhang, D.; Sun, M.; Wang, Y.; Chen, X. The Protective Effects of Corn Oligopeptides on Acute Alcoholic Liver Disease by Inhibiting the Activation of Kupffer Cells NF-κB/AMPK Signal Pathway. *Nutrients* **2022**, *14*, 4194. [CrossRef]
29. Park, J.E.; Ahn, C.H.; Lee, H.J.; Sim, D.Y.; Park, S.Y.; Kim, B.; Shim, B.S.; Lee, D.Y.; Kim, S.H. Antioxidant-Based Preventive Effect of Phytochemicals on Anticancer Drug-Induced Hepatotoxicity. *Antioxid. Redox Signal.* **2023**, *38*, 1101–1121. [CrossRef]
30. Xu, W.; Xiao, M.; Li, J.; Chen, Y.; Sun, Q.; Li, H.; Sun, W. Hepatoprotective effects of Di Wu Yang Gan: A medicinal food against CCl(4)-induced hepatotoxicity in vivo and in vitro. *Food Chem.* **2020**, *327*, 127093. [CrossRef]
31. Athuraliya, T.N.; Jones, A.L. Prolonged N-acetylcysteine therapy in late acetaminophen poisoning associated with acute liver failure—A need to be more cautious? *Crit. Care* **2009**, *13*, 144. [CrossRef] [PubMed]

32. Aubert, J.; Begriche, K.; Delannoy, M.; Morel, I.; Pajaud, J.; Ribault, C.; Lepage, S.; McGill, M.R.; Lucas-Clerc, C.; Turlin, B.; et al. Differences in early acetaminophen hepatotoxicity between obese ob/ob and db/db mice. *J. Pharmacol. Exp. Ther.* **2012**, *342*, 676–687. [CrossRef]
33. Lam, K.; Schreiner, P.; Leung, A.; Stainton, P.; Reid, S.; Yaworski, E.; Lutwyche, P.; Heyes, J. Optimizing Lipid Nanoparticles for Delivery in Primates. *Adv. Mater.* **2023**, *35*, e2211420. [CrossRef] [PubMed]
34. Qian, H.; Bai, Q.; Yang, X.; Akakpo, J.Y.; Ji, L.; Yang, L.; Rülicke, T.; Zatloukal, K.; Jaeschke, H.; Ni, H.M.; et al. Dual roles of p62/SQSTM1 in the injury and recovery phases of acetaminophen-induced liver injury in mice. *Acta Pharm. Sin. B* **2021**, *11*, 3791–3805. [CrossRef] [PubMed]
35. Ni, H.M.; McGill, M.R.; Chao, X.; Du, K.; Williams, J.A.; Xie, Y.; Jaeschke, H.; Ding, W.X. Removal of acetaminophen protein adducts by autophagy protects against acetaminophen-induced liver injury in mice. *J. Hepatol.* **2016**, *65*, 354–362. [CrossRef]
36. McGill, M.R.; Jaeschke, H. Metabolism and disposition of acetaminophen: Recent advances in relation to hepatotoxicity and diagnosis. *Pharm. Res.* **2013**, *30*, 2174–2187. [CrossRef]
37. Adas, F.; Salaün, J.P.; Berthou, F.; Picart, D.; Simon, B.; Amet, Y. Requirement for omega and (omega;-1)-hydroxylations of fatty acids by human cytochromes P450 2E1 and 4A11. *J. Lipid Res.* **1999**, *40*, 1990–1997. [CrossRef]
38. Ip, E.; Farrell, G.C.; Robertson, G.; Hall, P.; Kirsch, R.; Leclercq, I. Central role of PPARalpha-dependent hepatic lipid turnover in dietary steatohepatitis in mice. *Hepatology* **2003**, *38*, 123–132. [CrossRef]
39. Leclercq, I.A.; Farrell, G.C.; Field, J.; Bell, D.R.; Gonzalez, F.J.; Robertson, G.R. CYP2E1 and CYP4A as microsomal catalysts of lipid peroxides in murine nonalcoholic steatohepatitis. *J. Clin. Investig.* **2000**, *105*, 1067–1075. [CrossRef]
40. Hardwick, J.P.; Osei-Hyiaman, D.; Wiland, H.; Abdelmegeed, M.A.; Song, B.J. PPAR/RXR Regulation of Fatty Acid Metabolism and Fatty Acid omega-Hydroxylase (CYP4) Isozymes: Implications for Prevention of Lipotoxicity in Fatty Liver Disease. *PPAR Res.* **2009**, *2009*, 952734. [CrossRef]
41. Zhang, Q.; Wu, S.; Chen, Q.; Zhang, Y.; Zhang, C.; Yin, R.; Ouyang, Z.; Wei, Y. Reducing Oxidative Stress-Mediated Alcoholic Liver Injury by Multiplexed RNAi of Cyp2e1, Cyp4a10, and Cyp4a14. *Biomedicines* **2024**, *12*, 1505. [CrossRef] [PubMed]
42. Wang, J.; Li, Y. CD36 tango in cancer: Signaling pathways and functions. *Theranostics* **2019**, *9*, 4893–4908. [CrossRef] [PubMed]
43. Lebeau, P.F.; Byun, J.H.; Platko, K.; Al-Hashimi, A.A.; Lhoták, Š.; MacDonald, M.E.; Mejia-Benitez, A.; Prat, A.; Igdoura, S.A.; Trigatti, B.; et al. Pcsk9 knockout exacerbates diet-induced non-alcoholic steatohepatitis, fibrosis and liver injury in mice. *JHEP Rep.* **2019**, *1*, 418–429. [CrossRef]
44. Ma, L.; Wang, Y.; Chen, X.; Zhao, L.; Guo, Y. Involvement of CYP2E1-ROS-CD36/DGAT2 axis in the pathogenesis of VPA-induced hepatic steatosis in vivo and in vitro. *Toxicology* **2020**, *445*, 152585. [CrossRef]
45. Zhang, C.; Shi, X.; Su, Z.; Hu, C.; Mu, X.; Pan, J.; Li, M.; Teng, F.; Ling, T.; Zhao, T.; et al. CD36 deficiency ameliorates drug-induced acute liver injury in mice. *Mol. Med.* **2021**, *27*, 57. [CrossRef] [PubMed]
46. Liao, M.; Sun, C.; Li, R.; Li, W.; Ge, Z.; Adu-Frimpong, M.; Xu, X.; Yu, J. Amelioration action of gastrodigenin rhamno-pyranoside from Moringa seeds on non-alcoholic fatty liver disease. *Food Chem.* **2022**, *379*, 132087. [CrossRef]
47. Zhang, Y.; Yan, T.; Wang, T.; Liu, X.; Hamada, K.; Sun, D.; Sun, Y.; Yang, Y.; Wang, J.; Takahashi, S.; et al. Crosstalk between CYP2E1 and PPARα substrates and agonists modulate adipose browning and obesity. *Acta Pharm. Sin. B* **2022**, *12*, 2224–2238. [CrossRef]
48. Wan, Y.Y.; Cai, Y.; Li, J.; Yuan, Q.; French, B.; Gonzalez, F.J.; French, S. Regulation of peroxisome proliferator activated receptor alpha-mediated pathways in alcohol fed cytochrome P450 2E1 deficient mice. *Hepatol. Res.* **2001**, *19*, 117–130. [CrossRef]
49. Manautou, J.E.; Tveit, A.; Hoivik, D.J.; Khairallah, E.A.; Cohen, S.D. Protection by clofibrate against acetaminophen hepatotoxicity in male CD-1 mice is associated with an early increase in biliary concentration of acetaminophen-glutathione adducts. *Toxicol. Appl. Pharmacol.* **1996**, *140*, 30–38. [CrossRef]
50. Chen, C.; Hennig, G.E.; Whiteley, H.E.; Corton, J.C.; Manautou, J.E. Peroxisome proliferator-activated receptor alpha-null mice lack resistance to acetaminophen hepatotoxicity following clofibrate exposure. *Toxicol. Sci.* **2000**, *57*, 338–344. [CrossRef]
51. Mukai, T.; Egawa, M.; Takeuchi, T.; Yamashita, H.; Kusudo, T. Silencing of FABP1 ameliorates hepatic steatosis, inflammation, and oxidative stress in mice with nonalcoholic fatty liver disease. *FEBS Open Bio* **2017**, *7*, 1009–1016. [CrossRef] [PubMed]
52. Yoshioka, H.; Aoyagi, Y.; Fukuishi, N.; Gui, M.Y.; Jin, Y.R.; Li, X.W.; Adachi, Y.; Ohno, N.; Takeya, K.; Hitotsuyanagi, Y.; et al. Suppressive effect of kamebakaurin on acetaminophen-induced hepatotoxicity by inhibiting lipid peroxidation and inflammatory response in mice. *Pharmacol. Rep.* **2017**, *69*, 903–907. [CrossRef] [PubMed]
53. Wang, B.; Wang, Y.; Zhang, J.; Hu, C.; Jiang, J.; Li, Y.; Peng, Z. ROS-induced lipid peroxidation modulates cell death outcome: Mechanisms behind apoptosis, autophagy, and ferroptosis. *Arch. Toxicol.* **2023**, *97*, 1439–1451. [CrossRef] [PubMed]

current issues in molecular biology

MDPI

Article

Histopathological Analysis of Lipopolysaccharide-Induced Liver Inflammation and Thrombus Formation in Mice: The Protective Effects of Aspirin

Hayate Saitoh, Miina Sakaguchi, Fumito Miruno, Naoto Muramatsu, Nozomi Ito, Kanako Tadokoro, Kiyoharu Kawakami and Kazuhiko Nakadate *

Department of Functional Morphology, Meiji Pharmaceutical University, 2-522-1 Noshio, Kiyose 204-8588, Japan; y201111@std.my-pharm.ac.jp (H.S.); y201117@std.my-pharm.ac.jp (M.S.); y201290@std.my-pharm.ac.jp (F.M.); y201295@std.my-pharm.ac.jp (N.M.); s212009@std.my-pharm.ac.jp (N.I.); s212052@std.my-pharm.ac.jp (K.T.); k-kawakami@my-pharm.ac.jp (K.K.)

* Correspondence: nakadate@my-pharm.ac.jp

Abstract: Hepatitis, a significant medical concern owing to its potential to cause acute and chronic liver disease, necessitates early intervention. In this study, we aimed to elucidate the histopathological features of lipopolysaccharide-induced hepatitis in mice, focusing on tissue alterations. The results demonstrated that hepatocytes exhibited decreased eosin staining, indicating cellular shrinkage, whereas sinusoids were swollen with blood cells. Detailed electron microscope analysis identified these blood cells as leukocytes and erythrocytes, which confirmed a thrombus formation within the liver. Pre-treatment with aspirin significantly attenuated these pathological changes, including reductions in inflammatory markers such as C-reactive protein, interleukin-1β, and tumor necrosis factor-alpha. These findings highlight aspirin's anti-inflammatory and antiplatelet effects in mitigating liver inflammation and thrombus formation. In this study, we highlighted the potential of aspirin as a therapeutic agent for liver inflammation, in addition to providing insights into hepatocyte alterations and sinusoidal blood cell aggregation in liver inflammation. Aspirin, through the protection of endothelial cells and reduction of cytokine levels, may have broader applications in managing liver disease and other systemic inflammatory conditions. This emphasizes its value in prevention and therapy.

Keywords: lipopolysaccharide; thrombus; blood clot; liver; aspirin; inflammation; correlative light and electron microscopy; scanning electron microscopy

1. Introduction

Hepatitis is characterized by an inflammatory response in the liver that may progress to acute or chronic liver disease, cirrhosis, and hepatocellular carcinoma [1,2]. The liver is an important organ for detoxification, metabolism, and immune regulation. In addition, the progression of conditions associated with hepatitis can have serious health consequences [3]. The causes of hepatitis are diverse, including viruses, alcohol, drugs, and metabolic and immune abnormalities; in addition, lipopolysaccharide (LPS)-induced inflammatory responses have been highlighted as an important contributing factor [4–6]. This inflammatory response is known as hepatocellular injury [7], which leads to liver damage, hepatocyte injury, sinusoidal endothelial destruction, and thrombus (blood clotting) formation.

In the pathogenesis of LPS-induced hepatitis, sinusoidal and vascular endothelial cells in the liver are damaged, leading to abnormal activation and aggregation of platelets and the formation of thrombi in the blood vessels [8,9]. The histopathological changes in the liver following LPS administration include hepatocyte necrosis, inflammatory cell infiltration, and thrombus formation, which may significantly affect the progression and prognosis of hepatitis. These lesions may have a significant impact on the progression and prognosis of hepatitis.

Thrombosis occurs in atherosclerotic and chronic inflammatory diseases and causes serious complications, including stroke, myocardial infarction, and pulmonary thromboembolism [10–12]. These diseases are the leading causes of death worldwide, and their risk of development is significantly high in elderly and chronically ill patients [13]. In addition, recent studies have indicated that thrombosis occurs in organs such as the liver and intestine, suggesting that visceral thrombosis may worsen liver disease and intestinal ischemia [14,15]. In clinical practice, thrombosis prevention and management are important, and antiplatelet and anticoagulant drugs play a central role in the treatment [16,17].

Thrombosis is an important pathogenesis in cardiovascular and chronic inflammatory diseases involving several reactions, including vascular endothelial damage and platelet aggregation [18,19]. In the liver, damage to sinusoidal endothelial cells leads to impaired blood flow and abnormal platelet aggregation, which can lead to thrombus formation [14]. These reactions promote hepatic dysfunction and the progression of inflammatory liver disease [20].

LPS, a cell wall component of gram-negative bacteria, is an exogenous trigger that strongly induces an immune response, and strongly activates the immune response upon entry into the bloodstream, causing the release of inflammatory cytokines and chemokines [7,21,22]. The LPS-stimulated inflammatory response damages sinusoidal and vascular endothelial cells in the liver, leading to platelet activation and aggregation, and increasing the risk of thrombus formation [23,24]. The levels of inflammatory markers, including the acute phase protein, C-reactive protein (CRP), increase rapidly in response to LPS stimulation, which increases the risk of thrombus formation with an increase in systemic inflammation. CRP, an acute-phase protein produced by the liver, is significantly increased during inflammation and tissue injury and is an important marker reflecting the level of systemic inflammation, due to its marked increase during inflammation and tissue damage [25,26]. In addition, LPS-induced inflammatory response enhances the production of pro-inflammatory cytokines, including tumor necrosis factor (TNF-α) and interleukins (IL-1β), which further increases vascular permeability and platelet aggregation, thereby increasing the risk of thrombus formation [27–29].

Aspirin is widely used as a general anti-platelet agent and inhibits the formation of prostaglandins and thromboxane A2 (TXA2) primarily through the irreversible inhibition of cyclooxygenase (COX) [30,31]. TXA2 causes platelet aggregation and vasoconstriction, whereas aspirin prevents platelet aggregation and thrombus formation by inhibiting its production [32,33]. Furthermore, aspirin reportedly has anti-inflammatory and anti-platelet effects and contributes to the suppression of inflammatory markers, such as CRP, TNF-α, and IL-1β [34]. Therefore, aspirin has gained attention as an effective agent for reducing LPS-induced inflammatory responses and preventing thrombus formation [34]. The mechanism by which aspirin suppresses LPS-induced inflammatory responses is multifaceted and may reduce vascular endothelial damage through suppression of inflammatory markers including CRP and TNF-α. Following LPS administration, a rapid rise in CRP and increases in TNF-α and IL-1β are observed [35,36], before aspirin expectedly suppresses the elevation of these inflammatory markers and reduces the risk of thrombus formation. In addition, aspirin may reduce thrombus formation by reducing oxidative stress in endothelial cells and preventing increased vascular permeability [37]. These actions indicate aspirin's potential as an anti-platelet agent and a part of thrombosis prevention through the suppression of systemic inflammation.

In this study, we aimed to analyze the histopathology of LPS-induced thrombus formation, especially in the liver, and the preventive effect of aspirin and its mechanism of action on inflammatory markers (CRP, TNF-α, and IL-1β).

2. Materials and Methods

2.1. Animals

We used 42 eight-week-old male ICR mice from Japan SLC in this study. All experimental procedures adhered to the guidelines in the National Institutes of Health Guide for the Care and Use of Laboratory Animals. The Laboratory Animal Ethics Committee of Meiji Pharmaceutical University approved the experiments (approval number: 2704, 1 April 2023 and 2024). Finally, we took measures to minimize animal suffering and the number of animals.

2.2. LPS Induction and Aspirin Administration

LPS was administered as previously described [38]. Twenty-three mice were administered LPS (Wako Pure Chemical Industries, Tokyo, Japan) intraperitoneally at a dose of 1 mg/kg body weight. An equal volume of saline was administered intraperitoneally to mice who did not develop hepatitis. The data of four animals from each group were included in the histopathological analysis.

Aspirin (Bayer Aspirin Tablets, Sato Pharmaceutical Co. Ltd., Tokyo. Japan) was administered orally to 15 animals, once daily at 2 mg/0.1 mL saline per kg body weight. The controls (15 animals) received the same dose of saline, and after 7 days, LPS or saline was administered to 12 animals.

2.3. Tissue Preparation, Histopathology, and Scanning Electron Microscope (SEM) Analysis

The tissues were prepared for histochemical and SEM analyses according to a previously described protocol [38–40]. Mice were deeply anesthetized 24 h after LPS (4 mice) or saline (4 mice), or LPS after 7 days of aspirin (4 mice) administration for histopathological analysis. The liver was removed and postfixed in the same fixative overnight at 4 °C.

Liver sections were prepared for hematoxylin and eosin (HE) staining analysis by sectioning the tissue to 5-mm-thick sections using a slicer. Subsequently, the sections were subjected to several procedures. The sections were initially exposed to increasing concentrations of ethanol, cleared with Lemosol A (Wako Pure Chemical Industries, Ltd., Japan), and finally embedded in paraffin. The paraffin-embedded liver blocks were sectioned into 5-μm thick slices using a sliding microtome (REM-710, Yamato Kohki Industrial, Tokyo, Japan). The sections were mounted on glass slides, deparaffinized using Lemosol A, and immersed in a gradient of ethanol and distilled water. Furthermore, they were subjected to HE staining, followed by dehydration using increasing ethanol concentrations. Finally, the sections were cleared using Lemosol A and covered with coverslips. The stained sections were captured using a CCD camera (BZ-X700, Keyence, Osaka, Japan).

The 5-mm-thick sections were cut for SEM observation. Following fixation with 1% osmium tetroxide (OsO_4) solution for 2 h, the sections were dehydrated and replaced with t-butyl alcohol solution. Subsequently, after lyophilization, the liver tissue was placed on an aluminum table for SEM (IT-800, JEOL, Tokyo. Japan).

2.4. Correlative Light and Electron Microscope Analysis

The paraffin-embedded liver blocks were sectioned into 5-μm-thick slices for correlative light and electron microscopy analyses. Sections were mounted on glass slides, deparaffinized using Lemosol A, and immersed in a gradient of ethanol and distilled water. The sections were stained with HE and washed severally with distilled water. The sections were coverslipped with distilled water, and images were captured using a CCD camera (BZ-X700, Keyence, Japan). Following photography, the cover glass was removed, and sections were dehydrated with ethanol, dried, and treated with an OsO_4 coater. The field of view was observed using SEM and optical microscopy.

2.5. Measurement of CRP, IL-1β, and TNF-α in Blood

Mice serum CRP, IL-1β, and TNF-α concentrations were determined using an ELISA kit (ab157712, ab197742, ab208348, Abcam) following the manufacturer's protocol. To examine the effect of pre-administered aspirin on each value, the aspirin and saline groups were compared. Aspirin was administered for 7 days, and whole blood samples from heart were taken after deep anesthesia without LPS (0H, 3 animals), 2 h (2H, 3 animals), 6 h (6H, 3 animals), 12 h (12H, 3 animals) and 24 h (24H, 3 animals) after LPS administration. Subjects were pre-treated with saline for 7 days, and whole blood samples were collected without LPS administration (0H, 3 animals), 2 h after LPS administration (2H, 3 animals), 6 h (6H, 3 animals), 12 h (12H, 3 animals), and 24 h (24H, 3 animals) after deep anesthesia.

Blood samples were centrifuged (1500 rpm, 10 min, 4 °C) to obtain serum, which was stored at −20 °C until analysis. Serum samples were brought to room temperature and diluted appropriately before analysis. The ELISA procedure was as follows: standard solutions and diluted serum samples were added to the ELISA plate and incubated at room temperature for 2 h. The wells were washed four times using a wash solution. Subsequently, the detection antibody was added, and samples were incubated for 10 min at room temperature. The substrate solution (TMB) was added after washing, and the reaction proceeded in the dark for 5 min. The reaction was stopped by the addition of stop solution and the absorbance was measured at 450 nm using a microplate reader. Standard curves were generated to calculate the CRP, IL-1β, and TNF-α concentrations in each sample. The measurements were repeated in triplicate for each sample, and the results are expressed as mean ± standard deviation.

2.6. Data Analysis

The investigators evaluated 200 distinct cells per mouse across the four mice in each group to assess the hepatocyte area. The number of hematoxylin-positive cells within the sinusoids was determined by examining 10 separate images per mouse, with four mice per group. Statistical analysis was conducted using analysis of variance and statistical significance set at $p < 0.05$.

3. Results

3.1. Histopathological Analysis

First, acute LPS-induced hepatitis was analyzed histopathologically. Mice were injected with saline or LPS, and after 24 h, their livers were removed, and HE-stained specimens were analyzed (Figure 1). The livers of the controls (normal) were healthy (Figure 1A). However, hepatocytes in the hepatitis-induced liver (Figure 1B) showed reduced eosin staining. The area of hepatocytes was quantified because the sinusoids appeared to be vastly enlarged (Figure 1C). The results revealed significantly shrunken hepatocytes in mice with LPS-induced liver hepatitis. In addition, a large number of hematoxylin-positive cells were found in the sinusoids of mice with LPS-induced liver hepatitis (yellow arrows in Figure 1A,B). Quantitative results showed a significant increase in the number of hematoxylin-positive cells (Figure 1D). In contrast, there was an increasing trend in the number of hematoxylin-negative cells (black arrows in Figure 1B) in the sinusoids of mice with LPS-induced liver hepatitis; however, the increase was not significant.

3.2. Correlative Light and Electron Microscopy Analysis

A correlative light and electron microscopic method was used for detailed analysis to clarify the hematoxylin-positive cells observed in the sinusoids (Figure 2). HE staining, used in histopathological analysis, as shown in Figure 1, was not performed rather only hematoxylin staining was conducted (Figure 2A,A′). Hematoxylin-positive cells were identified under an optical microscope, treated with an osmium coater, and observed in the same field of view using a SEM (Figure 2B–B″). The hematoxylin-positive cells (yellow arrows) identified under an optical microscope were observed under an SEM. They were recognized as leukocytes using their morphology. In contrast, hematoxylin-negative cells (red arrows) were recognized as erythrocytes based on their morphology. Further analyses of these cells are difficult to perform using SEM, which can observe the cell surface; however, these analyses indicate that LPS treatment induces a significant accumulation of leukocytes in the sinusoids of the liver, and the aggregation of erythrocytes.

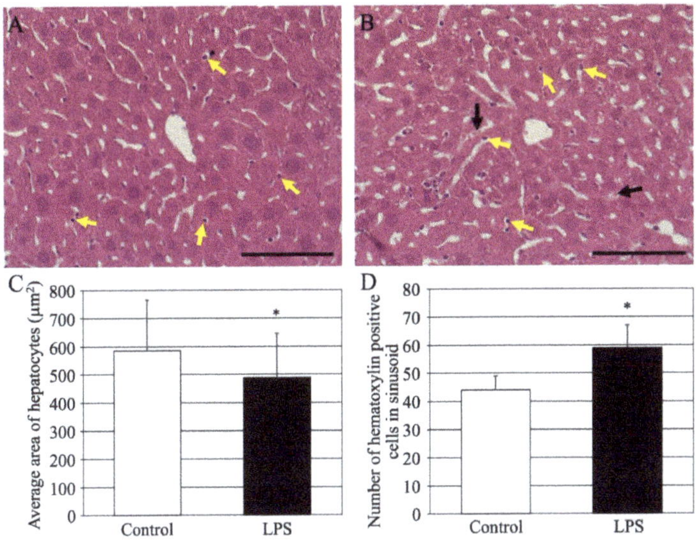

Figure 1. HE-stained images of livers from normal mice (**A**) and LPS-treated mice (**B**). Yellow arrows indicate hematoxylin-positive cells in the sinusoids. Black arrows indicate hematoxylin-negative cells within the sinusoids. Scale bars in (**A**,**B**) are 100 μm. (**C**) Average area in control (white bar) and LPS-treated mice (black bar) of hepatocytes (200 cells/animal, four animals from each group). (**D**) Number of hematoxylin-positive cells within sinusoids in control (white bar) and LPS-treated mice (black bar) of (numbers/200,000 μm^2, 10 pictures/animal, four animals from each group). Each graph shows the mean ± standard deviation. *; $p < 0.05$ compared to control value. HE, hematoxylin-eosin; LPS, liposaccharide. All sale bars are 100 μm.

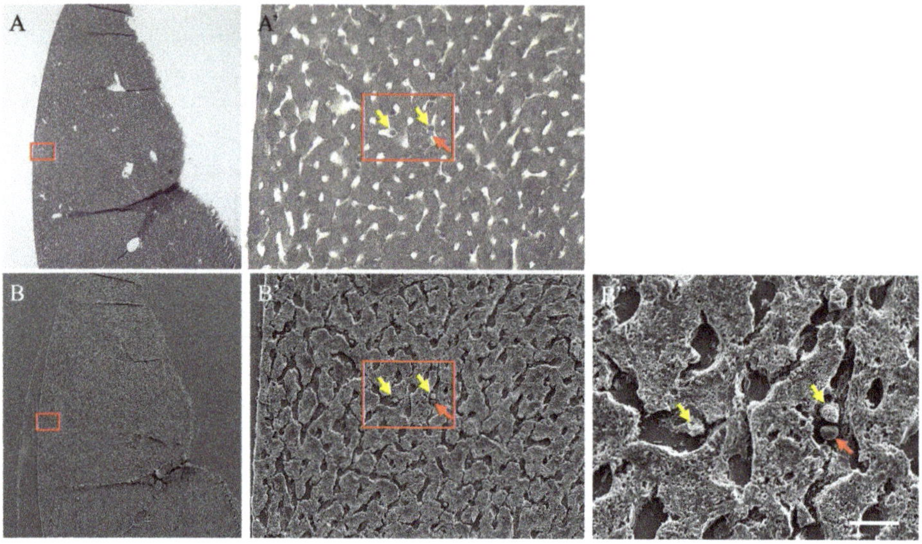

Figure 2. Correlative light and electron microscopical (CLEM) images of livers from liposaccharide-treated mice. (**A**,**A′**) Hematoxylin-stained images. (**B–B″**) Scanning electron microscope images of the same tissue. (**A′**) is a magnified image of the red frame of A. (**B′**) is a magnified image of the red frame of (**B**). (**B″**) is a magnified image of the red frame of (**B′**). Yellow arrows indicate hematoxylin-positive cells in the sinusoids. Red arrows indicate hematoxylin-negative cells. The scale bar in (**B″**) is 10 µm.

3.3. Histopathological and SEM Analysis of the Blood Vessels in the Liver

The above correlative light and electron microscopical analysis indicated that blood accumulation (aggregation) may occur in the sinusoids; however, histopathological analysis was performed on the blood vessels present in the liver (Figure 3). HE-stained images of livers with acute LPS-induced hepatitis showed many blood cells remaining in the blood vessels (Figure 3A). We observed the morphology of the accumulated cells using SEM Figure 3B,B′ based on the staining characteristics, as they may accumulate alongside erythrocytes and leukocytes, and possibly thrombi. Morphological analysis revealed that the accumulated cells were mostly erythrocytes, composed of leukocytes and platelets.

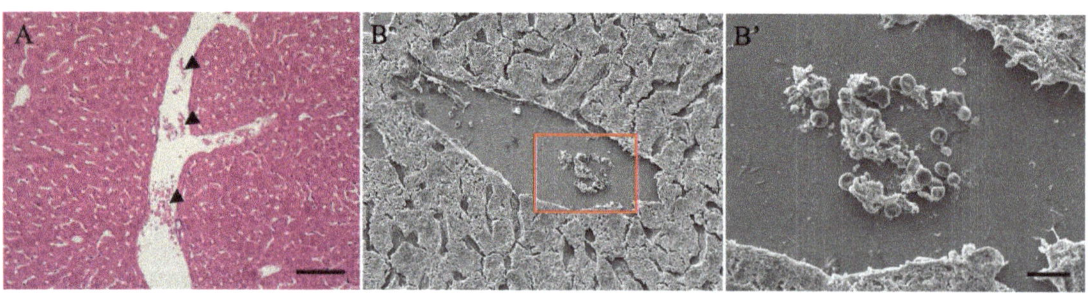

Figure 3. HE-stained (**A**) and SEM (**B**,**B′**) images of liver from LPS-treated mice. (**B′**) is a magnified image of the red frame of (**B**). Black arrowheads indicate hematocytes in the blood vessel. The scale bar in (**A**) is 10 µm, in (**B′**) is 10 µm. HE, hematoxylin-eosin; LPS, liposaccharide; SEM, scanning electron microscope.

3.4. SEM Analysis of the Blood Vessels and Hepatocytes in the Liver

In 5-µm-thick sections, an accumulation of blood cells was observed in blood vessels in the liver in LPS-induced liver hepatitis; however, it was impossible to determine if the accumulation was a thrombus. Therefore, SEM analysis was performed on thick sections (Figure 4). We identified numerous thrombi in the vessels of patients with LPS-induced liver hepatitis. These results indicate that thrombi are induced in blood vessels within the liver in LPS-induced liver hepatitis; however, they could not be determined in thin sections.

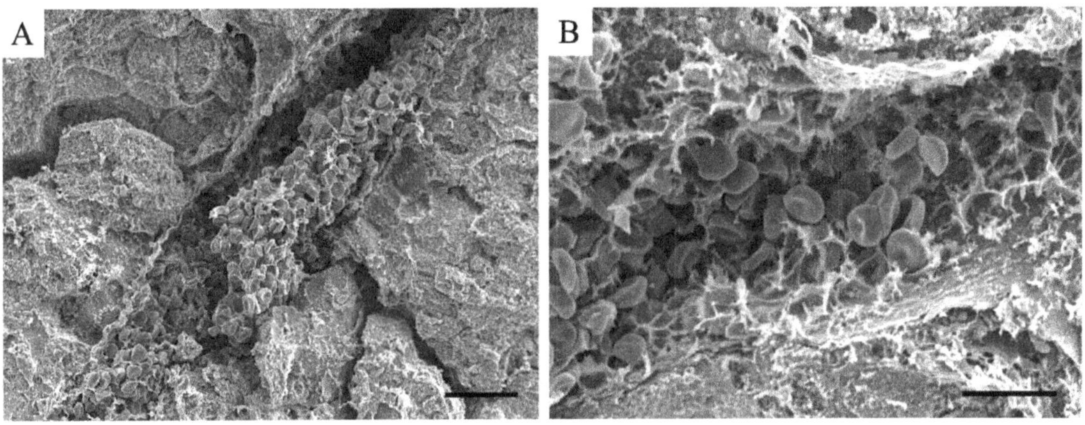

Figure 4. SEM images of livers from LPS-treated mice. (**A**) lower magnified image and (**B**) higher magnified image. The scale bar in (**A**) is 20 µm, (**B**) is 10 µm. LPS, liposaccharide; SEM, scanning electron microscope.

Based on these histopathological analyses and a series of SEM analyses, the accumulation of blood cells in the blood vessels observed under an optical microscope was determined to be a thrombus. Therefore, the presence or absence of a thrombus can be determined by the presence or absence of blood cell accumulation under an optical microscope.

Using SEM, a detailed observation of the inner surface of the vessel and the surface of the hepatocytes was possible. Therefore, we observed the inner surface of the blood vessels, in which no thrombus was observed, for confirmation (Figure 5A,B). The inner surface of endothelial cells within the liver of the controls was healthy (Figure 5A). In contrast, in the blood vessels within the liver in LPS-induced liver hepatitis, some areas showed rough inner surfaces (yellow arrows in Figure 5B), whereas others showed intact surfaces. A large thrombus was not noticed in the observed areas of the endothelium in this analysis; however, it is highly possible that the thrombus was caused by the disruption of the endothelium. Furthermore, capillary bile ducts were observed on the surface of the hepatocytes (Figure 5C,D). Using this analysis, no significant difference was observed in the capillary bile ducts of LPS-induced liver hepatitis (red arrow in Figure 5D) compared to the controls (red arrow in Figure 5C). This result indicates a normal bile transport function in the hepatocytes of LPS-induced liver hepatitis mice.

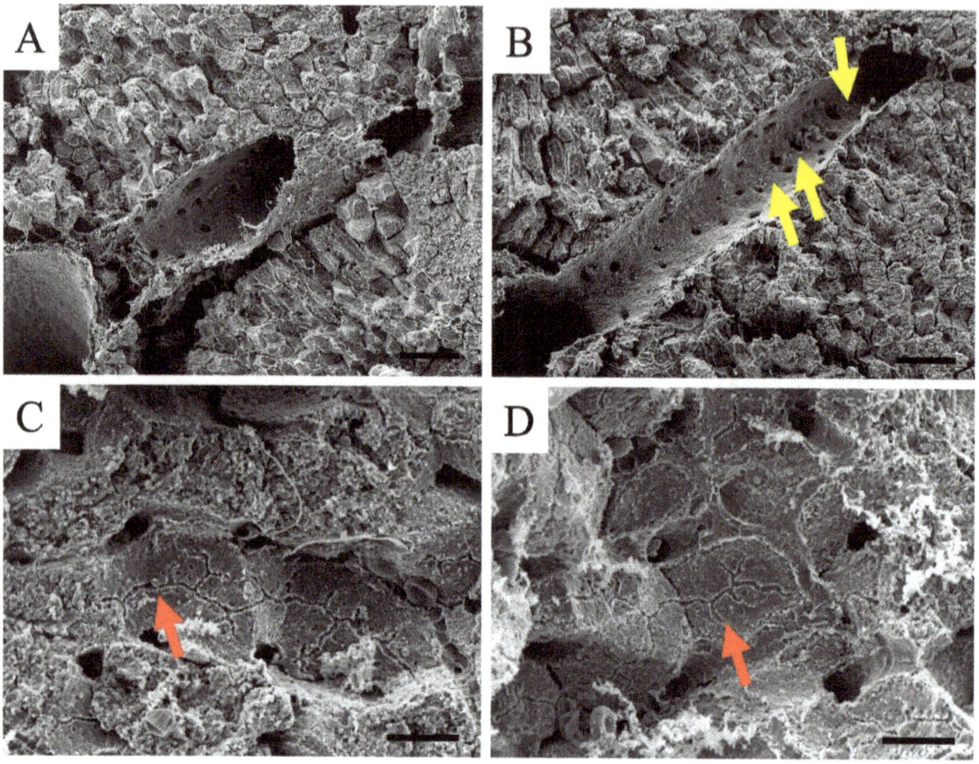

Figure 5. SEM images of livers from normal (**A**,**C**) and LPS-treated (**B**,**D**) mice. (**A**,**B**) The images of the inner surface of endothelial cells. (**C**,**D**) The images of capillary bile ducts. Yellow arrows in B indicate the rough inner surfaces of endothelial cells. Red arrows in (**C**,**D**) indicate the capillary bile ducts. LPS, liposaccharide; SEM, scanning electron microscope. Scale bars in A and B are 50 μm, and in C and D are 10 μm.

3.5. Thromboprophylaxis Effect of Aspirin Pre-Administration

Aspirin is believed to have a prophylactic effect on blood clots; however, there are no reports on its preventive effects on hepatocytes and liver thrombosis. Therefore, aspirin was administered for one week, followed by LPS-induced inflammation. Histopathological analysis showed that the size (area) of hepatocytes was 560.28 ± 167.2 (Figure 6A), which was not significantly different from that of normal animals (586.05 ± 180.1; Figure 1C) and improved. In addition, the number of hematoxylin-positive cells in the sinusoids was 47.12 ± 6.5 (Figure 6A), which was not significantly different from that in normal animals (44.19 ± 4.9; Figure 1D), indicating improvement. However, eosin staining of hepatocytes showed a slightly lighter tone, suggesting that the effect of LPS was reflected in the hepatocytes. In contrast, no accumulation of blood cells, considered a thrombus, was observed in the blood vessels of the liver. This indicates that aspirin is effective on the inflammation induced by LPS in the liver; however, its preventive effect on hepatocytes may not be complete.

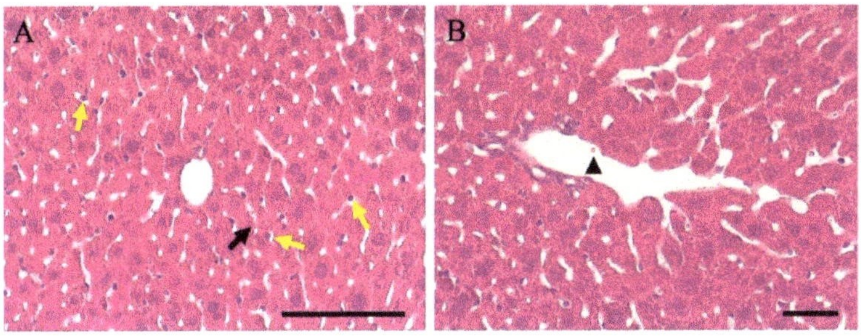

Figure 6. HE-stained image of hepatitis-induced mouse liver by LPS after pre-administration of aspirin. Yellow arrows indicate hematoxylin-positive cells in the sinusoids. Black arrows indicate hematoxylin-negative cells within the sinusoids. Black arrowheads indicate hematocytes in the blood vessel. Scale bars in (**A**,**B**) are 100 μm. HE, hematoxylin-eosin; LPS, liposaccharide; SEM.

Next, we examined whether aspirin's reduction of liver inflammation was effective in preventing inflammation in the blood (Figure 7). First, CRP levels increased with time after LPS administration; however, they were significantly reduced in the pre-aspirin group (Figure 7A). IL-1b levels peaked at 6 to 12 h after LPS administration; nevertheless, they were significantly reduced by pre-aspirin administration (Figure 7B). TNF-a peaked at 2 h after LPS administration; however, this increase was significantly reduced by pre-administration of aspirin (Figure 7C). These findings suggest that aspirin pre-administration significantly decreases liver and blood inflammation.

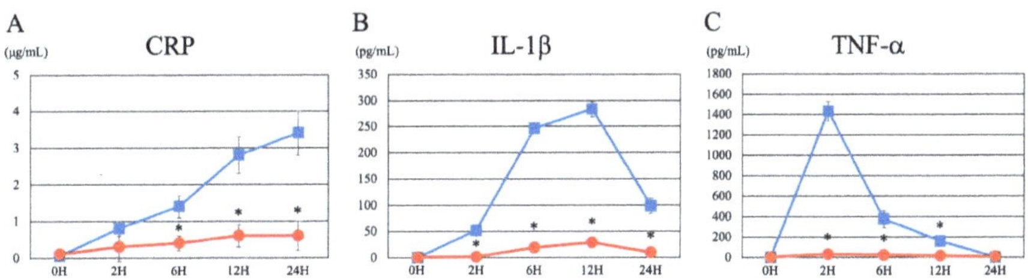

Figure 7. Changes in each inflammatory marker over time after saline pre-treatment and aspirin pre-treatment. (**A**) CRP (**B**) IL-1β (**C**) TNF-α. The blue line represents the control group (saline pre-treatment). The red line represents the aspirin group. 0H indicates values without LPS. 2H, 6H, 12H, and 24H indicate values 2 h, 6 h, 12 h, and 24 h after LPS administration, respectively. Each graph shows the mean ± standard deviation. *; $p < 0.05$ compared to control value. CRP, C-reactive protein; IL-1β, interleukin 1β; TNF-α; tumor necrosis factor alpha; LPS, liposaccharide.

4. Discussion

In this study, we investigated LPS-induced histopathological changes in the liver and the preventive effect of aspirin. We investigated the effects of LPS on hepatocytes, vascular endothelial damage, thrombus formation, and inflammatory cell infiltration. Hepatocyte shrinkage, sinusoidal swelling, and vascular endothelial damage were observed after LPS administration (Figure 1). Histopathological analysis revealed abnormal accumulation of leukocytes and erythrocytes in the sinusoids and blood vessels of the liver. Electron microscopy demonstrated that this accumulation was responsible for inflammation, induction of platelet aggregation, and thrombus formation, suggesting its major role in the progression of inflammation (Figures 2 and 3).

Furthermore, we discovered that the pre-administration of aspirin before LPS administration alleviated these pathological changes (Figure 6). In addition, aspirin reduced the levels of LPS-induced inflammatory markers (CRP, IL-1β, and TNF-α) and had an inhibitory effect on vascular endothelial damage and platelet aggregation (Figure 7). These results confirmed that the pre-administration of aspirin inhibited thrombus formation and the abnormal accumulation of inflammatory cells in the liver, effectively blocking the progression of LPS-induced liver inflammation. These findings suggested that aspirin's anti-inflammatory and antiplatelet effects may be useful as prophylaxis against LPS-induced liver injury.

This study is consistent with previous findings on LPS-induced liver inflammation, confirming that LPS produces a potent inflammatory response that causes the elevation of CRP, IL-1β, and TNF-α [41,42]. This demonstrated the process of vascular endothelial damage within the liver, the promotion of platelet aggregation, and the occurrence of thrombus formation. These inflammatory cascades occur when LPS invades the bloodstream, thereby stimulating the immune system and the production of multiple cytokines and chemokines, leading to endothelial cell damage and platelet activation [7,21,22]. We believe that these findings are consistent with the immune response mechanisms triggered when LPS functions as a cell wall component in gram-negative bacteria, and provide histopathological evidence for a previous study [22].

In addition, we gained a new perspective on the effects of LPS on the liver, particularly in the sinusoids. In this study, we demonstrated that LPS-induced liver tissue damage is accompanied by abnormal histopathological changes, including leukocyte and erythrocyte accumulation and endothelial roughness. The accumulation of blood cells in the sinusoids can cause thrombus formation and inflammatory cell infiltration in the blood vessels. Our results indicate that the release of pro-inflammatory cytokines is important in the primary mechanism by which LPS causes endothelial damage in the liver, and is consistent with previous studies [22,42,43]. In addition, pathological findings, including endothelial cell roughness and abnormal blood cell accumulation, are important indicators of LPS-induced microvascular abnormalities within the liver. Furthermore, they provide valuable information for understanding the progression of liver inflammation which causes vascular system damage.

In this study, we investigated the mechanism by which aspirin reduces LPS-induced inflammatory damage. Aspirin acts primarily by inhibiting TXA2 formation through inhibition of COX enzymes, particularly COX-1 and COX-2 [30–33]. TXA2 is an important mediator of platelet aggregation, aspirin prevents platelet aggregation and blood clot formation by suppressing its formation. This anti-platelet effect is expected to help maintain smooth blood flow and inhibit the accumulation of microthrombi in the liver's blood vessels. Furthermore, aspirin has potent anti-inflammatory effects that may suppress LPS-induced hepatic inflammation. The anti-inflammatory effects of aspirin are associated with the inhibition of inflammatory markers such as CRP, IL-1β, and TNF-α. These cytokines play important roles in the inflammatory cascade, causing endothelial damage and thrombus formation [34]. The results of this study indicate that aspirin suppresses the production of these inflammatory mediators, thereby protecting the liver endothelial cells from damage and inhibiting thrombus formation.

More interestingly, aspirin may reduce oxidative stress through its antioxidant properties [44]. Inflammatory reactions induced by LPS have been shown to increase oxidative stress and damage endothelial cells; however, aspirin may promote endothelial cell protection by reducing the level of oxidative stress. In addition, aspirin has been reported to inhibit vascular permeability, which may reduce excessive fluid leakage from the blood vessels and tissue swelling [45]. These mechanisms of action may be important for reducing liver inflammation, protecting endothelial integrity, and ameliorating histopathological changes. Collectively, these results indicate that aspirin's anti-platelet, anti-inflammatory, and antioxidant properties, may work together to reduce LPS-induced liver injury and contribute to tissue protection. This study suggests that aspirin is not just an anti-platelet agent, but has a multifaceted preventive and protective effects on liver inflammation.

This study's findings indicate that aspirin may be useful as an adjunct therapy to reduce LPS-induced liver injury. Pre-administration of aspirin significantly reduced inflammatory markers (CRP, IL-1β, TNF-α, etc.) and thrombus formation, suggesting its potential in reducing the effects of systemic inflammation on the liver. In particular, aspirin use may play an important role in protecting the liver and preventing vascular complications such as sepsis and acute liver failure, where systemic inflammation is amplified through LPS. The effects of aspirin extend beyond its antiplatelet action; aspirin promotes the reduction of vascular permeability and protection of endothelial cells within the liver through its anti-inflammatory effects and reduction of oxidative stress. This multifaceted effect may have great clinical significance in the management of liver inflammation and the prevention of thrombosis, especially in patients prone to severe LPS-induced inflammatory liver injury. Furthermore, because LPS-induced endothelial damage and vascular inflammation can cause serious thrombotic events such as myocardial infarction and pulmonary thromboembolism, aspirin use may have a dual benefit for patients, by simultaneously inhibiting platelet aggregation and inflammation. Prophylactic aspirin administration in patients with chronic liver disease and high-risk groups should be considered. Patients with chronic liver disease often have coexisting abnormalities in their coagulation and inflammatory systems, and the risk of thrombus formation increases with worsening liver function. In the context of preventive medicine and long-term disease management, a comprehensive treatment strategy that includes aspirin should be considered, as it may reduce the risk of thrombosis in these patients and help maintain liver function.

This study had some limitations. First, an animal model, which may not fully replicate the LPS-induced inflammatory response in humans, was used. In addition, aspirin significantly reduced inflammatory marker levels; however, residual hepatocellular damage was observed, suggesting that its protective effect may not be complete at high doses of LPS. Therefore, future studies should investigate the long-term effects of aspirin on liver tissue repair and its combined effects with other anti-inflammatory drugs at different stages of liver inflammation. In addition, optimizing aspirin dosing schedules and examining its effects on other inflammatory pathways may suggest the possibility of enhancing therapeutic efficacy.

In this study, we demonstrated LPS-induced liver inflammation and thrombus formation through histopathological and micro-morphological examinations using electron microscopy. These results indicated that the accumulation of blood cells observed in specimens used in general pathology was likely a thrombus, and may provide useful information for clinical diagnosis. Furthermore, we discovered that aspirin pre-administration effectively reduced LPS-induced liver inflammation and thrombus formation, and exerted a protective effect on the liver through its antiplatelet and anti-inflammatory effects. These results indicated that aspirin contributed to the maintenance of liver health by suppressing inflammatory cytokines, protecting endothelial cells, and preventing platelet aggregation. In addition, aspirin can be applied as a therapeutic option in the liver and systemic inflammatory diseases. It is important to review its position in prevention and treatment strategies, including its cost-effectiveness and versatility. Further clinical studies are required to investigate the range of applications and optimal administration of aspirin, thereby contributing to the establishment of a comprehensive treatment for inflammatory diseases and thrombosis.

5. Conclusions

In this study, we elucidated LPS-induced liver inflammation and thrombus formation through histopathological and electron microscopy and micromorphological examinations, revealing that the accumulation of blood cells observed on routine pathology examination could be a thrombus. Furthermore, aspirin pre-administration effectively inhibited LPS-induced hepatitis and thrombus formation and conferred hepatoprotection through its antiplatelet and anti-inflammatory effects. Notably, aspirin inhibited inflammatory cytokines, protected endothelial cells, and prevented platelet aggregation, indicating its potential as a therapeutic agent for hepatic and systemic inflammatory diseases. Future clinical studies are expected to contribute to the establishment of a comprehensive treatment for inflammatory diseases and thrombosis by investigating the range of applications and optimal aspirin administration methods.

Author Contributions: Conceptualization, H.S., K.K. and K.N.; methodology, H.S., M.S., F.M., N.M., N.I., K.T., K.K. and K.N.; formal analysis, H.S., M.S., F.M., N.M., N.I. and K.T.; writing—original draft preparation, K.K. and K.N.; writing—review and editing, K.K. and K.N.; funding acquisition, K.N. All authors have read and agreed to the published version of the manuscript.

Funding: This work was supported by JSPS KAKENHI Grant Number 23K08620 (K.N.).

Institutional Review Board Statement: This study was reviewed and approved by the relevant committees at Meiji Pharmaceutical University (No. 2704, 1 April 2023–2024).

Informed Consent Statement: Not applicable.

Data Availability Statement: The datasets used and/or analyzed during this study are available from the corresponding author upon reasonable request.

Conflicts of Interest: The authors declare no conflicts of interest.

References

1. Ohtsuki, T.; Kimura, K.; Tokunaga, Y.; Tsukiyama-Kohara, K.; Tateno, C.; Hayashi, Y.; Hishima, T.; Kohara, M. M2 Macrophages Play Critical Roles in Progression of Inflammatory Liver Disease in Hepatitis C Virus Transgenic Mice. *J. Virol.* **2016**, *90*, 300–307. [CrossRef]
2. Mohan, P.; Chandra, R.S.; Escolar, D.M.; Luban, N.L. Inflammatory myopathy and hepatitis C in a pediatric patient: Role of liver biopsy in evaluating the severity of liver disease. *Hepatology* **2001**, *34*, 851–852. [CrossRef]
3. Yoshihara, S.; Harada, K.; Ohta, S. Metabolism of 1-methyl-4-phenyl-1,2,3,6-tetrahydropyridine (MPTP) in perfused rat liver: Involvement of hepatic aldehyde oxidase as a detoxification enzyme. *Drug Metab. Dispos.* **2000**, *28*, 538–543.
4. Shin, M.S.; Kang, E.H.; Lee, Y.I. A flavonoid from medicinal plants blocks hepatitis B virus-e antigen secretion in HBV-infected hepatocytes. *Antiviral Res.* **2005**, *67*, 163–168. [CrossRef] [PubMed]
5. Hayashi, M.; Kanda, T.; Nakamura, M.; Miyamura, T.; Yasui, S.; Nakamoto, S.; Wu, S.; Arai, M.; Imazeki, F.; Yokosuka, O. Acute liver injury in a patient with alcohol dependence: A case resembling autoimmune hepatitis or drug-induced liver injury. *Case Rep. Gastroenterol.* **2014**, *8*, 129–133. [CrossRef] [PubMed]
6. Odena, G.; Chen, J.; Lozano, J.J.; Altamirano, J.; Rodrigo-Torres, D.; Affo, S.; Morales-Ibanez, O.; Matsushita, H.; Zou, J.; Dumitru, R.; et al. LPS-TLR4 Pathway Mediates Ductular Cell Expansion in Alcoholic Hepatitis. *Sci. Rep.* **2016**, *6*, 35610. [CrossRef] [PubMed]
7. Maldonado, R.F.; Sa-Correia, I.; Valvano, M.A. Lipopolysaccharide modification in Gram-negative bacteria during chronic infection. *FEMS Microbiol. Rev.* **2016**, *40*, 480–493. [CrossRef]
8. Li, M.; Zhao, Y.; Liu, X.; Dang, Z.; Wang, X.; Jiang, Y.; Yang, Z. Association and interaction between model for end-stage liver disease score and minimally invasive treatment with regard to mortality of patients with hepatitis B virus-associated hepatocellular carcinoma and portal vein tumor thrombi. *Oncol. Lett.* **2019**, *17*, 119–126. [CrossRef]
9. Kohler, J.; Ehler, J.; Kreikemeyer, B.; Bajorath, R.; Schurholz, T.; Oehmcke-Hecht, S. The synthetic LPS binding peptide 19-2.5 interferes with clotting and prevents degradation of high molecular weight kininogen in plasma. *Sci. Rep.* **2020**, *10*, 7142. [CrossRef] [PubMed]
10. Schafer, A.I.; Mann, D.L. Thrombotic, Cardiovascular, and Microvascular Complications of Myeloproliferative Neoplasms and Clonal Hematopoiesis (CHIP): A Narrative Review. *J. Clin. Med.* **2024**, *13*, 6084. [CrossRef]

11. Feigin, V.L.; Abate, M.D.; Abate, Y.H.; Abd ElHafeez, S.; Abd-Allah, F.; Abdelalim, A.; Abdelkader, A.; Abdelmasseh, M.; Abd-Elsalam, S.; Abdi, P.; et al. Global, regional, and national burden of stroke and its risk factors, 1990–2021: A systematic analysis for the Global Burden of Disease Study 2021. *Lancet Neurol.* **2024**, *23*, 973–1003. [CrossRef]
12. Gao, X.; Chen, H.; Huang, Z.; Lin, J.; Huang, J.; Chen, Q. Correlation Between Neutrophil-to-Lymphocyte Ratio and Platelet-to-Lymphocyte Ratio with Risk Stratification Indicators and Thrombus Burden in Patients with Moderate-to-High Risk Acute Pulmonary Embolism, and Changes After Treatment. *Clin. Appl. Thromb. Hemost.* **2024**, *30*, 10760296241285446. [CrossRef]
13. Qiu, J.; Yu, Y.; Wang, Z.; Hong, L.; Shao, L.; Wu, J. Developing Individualized Follow-Up Strategies Based on High-Risk Recurrence Factors and Dynamic Risk Assessment for Locally Advanced Rectal Cancer. *Cancer Med.* **2024**, *13*, e70323. [CrossRef] [PubMed]
14. Chen, Z.H.; Zhang, X.P.; Feng, S.; Feng, J.K.; Chai, Z.T.; Guo, W.X.; Shi, J.; Lau, W.Y.; Meng, Y.; Cheng, S.Q. Liver resection versus intensity-modulated radiation therapy for treatment of hepatocellular carcinoma with hepatic vein tumor thrombus: A propensity score matching analysis. *Hepatobiliary Surg. Nutr.* **2021**, *10*, 646–660. [CrossRef]
15. Ichida, A.; Kokudo, T.; Shimada, S.; Hatano, E.; Kubo, S.; Kato, Y.; Ishikawa, Y.; Mori, A.; Baba, H.; Matsuyama, Y.; et al. Liver Resection for Hepatocellular Carcinoma With Tumor Thrombus in the Inferior Vena Cava or Right Atrium: A Large-scale Multicenter Survey Conducted in Japan. *Ann. Surg.* **2023**, *278*, e549–e555. [CrossRef]
16. Bultink, I.E.M. Dialogue: Antiplatelet effects of hydroxychloroquine in patients with systemic lupus erythematosus evaluated by the total thrombus-formation analysis system (T-TAS). *Lupus Sci. Med.* **2024**, *11*, e001223. [CrossRef] [PubMed]
17. Tourdot, B.E.; Stoveken, H.; Trumbo, D.; Yeung, J.; Kanthi, Y.; Edelstein, L.C.; Bray, P.F.; Tall, G.G.; Holinstat, M. Genetic Variant in Human PAR (Protease-Activated Receptor) 4 Enhances Thrombus Formation Resulting in Resistance to Antiplatelet Therapeutics. *Arterioscler. Thromb. Vasc. Biol.* **2018**, *38*, 1632–1643. [CrossRef] [PubMed]
18. Kellermair, L.; Hofer, C.; Zeller, M.W.G.; Kubasta, C.; Bandke, D.; Weis, S.; Kellermair, J.; Forstner, T.; Helbok, R.; Vosko, M.R. Endothelial receptor proteins in acute venous thrombosis and delayed thrombus resolution in cerebral sinus vein thrombosis. *J. Neurol.* **2024**, *271*, 4095–4104. [CrossRef] [PubMed]
19. Rossi, E.; Pericacho, M.; Kauskot, A.; Gamella-Pozuelo, L.; Reboul, E.; Leuci, A.; Egido-Turrion, C.; El Hamaoui, D.; Marchelli, A.; Fernandez, F.J.; et al. Soluble endoglin reduces thrombus formation and platelet aggregation via interaction with alphaIIbbeta3 integrin. *J. Thromb. Haemost.* **2023**, *21*, 1943–1956. [CrossRef]
20. Ma, J.; Zhu, J.; Ding, T.; Cai, L.; Zhou, C.; Zhang, Y. Thrombus formation in the suprahepatic inferior vena cava after microwave ablation in patients with hepatic metastasis: A case report. *Thromb. J.* **2023**, *21*, 36. [CrossRef] [PubMed]
21. Zhang, J.; Oueslati, R.; Cheng, C.; Zhao, L.; Chen, J.; Almeida, R.; Wu, J. Rapid, highly sensitive detection of Gram-negative bacteria with lipopolysaccharide based disposable aptasensor. *Biosens. Bioelectron.* **2018**, *112*, 48–53. [CrossRef] [PubMed]
22. Gonzalez-Fernandez, C.; Bringas, E.; Oostenbrink, C.; Ortiz, I. In silico investigation and surmounting of Lipopolysaccharide barrier in Gram-Negative Bacteria: How far has molecular dynamics Come? *Comput. Struct. Biotechnol. J.* **2022**, *20*, 5886–5901. [CrossRef] [PubMed]
23. Hsu, C.S.; Chang, S.H.; Yang, R.C.; Lee, C.H.; Lee, M.S.; Kao, J.K.; Shieh, J.J. Lipopolysaccharide-Induced Lysosomal Cell Death Through Reactive Oxygen Species in Rat Liver Cell Clone 9. *Environ. Toxicol.* **2024**, *39*, 5008–5018. [CrossRef]
24. Guo, L.; Liu, Y.; Chen, Y.; Xu, J.; Liu, Y. Liver macrophages show an immunotolerance phenotype in nonalcoholic fatty liver combined with Porphyromonas gingivalis-lipopolysaccharide infection. *Hua Xi Kou Qiang Yi Xue Za Zhi* **2023**, *41*, 385–394. [CrossRef]
25. Tiirola, T.; Sinisalo, J.; Nieminen, M.S.; Silvennoinen-Kassinen, S.; Paldanius, M.; Saikku, P.; Jauhiainen, M.; Leinonen, M. Chlamydial lipopolysaccharide is present in serum during acute coronary syndrome and correlates with CRP levels. *Atherosclerosis* **2007**, *194*, 403–407. [CrossRef] [PubMed]
26. Tenforde, M.W.; Gupte, N.; Dowdy, D.W.; Asmuth, D.M.; Balagopal, A.; Pollard, R.B.; Sugandhavesa, P.; Lama, J.R.; Pillay, S.; Cardoso, S.W.; et al. C-reactive protein (CRP), interferon gamma-inducible protein 10 (IP-10), and lipopolysaccharide (LPS) are associated with risk of tuberculosis after initiation of antiretroviral therapy in resource-limited settings. *PLoS ONE* **2015**, *10*, e0117424. [CrossRef] [PubMed]
27. Yang, Y.; Wang, Y.; Wang, C.; Wu, S.; Yao, D. Macrophages and derived-TNF-alpha promote lipopolysaccharide-induced upregulation of endogenous beta-glucuronidase in the epithelial cells of the bile duct: A possible facilitator of hepatolithiasis formation. *Clin. Res. Hepatol. Gastroenterol.* **2023**, *47*, 102062. [CrossRef]
28. Komegae, E.N.; Fonseca, M.T.; Steiner, A.A. Diet-induced obesity attenuates the hypothermic response to lipopolysaccharide independently of TNF-alpha production. *Temperature* **2020**, *7*, 270–276. [CrossRef]
29. Lee, Y.; Ju, X.; Cui, J.; Zhang, T.; Hong, B.; Kim, Y.H.; Ko, Y.; Park, J.; Choi, C.H.; Heo, J.Y.; et al. Mitochondrial dysfunction precedes hippocampal IL-1beta transcription and cognitive impairments after low-dose lipopolysaccharide injection in aged mice. *Heliyon* **2024**, *10*, e28974. [CrossRef] [PubMed]
30. Tacconelli, S.; Contursi, A.; Falcone, L.; Mucci, M.; D'Agostino, I.; Fullone, R.; Sacco, A.; Zucchelli, M.; Bruno, A.; Ballerini, P.; et al. Characterization of cyclooxygenase-2 acetylation and prostanoid inhibition by aspirin in cellular systems. *Biochem. Pharmacol.* **2020**, *178*, 114094. [CrossRef]

31. Bastaki, S.M.A.; Padol, I.T.; Amir, N.; Hunt, R.H. Effect of Aspirin and ibuprofen either alone or in combination on gastric mucosa and bleeding time and on serum prostaglandin E(2) and thromboxane A(2) levels in the anaesthetized rats in vivo. *Mol. Cell Biochem.* **2018**, *438*, 25–34. [CrossRef]
32. Yin, Y.; Geng, H.; Cui, S.; Hua-Ji, L. Correlation of platelet-derived growth factor and thromboxane A2 expression with platelet parameters and coagulation indices in chronic altitude sickness patients. *Exp. Physiol.* **2022**, *107*, 807–812. [CrossRef] [PubMed]
33. Mizurini, D.M.; Aslan, J.S.; Gomes, T.; Ma, D.; Francischetti, I.M.; Monteiro, R.Q. Salivary Thromboxane A2-Binding Proteins from Triatomine Vectors of Chagas Disease Inhibit Platelet-Mediated Neutrophil Extracellular Traps (NETs) Formation and Arterial Thrombosis. *PLoS Negl. Trop. Dis.* **2015**, *9*, e0003869. [CrossRef] [PubMed]
34. Feng, Y.; Zhang, H.; Dai, S.; Li, X. Aspirin treatment for unruptured intracranial aneurysms: Focusing on its anti-inflammatory role. *Heliyon* **2024**, *10*, e29119. [CrossRef]
35. Li, M.; Wang, J.; Wang, X.; Li, G. Clinical efficacy of aspirin combined with clopidogrel in treating cerebral infarction and its effect on serum hs-CRP, sICAM-1 and TNF-alpha. *Exp. Ther. Med.* **2020**, *19*, 939–944. [CrossRef]
36. Wu, Z.; Wu, Y.; Zhong, W.; Zhong, Q.; Rao, S.; Yu, D.; Luo, X.; Qiu, F.; Song, Z.; Jin, D.; et al. The hepatoprotective effect of aspirin on carbon tetrachloride-induced hepatic fibrosis via inhibition of TGFbeta-1 pathway and pro-inflammatory cytokines IL-1beta and COX-2 in rats. *Exp. Ther. Med.* **2023**, *25*, 232. [CrossRef] [PubMed]
37. Rood, K.M.; Patel, N.; DeVengencie, I.M.; Quinn, J.P.; Gowdy, K.M.; Costantine, M.M.; Kniss, D.A. Aspirin modulates production of pro-inflammatory and pro-resolving mediators in endothelial cells. *PLoS ONE* **2023**, *18*, e0283163. [CrossRef]
38. Hikita, M.; Motojima, K.; Kamata, S.; Yoshida, T.; Tanaka-Nakadate, S.; Nakadate, K. Protective Efficacy of the Ingestion of Mandarin Orange Containing beta-Cryptoxanthin on Lipopolysaccharide-induced Acute Nephritis. *Yakugaku Zasshi* **2016**, *136*, 1031–1040. [CrossRef]
39. Itoh, M.; Nakadate, K.; Horibata, Y.; Matsusaka, T.; Xu, J.; Hunziker, W.; Sugimoto, H. The structural and functional organization of the podocyte filtration slits is regulated by Tjp1/ZO-1. *PLoS ONE* **2014**, *9*, e106621. [CrossRef]
40. Nakadate, K.; Sono, C.; Mita, H.; Itakura, Y.; Kawakami, K. Severe Acute Liver Dysfunction Induces Delayed Hepatocyte Swelling and Cytoplasmic Vacuolization, and Delayed Cortical Neuronal Cell Death. *Int. J. Mol. Sci.* **2023**, *24*, 7351. [CrossRef]
41. Huang, H.; Wu, S.; Yan, Y. Growth differentiation factor 11 suppresses intrahepatic inflammation via restricting NLRP3 inflammasome activation in LPS-induced liver injury. *Cell Mol. Biol.* **2023**, *69*, 250–255. [CrossRef] [PubMed]
42. Khan, H.U.; Aamir, K.; Jusuf, P.R.; Sethi, G.; Sisinthy, S.P.; Ghildyal, R.; Arya, A. Lauric acid ameliorates lipopolysaccharide (LPS)-induced liver inflammation by mediating TLR4/MyD88 pathway in Sprague Dawley (SD) rats. *Life Sci.* **2021**, *265*, 118750. [CrossRef] [PubMed]
43. Sae-Tan, S.; Kumrungsee, T.; Yanaka, N. Mungbean seed coat water extract inhibits inflammation in LPS-induced acute liver injury mice and LPS-stimulated RAW 246.7 macrophages via the inhibition of TAK1/IkappaBalpha/NF-kappaB. *J. Food Sci. Technol.* **2020**, *57*, 2659–2668. [CrossRef] [PubMed]
44. Liang, H.; Shi, H.; Li, Y.; Wang, D.; Zhang, Y. Mechanism of Aspirin oxidative stress regulating interleukin-induced apoptosis in nucleus pulposus cells in a rat model of intervertebral disc degeneration. *Ann. Transl. Med.* **2023**, *11*, 124. [CrossRef] [PubMed]
45. Serhan, C.N.; Takano, T.; Chiang, N.; Gronert, K.; Clish, C.B. Formation of endogenous "antiinflammatory" lipid mediators by transcellular biosynthesis. Lipoxins and aspirin-triggered lipoxins inhibit neutrophil recruitment and vascular permeability. *Am. J. Respir. Crit. Care Med.* **2000**, *161*, S95–S101. [CrossRef]

current issues in molecular biology

MDPI

Article

Long-Term Outcomes in Patients with EGFR Positive Lung Adenocarcinoma and Subgroup Analysis Based on Presence of Liver Metastases

Vesna Ćeriman Krstić [1,2,*], Ivan Soldatović [3], Natalija Samardžić [2], Milija Gajić [2], Milica Kontić [1,2], Aleksandar Reljić [4], Milan Savić [1,5], Marina Roksandić Milenković [6] and Dragana Jovanović [7]

1 Faculty of Medicine, University of Belgrade, 11000 Belgrade, Serbia; milicakontic@yahoo.com (M.K.); drmilansavic@gmail.com (M.S.)
2 Clinic for Pulmonology, University Clinical Center of Serbia, 11000 Belgrade, Serbia; natalis.dm@gmail.com (N.S.); milijagajic@gmail.com (M.G.)
3 Institute of Medical Statistics, Faculty of Medicine, University of Belgrade, 11000 Belgrade, Serbia; soldatovic.ivan@gmail.com
4 Clinic for Ortopedics, University Clinical Center of Serbia, 11000 Belgrade, Serbia; reljic1993@gmail.com
5 Clinic for Thoracic Surgery, University Clinical Center of Serbia, 11000 Belgrade, Serbia
6 Municipal Institute for Lung Diseases and TB, 11000 Belgrade, Serbia; dr.marr@gmail.com
7 Internal Medicine Clinic "Akta Medica", 11000 Belgrade, Serbia; draganajv@yahoo.com
* Correspondence: ceriman.vesna@gmail.com

Abstract: Lung cancer represents the most common cause of cancer related death. Patients with non-small cell lung cancer (NSCLC) and liver metastases (LM) have worse prognosis with an overall survival (OS) of three to six months. The aim of this study was to investigate long-term outcomes in patients with EGFR mutated (EGFRmut) lung adenocarcinoma as well as the presence of LM. (A total of 105 patients were included in the analysis). They were divided into two groups based on the presence of LM. OS was 13 months for the whole group and also 13 months for patients with and without LM. The 9-year survival rate for patients with and without LM was 12.5% and 3.4%, respectively. Further, the 9-year survival rate for the whole group of patients was 4.8%. There are few data about survival rates beyond 5 years for patients with locally advanced and metastatic EGFRmut NSCLC, mainly because patients with lung cancer rarely live for such a long time. Regarding patients with liver metastases, the results of our study showed similar outcomes compared to patients without LM. As these patients represent a significant number of patients, we need a wider range of therapeutic options. It might be that combination therapies represent a better therapeutic option.

Keywords: NSCLC; EGFR; liver metastases; immunotherapy; combination therapies; radiotherapy

1. Introduction

Lung cancer is the most common cause of death among all cancers [1]. Non-small cell lung cancer (NSCLC) represents around 85% of all lung cancers [1]. In the majority of patients, the disease is diagnosed at the locally advanced or metastatic stage [1]. Standard treatment for these patients, in the past, was platinum doublets, with a 5-year survival rate less than 10% [1]. The introduction of molecular therapy in the treatment of NSCLC has led to better clinical outcomes [2–4]. It has also been shown that female sex, never having smoked, Asian ethnicity and adenocarcinoma histology could be predictors of a better response to epidermal growth factor receptor (EGFR) tyrosine kinase inhibitor (TKIs) therapy [2–4]. More specifically, the type of driver mutation has been shown to have a greater impact on clinical outcomes. Accordingly, it was shown that the best outcomes were found in patients with common mutations, *exon 19 deletion* and *exon 21 L858R* mutation [5, 6]. Further, some studies have suggested that patients with *exon 19 deletion* have better outcomes than patients with *exon 21 L858R* mutation [7–9] and that patients with the other EGFR mutations—the rare EGFR mutations—have worse outcomes [6]. Recent data showed that patients with *insertion in exon 20* were resistant to EGFR TKIs, with the exception of insertions at amino acid A763 [10–12]. However, regardless of the initial response to first- and second-generation EGFR TKIs, it has been shown that acquired resistance has usually developed after 9–13 months [13–15]. Additionally, the most common mechanism of resistance, in approximately 60% of cases, is the development of secondary mutation *T790M in exon 20* [16,17].

Although the majority of patients with EGFR mutated (EGFRmut) NSCLC do have a response to EGFR TKIs, 20–30% of patients do not respond to EGFR TKIs or respond for a short time, usually less than 3 months, due to primary resistance to EGFR TKIs [13,15]. It has been suggested that the presence of other co-mutations is the main reason for primary resistance to EGFR TKIs [18–20].

Further, patients with EGFRmut NSCLC have poor response to mono-immunotherapy [21,22]. Some studies suggest that the upregulation of the programmed cell death protein 1 (PD-1)/programmed cell death-ligand 1 (PD-L1) pathway is associated with resistance to EGFR TKIs [23–26]. It has also been reported that high levels of membrane PD-L1 are correlated with a primary resistance and poor response to EGFR TKIs [23–26]. The results of one study suggested that PD-L1 expression was increased in patients who acquired resistance to EGFR TKIs [24], whereas several other studies have reported a correlation between high PD-L1 expression and primary resistance and poor response to EGFR TKIs [25,26].

Patients with NSCLC and liver metastases (LM) have worse prognosis compared to patients without LM, and they account for approximately one fifth of patients [27]. Studies have shown that these patients live approximately three to six months [27]. The results of one large study which involved more than 20,000 lung cancer patients showed that patients with liver metastases had a 53% higher risk of death compared to patients with brain metastases [28,29]. The majority of them have poor response to chemotherapy [27]. It has also been shown that patients with LM and driver mutations have a worse response to targeted therapies compared to patients without liver metastases [29–31], but response to immunotherapy is controversial.

The liver is considered to be an immune-tolerant organ, characterized by T cell anergy and immunosuppressive signals [32]. This could be the reason why patients with NSCLC and liver metastases have a worse response to PD-1/PD-L1 inhibitors compared to patients without LM [27].

2. Materials and Methods

2.1. Patients and Data Collection

In our study, we enrolled 105 patients with locally advanced and metastatic EGFRmut lung adenocarcinoma, who had been diagnosed at the Clinic for Pulmonology, University Clinical Center of Serbia, from February 2012 until January 2017. All of them were treated with first-line EGFR TKIs-gefitinib, erlotinib or afatinib. All patients met the inclusion criteria—histopathologically confirmed lung adenocarcinoma with confirmed EGFR mutations, locally advanced or metastatic stage of disease, measurable disease at baseline, ECOG PS 0 or 1. The exclusion criteria were as follows: histopathologically confirmed other subtypes of lung carcinoma, early stage of disease, prior treatment for lung adenocarcinoma, ECOG PS $\geq$ 2. All patients underwent a computed tomography of chest and head or magnetic resonance of the head at baseline to determine the stage of disease. The data about sex, age, smoking status [i.e., non-smokers, ex-smokers (patients who stopped smoking one year before treatment), smokers], stage of disease, response to therapy (ORR), time to disease progression (PFS) and overall survival (OS) were collected. Response to therapy included complete response (CR), partial response (PR), stable disease (SD) and progression of disease (PD). Patients were followed for a minimum of one year (the majority of them until death), and some of them for more than 10 years.

A responder is defined as a patient who had a complete response or partial response for at least 4 weeks (confirmed response). Disease control is defined as response—as defined above—or stable disease for at least 6 weeks. PFS is defined as time from the start of therapy until disease progression or death, whichever occurs first. OS is defined as time from the start of therapy until death.

2.2. Statistical Analysis

Results are presented as count (%) or means $\pm$ standard deviation, depending on data type. Groups were compared using parametric, t test and nonparametric Pearson Chi-square test. The Kaplan–Meier with Log-ranks test was used to assess survival and group differences regarding survival. Survival is presented using the median (95% CI) or percentage of participants without an event of interest in a specific time period. All p values less than 0.05 have been considered significant. All data were analyzed using SPSS 29.0 (IBM Corp. Released 2023. IBM SPSS Statistics for Windows, Version 20.0. Armonk, NY, USA: IBM Corp.) and R 3.4.2. (R Core Team (2017). R: A language and environment for statistical computing. R Foundation for Statistical Computing, Vienna, Austria. URL https://www.R-project.org/).

3. Results

One hundred and five patients with locally advanced and metastatic EGFRmut lung adenocarcinoma were included in the analysis. Patients were divided into two groups based on the presence of liver metastases. There were 16 patients with LM (15.2%) and 89 patients without LM (84.8%). There were no significant differences between groups by age, gender, smoking status or type of mutation.

The main baseline demographic characteristics for the whole group and for groups based on the presence of liver metastases are presented in Table 1.

Table 1. Comparative view of demographic characteristics for the whole group and for patients with and without liver metastases.

Demographic Characteristics	All Patients $N = 105$	Patients with Liver Metastases $N = 16$	Patients Without Liver Metastases $N = 89$	p Value
Age, mean ± sd	63.7 ± 10.6	62.4 ± 8.7	63.9 ± 10.9	$p = 0.603$ [a]
Gender				$p = 1.000$ [b]
Female sex, *n* (%)	74 (70.5%)	11 (69%)	63 (71%)	
Male sex, *n* (%)	31 (29.5%)	5 (31%)	26 (29%)	
Smoking, *n* (%)				$p = 0.487$ [b]
Never smokers	60 (57.7%)	11 (69%)	49 (55%)	
Former smokers	17 (16.3%)	1 (6%)	16 (18%)	
Current smokers	27 (26.0%)	4 (25%)	23 (26%)	
EGFR mutation				$p = 0.248$ [b]
Del 19	54 (51.4%)	7 (44%)	47 (53%)	
L858R point mutation exon 21	35 (33.3%)	8 (50%)	27 (30%)	
Rare mutations	16 (15.2%)	1 (6%)	15 (17%)	
Del 19 and *T790M*	1	1		
Del 20	1		1	
Exon 18	1		1	
G719X exon 18	3		3	
G719X exon 18 and *S768I exon 20*	2		2	
Exon 20 insertion	7		7	
L828V	1		1	

[a] Independent samples t test; [b] Pearson chi square test.

Patients with LM achieved disease control in 68.7% of cases, and patients without LM in 77.5% of cases. Fifty percent of patients were responders in the group of patients with LM, and 41.6% in the group of patients without LM. There were no statistically significant differences between groups regarding response to EGFR TKIs (Chi-square test, $p = 0.206$) (Table 2).

Table 2. Response to EGFR TKIs by group ($p = 0.206$).

Variable	Patients with LM N = 16	Patients Without LM N = 89
Complete response	1/16 (6.3%)	2/89 (2.2%)
Partial response	7/16 (43.8%)	35/89 (39.3%)
Response	**8/16 (50%)**	**37/89 (41.6%)**
Stable disease	3/16 (18.8%)	32/89 (36.0%)
Disease control	**11/16 (68.7%)**	**69/89 (77.5%)**
Disease progression	**5/16 (31.3%)**	**20/89 (22.5%)**

Progression-free survival was 9 months for the whole group of patients, 95% CI (7.7–10.3). Further, PFS for patients with LM was 9 months, 95% CI (6.2–11.8), and also 9 months, 95% CI (7.6–10.4) for patients without LM. There was no statistically significant difference in PFS between these groups ($p = 0.927$). Figure 1 shows the Kaplan–Meier curves for PFS in patients with and without LM.

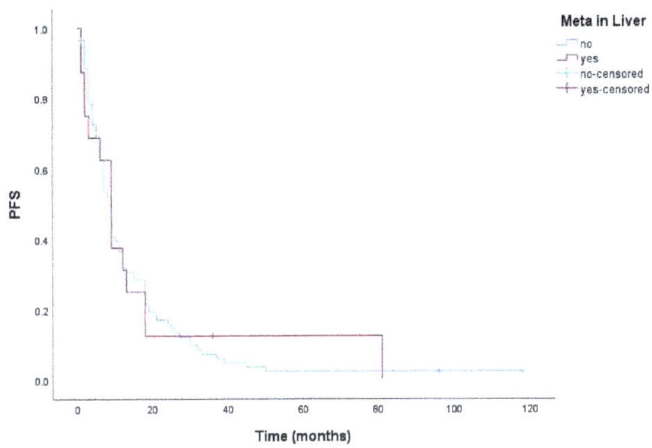

Figure 1. Kaplan–Meier curves for PFS in patients with and without LM ($p = 0.927$).

Overall survival was 13 months for the whole group of patients, 95% CI (10.3–15.7). Further, OS for patients with LM was 13 months, 95% CI (10.2–15.8) and also 13 months, 95% CI (5.2–20.8) in the group without LM. There was no statistically significant difference in OS between these groups ($p = 0.337$). Figure 2 shows Kaplan–Meier curves for OS in patients with and without LM.

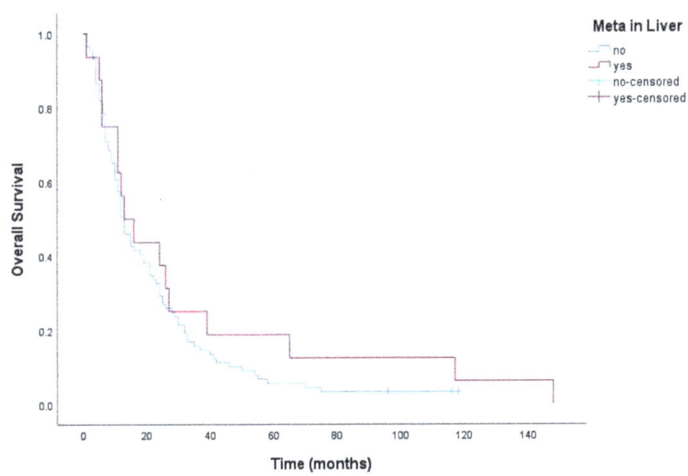

Figure 2. Kaplan–Meier curves for OS in patients with and without liver metastases ($p = 0.337$).

The 5-year survival rate for patients with and without LM was 12.5% and 4.5%, respectively. The 9-year survival rate for patients with and without LM was 12.5% and 3.4%, respectively. Further, the 5-year and 9-year survival rates for whole group of patients were 6.7% and 4.8%, respectively. The data on 5-year and 9-year OS rates for the whole group, for patients with and without LM and by gender are presented in Table 3.

Table 3. Comparative view of 5-year and 9-year OS rate.

Patients	5-Year OS Rate	9-Year OS Rate	Female No (%)	Female 5-Year OS Rate	Female 9-Year OS Rate	Male No (%)	Male 5-Year OS Rate	Male 9-Year OS Rate
Whole group of patients	6.7%	4.8%	74 (70.5%)	6.8%	4%	31 (29.5%)	9.7%	6.5%
Patients with LM	12.5%	12.5%	11 (31%)	18.2%	18.2%	5 (69%)	20%	0%
Patients without LM	4.5%	3.4%	63 (70.2%)	4.8%	1.6%	26 (29.2%)	7.7%	7.7%

4. Discussion

To the best of our knowledge, this is the second study to have reported this long survival rate for patients with EGFRmut NSCLC treated with first- and second-generation EGFR TKIs, and none of the previous studies reported long-term survival rates for patients with and without LM.

Hirsch et al. [33] reported 10-year and 15-year survival rates of 86% and 59%, respectively, in patients treated with gefitinib, but EGFR mutational status was known for only 22% of patients.

We also found a couple of case reports which reported long-term survival with erlotinib or gefitinib [34–39], but only one patient reported by Jovanovic et al. [40] had LM.

Lin et al. [40] reported a 5-year survival rate of 14.6% in patients with EGFRmut NSCLC treated with erlotinib and gefitinib. Another real-world study reported a 5-year survival rate of 21.5% [41].

Tompkins et al. [42] investigated whether patients who were treated with first- and second-generation EGFR TKIs and who lived for 5 years or longer had different demographics and co-mutations compared to patients who lived less than 5 years. The number of co-mutations and the frequency of at least one pathogenic co-mutation did not differ between groups [42]. The most frequent co-mutations were *TP53* and *PIK3CA*, which were found at a similar frequency in both groups [42]. Further, they found that patients with baseline brain metastases had higher odds of death within 5 years, as well as a non-significant trend toward worse OS [42]. Additionally, for patients with baseline liver metastases they found an association with worse 5-year OS but no higher odds of death within 5 years [42]. Patients with EGFRmut NSCLC but not *exon 19 deletion* or *exon 21 L858R* mutation had worse OS and higher odds of death within 5 years [42]. Patients with smoking history also had higher odds of death within 5 years, but smoking history was not associated with worse OS [42].

Several studies investigated the impact of *TP53* mutation on survival in patients with EGFRmut NSCLC. All those studies found that *TP53* was a negative prognostic factor in patients with EGFRmut NSCLC [43–46].

Jiang et al. [47] investigated whether the addition of local therapy to EGFR-TKIs could provide better outcomes for EGFRmut patients with oligometastatic or oligoprogressive liver metastases compared to EGFR-TKIs alone. The results showed significantly better PFS (13.8 vs. 8.6 months) and OS (31.2 vs. 18.5 months) in the whole group [47]. In the oligometastatic group, there were 23 patients who were treated with EGFR-TKIs and local therapy and 20 patients treated with EGFR-TKIs alone. Patients treated with a combination of EGFR-TKIs and local therapy had significantly better PFS (12.9 vs. 7.9 months) and OS (36.8 vs. 21.3 months) [47].

In another study by Jiang et al. [30], it was investigated whether liver metastasis could be a predictor of EGFR TKI efficacy. There were 598 patients with NSCLC in total, and 99 had liver metastases [30]. Two hundred and ninety-six of them were EGFRmut, and only 56 of EGFRmut received EGFR TKIs in a first-line setting. In the EGFRmut group, patients who had LM and who were treated with EGFR TKIs had shorter PFS (7.5 vs. 11.8 months) and OS (20.8 vs. 30.6 months) compared to patients without liver metastases [30]. Patients with liver metastases who were treated with first-line chemotherapy, regardless of EGFR status, had similar PFS and OS compared to patients without liver metastases [30].

Similar results were obtained in a study conducted by Wu et al. [31]. There were 148 patients included in the analysis; among them, 19 patients had LM (13%) [31]. Patients with liver metastases who had been treated with gefitinib had shorter PFS and OS compared to those without liver metastases [31]. It was shown by multivariable analysis that liver metastases were an independent poor prognostic factor for PFS and OS [31].

Chang et al. [48] conducted a study the results of which showed similar OS in patients with EGFRmut and EGFRwt NSCLC who had baseline liver metastases. Patients with EGFRmut NSCLC were all treated with TKIs in first line, and patients with EGFRwt NSCLC were treated with chemotherapy or best supportive care if they had poor performance status [48]. A total of 490 patients were included, and among them 39 had EGFRmut NSCLC and LM [48]. Liver metastases were a poor prognostic factor in patients with EGFRmut NSCLC, but it did not have an impact on OS in patients with EGFRwt NSCLC [48].

In another large study that included more than 2000 patients with EGFRmut NSCLC, 13.2% of patients had baseline liver metastases [49]. It was also shown that these patients had a lower response to treatment with first- and second-generation EGFR TKIs [49].

A retrospective analysis evaluated the clinical outcomes in 236 patients with metastatic NSCLC [50]. Patients with liver metastases accounted for 16.9% of all patients, and they had lower OS compared to patients without liver metastases (10 vs. 21 months, respectively) [50]. In a subgroup analysis, patients with EGFRmut NSCLC and liver metastases had better OS compared to patients with EGFRwt NSCLC (not reached vs. 13 months) [50].

A RELAY study showed a better HR of PFS with the addition of ramucirumab (anti-VEGF receptor antibody) to erlotinib in patients with liver metastases compared to erlotinib alone [51].

Further, Chiu et al. [52] investigated which factors were associated with prolonged survival (>36 months) in EGFRmut advanced NSCLC patients treated with first-line afatinib. They found that liver metastases were independently and negatively associated with prolonged PFS [52].

Gen et al. [53] investigated the efficacy of osimertinib in EGFRmut metastatic NSCLC compared to first-/second-generation EGFR TKIs. Clinical data for 388 patients were retrospectively evaluated, and 8.8% of them had liver metastases [53]. In the osimertinib group, patients with liver metastases had significantly lower PFS and OS compared to patients without (7.4 vs. 19.7 months, and 12.1 vs. not reached) [53]. In the gefitinib/erlotinib group and afatinib group, the presence of liver metastases was not associated with shorter PFS [53]. In addition, PFS in patients with liver metastases was not significantly different among three groups [53]. The ORR of osimertinib was significantly lower in patients with liver metastases compared to those without, and it was similar to the ORR of gefitinib, erlotinib and afatinib [53].

Similar results for osimertinib efficacy in patients with liver metastases were shown in a retrospective study by Sakata et al. [54].

Another study investigated the efficacy of osimertinib according to mutation subtypes and different sites of baseline metastases [55]. Fourteen percent of patients had liver metastases, and it was shown that those patients had shorter PFS compared to patients without liver metastases (10.3 vs. 24.1 months, respectively) [55].

A comparison of clinical outcomes for patients treated with EGFR TKIs in some studies is presented in Table 4.

Table 4. Comparison of clinical outcomes for patients treated with EGFR TKIs in some studies.

Study	No of Patients	OS (Months)	PFS (Months)	5-Year OS Rate	9-Year OS Rate
Hirsh et al. [33] *	79	/	/	/	10-year OS rate 86%, 15-year OS rate 59%
Lin et al. [40]	137	30.9	12.1	14.6%	/
Chiu et al. [52]	546	27.2	14.5	/	/
Gen et al. [53]	388				
Sakata et al. [54]	538	/	20.5	/	/
Takeyasu et al. [55]	229	33.7	23.3	/	/
Ćeriman Krstić et al.	105	13	9	6.7%	4.8%

* EGFR ststus was known for 17 patients (22%).

The results of the Mariposa study showed significantly better PFS for patients with advanced EGFRmut lung adenocarcinoma who were treated with amivantanab-lazertinib compared to osimertinib in a first-line setting [56]. Subgroup analysis also showed that patients with liver metastases had better PFS if they were treated with amivantanab-lazertinib compared to osimertinib (18.2 vs. 11 months) [56].

An ImPower150 study [57] investigated the efficacy of different therapeutic options in patients with metastatic NSCLC in first-line treatment. Patients with EGFRmut (*deletion 19* or *exon 21 L858R*) NSCLC and patients with ALK translocations were also included, but it was required that patients have disease progression or treatment intolerance with at least one TKI [57]. It was shown that patients with liver metastases had better outcomes if they were treated with bevacizumab (VEGF inhibitor), atezolizumab (PD-L1 inhibitor) and chemotherapy compared to those who were treated with bevacizumab and chemotherapy alone (PFS 7.4 vs. 4.9 months; OS 13.3 vs. 9.4 months) [58]. There was no difference in outcomes for patients who were treated with atezolizumab and chemotherapy compared to those treated with bevacizumab and chemotherapy [58]. In a group of patients with sensitizing EGFR mutations, OS was longer in a group of patients who were treated with bevacizumab, atezolizumab and chemotherapy compared to patients who were treated with bevacizumab and chemotherapy (OS 29.4 vs. 18.1 months, HR 0.60) [57]. Further, OS was similar in patients who were treated with atezolizumab and chemotherapy compared to those treated with bevacizumab and chemotherapy (OS 19.0 vs. 18.1 months, HR 1.00) [57]. The 3-year survival rate was 41.9% for patients treated with bevacizumab, atezolizumab and chemotherapy, 25.6% for patients treated with atezolizumab and chemotherapy and 24.6% for patients treated with bevacizumab and chemotherapy [57].

One retrospective study evaluated the efficacy of immunotherapy + chemotherapy, immunotherapy + chemotherapy + antiangiogenic therapy and chemotherapy + antiangiogenic therapy in EGFRmut patients after the failure of EGFR TKIs [58]. Liver metastases were found in 29% of patients [58]. The results showed the superior PFS of immunotherapy + chemotherapy + antiangiogenic therapy in this subgroup of patients compared to other treatment options (6.44 months vs. 0.95 months for patients treated with immunotherapy + chemotherapy and 3.48 months for patients treated with chemotherapy + antiangiogenic therapy) [58].

As was mentioned previously, the liver is considered to be an immune-tolerant organ, and it is likely that liver metastases will be resistant to immunotherapy [32]. Some studies suggested that liver metastases could actively induce the apoptosis of CD8+ T cells after their interaction with macrophages, which could result in liver metastases taking over mechanisms of peripheral tolerance and lead to an acquired resistance to immunotherapy through the removal of CD8+ T cells [32].

PD-L1 is an immune checkpoint that limits T cell activity by binding to its receptor PD-1 on activated T cells [32]. PD-L1 can also be found on tumor cells and that way blocks the activity of the immune system to fight against various cancers [32]. Furthermore, blocking PD-1/PD-L1 leads to a reactivation of exhausted T cells and to a reactivation of the immune system to fight against tumors [32]. The tumor as well as the organ specific microenvironment plays an important role in this process [32]. Vascular endothelial growth factor (VEGF) could block antitumor activity by promoting vascularization [59], which is often used by tumors for growth and metastasis, and interrupting T cell infiltration into the tumor [60,61]. It has been suggested that VEGF inhibitor in combination with chemotherapy could enhance PD-L1 inhibitor-induced T cell-mediated cancer cell death, by reversing VEGF-induced immunosuppression by promoting the infiltration of T cells into the tumor microenvironment and enabling T cell activity against tumor antigens, as well as chemotherapy-induced cancer cell death [57].

It could be assumed that susceptibility to VEGF inhibitor in patients with NSCLC and liver metastases arises from a reversal of immune suppression by VEGF inhibitor, which basically reduces the infiltration of CD8+ T cells, which may further enhance the T cell-mediated killing of cancer cells by PD-L1 inhibitor [57]. The efficacy of this has been shown regardless of PD-L1 expression [57]. It could be explained by the fact that activating both mechanisms—T cell activation by atezolizumab and the interferon γ-mediated induction of PD-L1 expression—throw the reprogramming tumor microenvironment into the active immune system by VEGFR inhibitor [57]. The efficacy of bevacizumab and atezolizumab in combination with chemotherapy in patients with NSCLC and liver metastases is supported by an investigation in patients with unresectable hepatocellular carcinoma who were treated with bevacizumab and atezolizumab in combination with chemotherapy [57]. It seems that patients with NSCLC and liver metastases respond to treatment similarly to patients with primary liver cancer [57].

In the past few years, there have been a large number of papers investigating the role of tumor-educated platelets as a biomarker for the early diagnosis of tumors, including NSCLC, as well as their potential role in the early detection of tumor progression. Platelet transcriptome is a product of parental megakaryocyte profile and local and systemic pathological conditions like cancer, which can modulate their transcriptome [62,63]. Thus, tumor cells can educate platelets and "leave their signature", which can lead to diagnosis [62,63]. In the interaction process with tumor cells, platelets can modify their RNA profile and thus derive tumor-educated platelets which produce a different microenvironment from the normal one [62,63].

Several studies were conducted which included different platelet RNA in patients with NSCLC. It was shown that tumor-educated platelet RNA may enable an early diagnosis of NSCLC and an early detection of progression and may serve as a potential biomarker of response to chemotherapy [62].

However, EGFR mutations could not be detected in the DNA of tumor-educated platelets isolated from patients with NSCLC [64].

5. Conclusions

There are few data about survival rates beyond 5 years for patients with locally advanced and metastatic EGFRmut NSCLC, mainly because patients with lung cancer rarely live for a very long time. Regarding patients with liver metastases, the results of other studies have shown that they have worse prognosis compared to other groups of patients regardless of the applied treatment. In our study, clinical outcomes were similar in patients with EGFRmut NSCLC with and without LM, but, in our study, there were a small number of patients with LM. Moreover, patients were treated with first-/second-generation TKIs, but in some countries these drugs are still first-line treatment options. As these patients represent a significant number of patients, we need a wider range of therapeutic options. It might be that combination therapies represent better a therapeutic option for this subset of patients.

Author Contributions: Conceptualization, V.Ć.K.; methodology, V.Ć.K., I.S. and D.J.; software, I.S.; validation, V.Ć.K.; formal analysis, V.Ć.K., I.S., N.S., M.G., M.K., A.R., M.S., M.R.M. and D.J.; investigation, V.Ć.K., I.S., N.S., M.G., M.K., A.R., M.S., M.R.M. and D.J.; resources, V.Ć.K., N.S., M.G., M.K., A.R., M.S. and M.R.M.; data curation, V.Ć.K. and I.S.; writing—original draft preparation, V.Ć.K.; writing—review and editing, V.Ć.K.; supervision, V.Ć.K., I.S. and D.J.; project administration, V.Ć.K. and D.J. All authors have read and agreed to the published version of the manuscript.

Funding: This research received no external funding.

Institutional Review Board Statement: The study was conducted in accordance with the Declaration of Helsinki, and it was approved by the Ethics committee of School of Medicine Belgrade University of 06.06.2017 (No 61206-2300/2-17).

Informed Consent Statement: All subjects gave their informed consent for participation in the study before any procedure was performed.

Data Availability Statement: The data presented in this study are available upon reasonable request from the corresponding author (accurately indicating status).

Conflicts of Interest: The authors declare no conflicts of interest.

References

1. Global Data. Non-Small Cell Lung Cancer (NSCLC)—Epidemiology Forecast to 2025. Available online: https://store.globaldata.com/report/gdhcer132-16--epicast-report-non-small-cell-lung-cancer-nsclc-epidemiology-forecast-to-2025/ (accessed on 15 May 2021).
2. Lynch, T.J.; Bell, D.W.; Sordella, R.; Gurubhagavatula, S.; Okimoto, R.A.; Brannigan, B.W.; Harris, P.L.; Haserlat, S.M.; Supko, J.G.; Haluska, F.G.; et al. Activating mutations in the epidermal growth factor receptor underlying responsiveness of non-small-cell lung cancer to gefitinib. *N. Engl. J. Med.* **2004**, *350*, 2129–2139. [CrossRef] [PubMed]
3. Mok, T.S.; Wu, Y.L.; Thongprasert, S.; Yang, C.H.; Chu, D.T.; Saijo, N.; Sunpaweravong, P.; Han, B.; Margono, B.; Ichinose, Y.; et al. Gefitinib or carboplatin-paclitaxel in pulmonary adenocarcinoma. *N. Engl. J. Med.* **2009**, *361*, 947–957. [CrossRef] [PubMed]
4. Maemondo, M.; Inoue, A.; Kobayashi, K.; Sugawara, S.; Oizumi, S.; Isobe, H.; Gemma, A.; Harada, M.; Yoshizawa, H.; Kinoshita, I.; et al. Gefitinib or chemotherapy for non-small-cell lung cancer with mutated EGFR. *N. Engl. J. Med.* **2010**, *362*, 2380–2388. [CrossRef]
5. Yang, C.H.; Yu, C.J.; Shih, J.Y.; Chang, Y.C.; Hu, F.C.; Tsai, M.C.; Chen, K.Y.; Lin, Z.Z.; Huang, C.J.; Shun, C.T.; et al. Specific EGFR mutations predict treatment outcome of stage IIIB/IV patients with chemotherapy-naive non-small-cell lung cancer receiving first-line gefitinib monotherapy. *J. Clin. Oncol.* **2008**, *26*, 2745–2753. [CrossRef]
6. Locatelli-Sanchez, M.; Couraud, S.; Arpin, D.; Riou, R.; Bringuier, P.P.; Souquet, P.J. Routine EGFR molecular analysis in non-small-cell lung cancer patients is feasible: Exons 18–21 sequencing results of 753 patients and subsequent clinical outcomes. *Lung* **2013**, *191*, 491–499. [CrossRef]
7. Won, Y.W.; Han, J.Y.; Lee, G.K.; Park, S.Y.; Lim, K.Y.; Yoon, K.A.; Yun, T.; Kim, H.T.; Lee, J.S. Comparison of clinical outcome of patients with non-small-cell lung cancer harbouring epidermal growth factor receptor exon 19 or exon 21 mutations. *J. Clin. Pathol.* **2011**, *64*, 947–952. [CrossRef]
8. Jackman, D.M.; Yeap, B.Y.; Sequist, L.V.; Lindeman, N.; Holmes, A.J.; Joshi, V.A.; Bell, D.W.; Huberman, M.S.; Halmos, B.; Rabin, M.S.; et al. Exon 19 deletion mutations of epidermal growth factor receptor are associated with prolonged survival in non-small cell lung cancer patients treated with gefitinib or erlotinib. *Clin. Cancer Res.* **2006**, *12*, 3908–3914. [CrossRef] [PubMed]
9. Riely, G.J.; Pao, W.; Pham, D.; Li, A.R.; Rizvi, N.; Venkatraman, E.S.; Zakowski, M.F.; Kris, M.G.; Ladanyi, M.; Miller, V.A. Clinical course of patients with non-small cell lung cancer and epidermal growth factor receptor exon 19 and exon 21 mutations treated with gefitinib or erlotinib. *Clin. Cancer Res.* **2006**, *12*, 839–844. [CrossRef] [PubMed]
10. Robichaux, J.P.; Elamin, Y.Y.; Tan, Z.; Carter, B.W.; Zhang, S.; Liu, S.; Li, S.; Chen, T.; Poteete, A.; Estrada-Bernal, A.; et al. Mechanisms and clinical activity of an EGFR and HER2 exon 20-selective kinase inhibitor in non-small cell lung cancer. *Nat. Med.* **2018**, *24*, 638–646; Erratum in *Nat. Med.* **2024**, *30*, 2694–2695. [CrossRef]
11. Vasconcelos, P.E.N.S.; Gergis, C.; Viray, H.; Varkaris, A.; Fujii, M.; Rangachari, D.; VanderLaan, P.A.; Kobayashi, I.S.; Kobayashi, S.S.; Costa, D.B. EGFR-A763_Y764insFQEA Is a Unique Exon 20 Insertion Mutation That Displays Sensitivity to Approved and In-Development Lung Cancer EGFR Tyrosine Kinase Inhibitors. *JTO Clin. Res. Rep.* **2020**, *1*, 100051. [CrossRef]

12. Yasuda, H.; Park, E.; Yun, C.H.; Sng, N.J.; Lucena-Araujo, A.R.; Yeo, W.L.; Huberman, M.S.; Cohen, D.W.; Nakayama, S.; Ishioka, K.; et al. Structural, biochemical, and clinical characterization of epidermal growth factor receptor (EGFR) exon 20 insertion mutations in lung cancer. *Sci. Transl. Med.* **2013**, *5*, 216ra177. [CrossRef] [PubMed]
13. Santoni-Rugiu, E.; Melchior, L.C.; Urbanska, E.M.; Jakobsen, J.N.; de Stricker, K.; Grauslund, M.; Sørensen, J.B. Intrinsic Resistance to EGFR-Tyrosine Kinase Inhibitors in EGFR-Mutant Non-Small Cell Lung Cancer: Differences and Similarities with Acquired Resistance. *Cancers* **2019**, *11*, 923. [CrossRef] [PubMed]
14. Camidge, D.R.; Pao, W.; Sequist, L.V. Acquired resistance to TKIs in solid tumours: Learning from lung cancer. *Nat. Rev. Clin. Oncol.* **2014**, *11*, 473–481. [CrossRef] [PubMed]
15. Jackman, D.; Pao, W.; Riely, G.J.; Engelman, J.A.; Kris, M.G.; Jänne, P.A.; Lynch, T.; Johnson, B.E.; Miller, V.A. Clinical definition of acquired resistance to epidermal growth factor receptor tyrosine kinase inhibitors in non-small-cell lung cancer. *J. Clin. Oncol.* **2010**, *28*, 357–360. [CrossRef] [PubMed]
16. Morgillo, F.; Della Corte, C.M.; Fasano, M.; Ciardiello, F. Mechanisms of resistance to EGFR-targeted drugs: Lung cancer. *ESMO Open* **2016**, *1*, e000060. [CrossRef]
17. Hochmair, M.J.; Buder, A.; Schwab, S.; Burghuber, O.C.; Prosch, H.; Hilbe, W.; Cseh, A.; Fritz, R.; Filipits, M. Liquid-Biopsy-Based Identification of EGFR T790M Mutation-Mediated Resistance to Afatinib Treatment in Patients with Advanced EGFR Mutation-Positive NSCLC, and Subsequent Response to Osimertinib. *Target. Oncol.* **2019**, *14*, 75–83. [CrossRef]
18. Blakely, C.M.; Watkins, T.B.K.; Wu, W.; Gini, B.; Chabon, J.J.; McCoach, C.E.; McGranahan, N.; Wilson, G.A.; Birkbak, N.J.; Olivas, V.R.; et al. Evolution and clinical impact of co-occurring genetic alterations in advanced-stage EGFR-mutant lung cancers. *Nat. Genet.* **2017**, *49*, 1693–1704. [CrossRef]
19. Guibert, N.; Barlesi, F.; Descourt, R.; Léna, H.; Besse, B.; Beau-Faller, M.; Mosser, J.; Pichon, E.; Merlio, J.P.; Ouafik, L.; et al. Characteristics and Outcomes of Patients with Lung Cancer Harboring Multiple Molecular Alterations: Results from the IFCT Study Biomarkers France. *J. Thorac. Oncol.* **2017**, *12*, 963–973. [CrossRef]
20. VanderLaan, P.A.; Rangachari, D.; Mockus, S.M.; Spotlow, V.; Reddi, H.V.; Malcolm, J.; Huberman, M.S.; Joseph, L.J.; Kobayashi, S.S.; Costa, D.B. Mutations in TP53, PIK3CA, PTEN and other genes in EGFR mutated lung cancers: Correlation with clinical outcomes. *Lung Cancer* **2017**, *106*, 17–21. [CrossRef]
21. Kuribayashi, T.; Ohashi, K.; Nishii, K.; Ninomiya, K.; Tsubata, Y.; Ishikawa, N.; Kodani, M.; Kanaji, N.; Yamasaki, M.; Fujitaka, K.; et al. Clinical characteristics of patients treated with immune checkpoint inhibitors in EGFR-mutant non-small cell lung cancer: CS-Lung-003 prospective observational registry study. *J. Cancer Res. Clin. Oncol.* **2024**, *150*, 89. [CrossRef]
22. Lee, C.K.; Man, J.; Lord, S.; Cooper, W.; Links, M.; Gebski, V.; Herbst, R.S.; Gralla, R.J.; Mok, T.; Yang, J.C. Clinical and Molecular Characteristics Associated With Survival Among Patients Treated with Checkpoint Inhibitors for Advanced Non-Small Cell Lung Carcinoma: A Systematic Review and Meta-analysis. *JAMA Oncol.* **2018**, *4*, 210–216. [CrossRef] [PubMed]
23. Jia, Y.; Li, X.; Zhao, C.; Ren, S.; Su, C.; Gao, G.; Li, W.; Zhou, F.; Li, J.; Zhou, C. Soluble PD-L1 as a Predictor of the Response to EGFR-TKIs in Non-small Cell Lung Cancer Patients With EGFR Mutations. *Front. Oncol.* **2020**, *10*, 1455. [CrossRef] [PubMed]
24. Han, J.J.; Kim, D.W.; Koh, J.; Keam, B.; Kim, T.M.; Jeon, Y.K.; Lee, S.H.; Chung, D.H.; Heo, D.S. Change in PD-L1 Expression After Acquiring Resistance to Gefitinib in EGFR-Mutant Non-Small-Cell Lung Cancer. *Clin. Lung Cancer* **2016**, *17*, 263–270.e2. [CrossRef] [PubMed]
25. Hsu, K.H.; Huang, Y.H.; Tseng, J.S.; Chen, K.C.; Ku, W.H.; Su, K.Y.; Chen, J.J.W.; Chen, H.W.; Yu, S.L.; Yang, T.Y.; et al. High PD-L1 expression correlates with primary resistance to EGFR-TKIs in treatment naïve advanced EGFR-mutant lung adenocarcinoma patients. *Lung Cancer* **2019**, *127*, 37–43. [CrossRef]
26. Su, S.; Dong, Z.Y.; Xie, Z.; Yan, L.X.; Li, Y.F.; Su, J.; Liu, S.Y.; Yin, K.; Chen, R.L.; Huang, S.M.; et al. Strong Programmed Death Ligand 1 Expression Predicts Poor Response and De Novo Resistance to EGFR Tyrosine Kinase Inhibitors Among NSCLC Patients With EGFR Mutation. *J. Thorac. Oncol.* **2018**, *13*, 1668–1675. [CrossRef]
27. Qin, B.D.; Jiao, X.D.; Liu, J.; Liu, K.; He, X.; Wu, Y.; Ling, Y.; Duan, X.P.; Qin, W.X.; Wang, Z.; et al. The effect of liver metastasis on efficacy of immunotherapy plus chemotherapy in advanced lung cancer. *Crit. Rev. Oncol. Hematol.* **2020**, *147*, 102893. [CrossRef]
28. Riihimäki, M.; Hemminki, A.; Fallah, M.; Thomsen, H.; Sundquist, K.; Sundquist, J.; Hemminki, K. Metastatic sites and survival in lung cancer. *Lung Cancer* **2014**, *86*, 78–84. [CrossRef]
29. Qu, F.J.; Zhou, Y.; Wu, S. Progress of immune checkpoint inhibitors therapy for non-small cell lung cancer with liver metastases. *Br. J. Cancer* **2024**, *130*, 165–175. [CrossRef]
30. Jiang, T.; Cheng, R.; Zhang, G.; Su, C.X.; Zhao, C.; Li, X.; Cai, W.; Zhou, F.; Zhao, J.; Xiong, A.; et al. Characterization of liver metastasis and its effect on targeted therapy in EGFR-mutant NSCLC: A Multicenter Study. *Clin. Lung Cancer* **2017**, *18*, 631–639.e2. [CrossRef]
31. Wu, K.L.; Tsai, M.J.; Yang, C.J.; Chang, W.A.; Hung, J.Y.; Yen, C.J.; Shen, C.H.; Kuo, T.Y.; Lee, J.Y.; Chou, S.H.; et al. Liver metastasis predicts poorer prognosis in stage IV lung adenocarcinoma patients receiving first-line gefitinib. *Lung Cancer* **2015**, *88*, 187–194. [CrossRef]

32. Deng, J.-Y.; Gou, Q.; Yang, L.; Chen, Z.H.; Yang, M.Y.; Yang, X.R.; Yan, H.H.; Wei, X.W.; Liu, J.Q.; Su, J.; et al. Immune suppressive microenvironment in liver metastases contributes to organ-specific response of immunotherapy in advanced non-small cell lung cancer. *J. ImmunoTher. Cancer* **2023**, *11*, e007218. [CrossRef] [PubMed]
33. Hirsch, F.R.; Sequist, L.V.; Gore, I.; Mooradian, M.; Simon, G.; Croft, E.F.; DeVincenzo, D.; Munley, J.; Stein, D.; Freivogel, K.; et al. Long-term safety and survival with gefitinib in select patients with advanced non-small cell lung cancer: Results from the US IRESSA Clinical Access Program (ICAP). *Cancer* **2018**, *124*, 2407–2414. [CrossRef]
34. Ricciuti, B.; Baglivo, S.; Ludovini, V.; Sidoni, A.; Metro, G.; Brambilla, M.; Siggillino, A.; Reda, M.S.; Rebonato, A.; Maiettini, D.; et al. Long-term survival with erlotinib in advanced lung adenocarcinoma harboring synchronous EGFR G719S and KRAS G12C mutations. *Lung Cancer* **2018**, *120*, 70–74. [CrossRef] [PubMed]
35. Zhang, M.X.; Tan, W.; Zhang, R.X.; Tian, Y.L.; Gao, H.M.; Gao, Z.; Zhang, N.; Zhao, J.Q.; Jia, Y.F.; Wang, Y.S. Long-term survival of non-small-cell lung cancer patients with EGFR inhibitor treatment. *Genet. Mol. Res.* **2014**, *13*, 8657–8660. [CrossRef]
36. Alkassis, S.; Alshare, B.; Ahmed, S. Maintained Complete Response and Long-Term Survival in Epidermal Growth Factor Receptor Mutated Metastatic Non-Small Cell Lung Cancer with Erlotinib. *Cureus* **2021**, *13*, e12451. [CrossRef]
37. Kanbe, M.; Sunaga, N.; Hara, K.; Sawada, H.; Wakamatsu, I.; Hara, K.; Muto, S.; Sawada, Y.; Masubuchi, H.; Sato, M.; et al. Durable response to afatinib rechallenge in a long-term survivor of non-small cell lung cancer harboring EGFR L858R and L747V mutations. *Thorac. Cancer* **2022**, *13*, 3225–3228. [CrossRef]
38. Matsuzaki, T.; Iwami, E.; Sasahara, K.; Kuroda, A.; Nakajima, T.; Terashima, T. A case report of metastatic lung adenocarcinoma with long-term survival for over 11 years. *Medicine* **2019**, *98*, e14100. [CrossRef] [PubMed]
39. Jovanovic, D.; Stevic, R.; Velinovic, M.; Kontic, M.; Maric, D.; Spasic, J.; Radosavljevic, D. Durable complete remission of poor performance status metastatic lung adenocarcinoma patient treated with second-line erlotinib: A case report. *Onco Targets Ther.* **2017**, *10*, 4347–4354. [CrossRef]
40. Lin, J.J.; Cardarella, S.; Lydon, C.A.; Dahlberg, S.E.; Jackman, D.M.; Jänne, P.A.; Johnson, B.E. Five-Year Survival in EGFR-Mutant Metastatic Lung Adenocarcinoma Treated with EGFR-TKIs. *J. Thorac. Oncol.* **2016**, *11*, 556–565. [CrossRef]
41. Okamoto, I.; Morita, S.; Tashiro, N.; Imamura, F.; Inoue, A.; Seto, T.; Yamamoto, N.; Ohe, Y.; Nakagawa, K.; Fukuoka, M. Real world treatment and outcomes in EGFR mutation-positive non-small cell lung cancer: Long-term follow-up of a large patient cohort. *Lung Cancer* **2018**, *117*, 14–19. [CrossRef]
42. Tompkins, W.; Grady, C.B.; Hwang, W.T.; Chandrasekhara, K.; McCoach, C.; Sun, F.; Liu, G.; Patel, D.; Nieva, J.; Herrmann, A.; et al. Characteristics of Long-Term Survivors With EGFR-Mutant Metastatic NSCLC. *JTO Clin. Res. Rep.* **2024**, *5*, 100669. [CrossRef] [PubMed]
43. Xu, Y.; Tong, X.; Yan, J.; Wu, X.; Shao, Y.W.; Fan, Y. Short-Term Responders of Non-Small Cell Lung Cancer Patients to EGFR Tyrosine Kinase Inhibitors Display High Prevalence of TP53 Mutations and Primary Resistance Mechanisms. *Transl. Oncol.* **2018**, *11*, 1364–1369. [CrossRef] [PubMed]
44. Canale, M.; Petracci, E.; Delmonte, A.; Chiadini, E.; Dazzi, C.; Papi, M.; Capelli, L.; Casanova, C.; De Luigi, N.; Mariotti, M.; et al. Impact of TP53 Mutations on Outcome in EGFR-Mutated Patients Treated with First-Line Tyrosine Kinase Inhibitors. *Clin. Cancer Res.* **2017**, *23*, 2195–2202. [CrossRef] [PubMed]
45. Aggarwal, C.; Davis, C.W.; Mick, R.; Thompson, J.C.; Ahmed, S.; Jeffries, S.; Bagley, S.; Gabriel, P.; Evans, T.L.; Bauml, J.M.; et al. Influence of TP53 Mutation on Survival in Patients With Advanced EGFR-Mutant Non-Small-Cell Lung Cancer. *JCO Precis. Oncol.* **2018**, *2018*, PO.18.00107. [CrossRef] [PubMed]
46. Aisner, D.L.; Sholl, L.M.; Berry, L.D.; Rossi, M.R.; Chen, H.; Fujimoto, J.; Moreira, A.L.; Ramalingam, S.S.; Villaruz, L.C.; Otterson, G.A.; et al. The Impact of Smoking and TP53 Mutations in Lung Adenocarcinoma Patients with Targetable Mutations-The Lung Cancer Mutation Consortium (LCMC2). *Clin. Cancer Res.* **2018**, *24*, 1038–1047. [CrossRef]
47. Jiang, T.; Chu, Q.; Wang, H.; Zhou, F.; Gao, G.; Chen, X.; Li, X.; Zhao, C.; Xu, Q.; Li, W.; et al. EGFR-TKIs plus local therapy demonstrated survival benefit than EGFR-TKIs alone in EGFR-mutant NSCLC patients with oligometastatic or oligoprogressive liver metastases. *Int. J. Cancer* **2019**, *144*, 2605–2612. [CrossRef]
48. Chang, Y.P.; Chen, Y.M.; Lai, C.H.; Lin, C.Y.; Fang, W.F.; Huang, C.H.; Li, S.H.; Chen, H.C.; Wang, C.C.; Lin, M.C. The impact of de novo liver metastasis on clinical outcome in patients with advanced non-small-cell lung cancer. *PLoS ONE* **2017**, *12*, e0178676. [CrossRef]
49. Chang, J.W.; Huang, C.Y.; Fang, Y.F.; Chang, C.F.; Yang, C.T.; Kuo, C.S.; Hsu, P.C.; Wu, C.E. Risk Stratification Using a Novel Nomogram for 2190 EGFR-Mutant NSCLC Patients Receiving the First or Second Generation EGFR-TKI. *Cancers* **2022**, *14*, 977. [CrossRef]
50. Castañón, E.; Rolfo, C.; Viñal, D.; López, I.; Fusco, J.P.; Santisteban, M.; Martin, P.; Zubiri, L.; Echeveste, J.I.; Gil-Bazo, I. Impact of epidermal growth factor receptor (EGFR) activating mutations and their targeted treatment in the prognosis of stage IV non-small cell lung cancer (NSCLC) patients harboring liver metastasis. *J. Transl. Med.* **2015**, *13*, 257. [CrossRef]

51. Nakagawa, K.; Garon, E.B.; Seto, T.; Nishio, M.; Ponce Aix, S.; Paz-Ares, L.; Chiu, C.H.; Park, K.; Novello, S.; Nadal, E.; et al. Ramucirumab plus erlotinib in patients with untreated, EGFR-mutated, advanced non-small-cell lung cancer (RELAY): A randomised, double-blind, placebo-controlled, phase 3 trial. *Lancet Oncol.* **2019**, *20*, 1655–1669. [CrossRef]
52. Chiu, L.C.; Hsu, P.C.; Wang, C.C.; Ko, H.W.; Kuo, S.C.; Ju, J.S.; Tung, P.H.; Huang, A.C.; Yang, C.T. Factors associated with prolonged progression-free survival of patients treated with first-line afatinib for advanced epidermal growth factor receptor-mutated non-small cell lung cancer. *Thorac. Cancer* **2024**, *15*, 529–537. [CrossRef] [PubMed]
53. Gen, S.; Tanaka, I.; Morise, M.; Koyama, J.; Kodama, Y.; Matsui, A.; Miyazawa, A.; Hase, T.; Hibino, Y.; Yokoyama, T.; et al. Clinical efficacy of osimertinib in EGFR-mutant non-small cell lung cancer with distant metastasis. *BMC Cancer* **2022**, *22*, 654. [CrossRef] [PubMed]
54. Sakata, Y.; Sakata, S.; Oya, Y.; Tamiya, M.; Suzuki, H.; Shibaki, R.; Okada, A.; Kobe, H.; Matsumoto, H.; Yokoi, T.; et al. Osimertinib as first-line treatment for advanced epidermal growth factor receptor mutation-positive non-small-cell lung cancer in a real-world setting (OSI-FACT). *Eur. J. Cancer* **2021**, *159*, 144–153. [CrossRef] [PubMed]
55. Takeyasu, Y.; Yoshida, T.; Masuda, K.; Matsumoto, Y.; Shinno, Y.; Okuma, Y.; Goto, Y.; Horinouchi, H.; Yamamoto, N.; Ohe, Y. Distinct Progression and Efficacy of First-Line Osimertinib Treatment According to Mutation Subtypes in Metastatic NSCLC Harboring EGFR Mutations. *JTO Clin. Res. Rep.* **2024**, *5*, 100636. [CrossRef] [PubMed]
56. Felip, E.; Cho, B.C.; Gutiérrez, V.; Alip, A.; Besse, B.; Lu, S.; Spira, A.I.; Girard, N.; Califano, R.; Gadgeel, S.M.; et al. Amivantamab plus lazertinib versus osimertinib in first-line EGFR-mutant advanced non-small-cell lung cancer with biomarkers of high-risk disease: A secondary analysis from MARIPOSA. *Ann. Oncol.* **2024**, *35*, 805–816. [CrossRef]
57. Nogami, N.; Barlesi, F.; Socinski, M.A.; Reck, M.; Thomas, C.A.; Cappuzzo, F.; Mok, T.S.K.; Finley, G.; Aerts, J.G.; Orlandi, F.; et al. IMpower150 Final Exploratory Analyses for Atezolizumab Plus Bevacizumab and Chemotherapy in Key NSCLC Patient Subgroups With EGFR Mutations or Metastases in the Liver or Brain. *J. Thorac. Oncol.* **2022**, *17*, 309–323. [CrossRef]
58. Cai, R.; Liu, Y.; Yu, M.; Sha, H.; Guo, M.; Chen, Y.; Ye, J.; Zhou, G.; Fang, Y.; Shen, B. A retrospective real-world study: The efficacy of immune-related combination therapies in advanced non-small cell lung cancer after resistance to EGFR-TKIs. *Cancer Immunol. Immunother.* **2023**, *72*, 4355–4365. [CrossRef]
59. Hicklin, D.J.; Ellis, L.M. Role of the vascular endothelial growth factor pathway in tumor growth and angiogenesis. *J. Clin. Oncol.* **2005**, *23*, 1011–1027. [CrossRef]
60. Kim, J.M.; Chen, D.S. Immune escape to PD-L1/PD-1 blockade: Seven steps to success (or failure). *Ann. Oncol.* **2016**, *27*, 1492–1504. [CrossRef]
61. Terme, M.; Colussi, O.; Marcheteau, E.; Tanchot, C.; Tartour, E.; Taieb, J. Modulation of immunity by antiangiogenic molecules in cancer. *Clin. Dev. Immunol.* **2012**, *2012*, 492920. [CrossRef]
62. Walke, V.; Das, S.; Mittal, A.; Agrawal, A. Tumor Educated Platelets as a Biomarker for Diagnosis of Lung cancer: A Systematic Review. *Asian Pac. J. Cancer Prev.* **2024**, *25*, 1911–1920. [CrossRef] [PubMed]
63. Roweth, H.G.; Battinelli, E.M. Lessons to learn from tumor-educated platelets. *Blood* **2021**, *137*, 3174–3180. [CrossRef] [PubMed]
64. Luo, C.L.; Xu, Z.G.; Chen, H.; Ji, J.; Wang, Y.H.; Hu, W.; Wang, K.; Zhang, W.W.; Yuan, C.H.; Wang, F.B. LncRNAs and EGFRvIII sequestered in TEPs enable blood-based NSCLC diagnosis. *Cancer Manag. Res.* **2018**, *10*, 1449–1459. [CrossRef] [PubMed]

current issues in molecular biology

MDPI

Article

The Pre-/Post-Transplant Hepatitis C Antibody Associated with the IL-28B RS8099917 TT Genotype and miRNA-122 Expression May Protect Acute Cellular Rejection After LDLT

King-Wah Chiu [1,2,3], Yu-Cheng Lin [2,4], Wei-Feng Li [2,4], Kuang-Tzu Huang [5], Li-Wen Hsu [5] and Chih-Chi Wang [2,3,4,*]

1 Division of Hepato-Gastroenterology, Department of Internal Medicine, Kaohsiung Chang Gung Memorial Hospital, Kaohsiung 83301, Taiwan; c471026@ms6.hinet.net
2 Liver Transplantation Program, Kaohsiung Chang Gung Memorial Hospital, Kaohsiung 83301, Taiwan; dr.linyucheng@gmail.com (Y.-C.L.); webphone0613@yahoo.com.tw (W.-F.L.)
3 College of Medicine, Chang Gung University, Tao-Yuen 33302, Taiwan
4 Department of Surgery, Kaohsiung Chang Gung Memorial Hospital, Kaohsiung 83301, Taiwan
5 Institute for Translational Research in Biomedicine, Kaohsiung Chang Gung Memorial Hospital, Kaohsiung 83301,Taiwan; huangkt@cgmh.org.tw (K.-T.H.); hsuliwen@ms55.hinet.net (L.-W.H.)
* Correspondence: ufel4996@ms26.hinet.net; Tel.: +886-7-731-7123 (ext. 8301); Fax: +886-7-733-6856

Abstract: This study aimed to investigate the relationship between the IL-28B SNP rs8099917 genotype, miRNA-122 expression, and the immune mechanism of ACR after LT using anti-HCV antibody calibration. A total of 45 patients with HCV received LT. IL-28B SNP rs8099917 genotyping was used to divide patients into TT and GT groups. The relative expression levels of miRNA-122 were calculated by quantitative PCR. Anti-HCV titers before and after LT were tracked to observe the relationship with ACR. The ACR rates were 27.6% for genotype TT and 62.5% for genotype GT, indicating a significantly higher rate in the GT group compared to the TT group ($p = 0.024$). In the rs8099917 genotype, TT was significantly associated with higher serum miRNA-122 levels than GT ($p < 0.001$). The TT group had significantly better outcomes than the GT group ($p = 0.005$). The Mann–Whitney U test showed significant differences in pre-LT and post-LT anti-HCV titers between the IL-28B genotypes (TT and GT) (p values of 0.006 and 0.027, respectively). These results suggested that the IL-28B rs8099917 genotype TT may play a significant role in modulating immune responses, both in terms of anti-HCV titers and the risk of ACR, possibly mediated through miRNA-122 levels.

Keywords: acute cellular rejection; hepatitis C virus; IL-28B; liver transplantation; microRNA-122; single-nucleotide polymorphism

1. Introduction

The immune mechanism of hepatitis C virus (HCV) infection is indeed complex and involves multiple interactions between host and viral factors [1]. When a patient is infected with HCV, the immune system produces antibodies against HCV, but these antibodies are not protective and cannot eliminate the virus from the body. Non-protective antibodies do not provide immune protection against HCV, possibly due to the high mutation rate of HCV, which makes it difficult for antibodies to effectively neutralize the virus [2,3]. These antibodies can only be used as indicators for diagnosing HCV infection. Anti-HCV antibodies can be used for clinical diagnosis to determine whether a person has been infected or is currently infected with HCV. As we know, HCV can evade the immune system by impairing the signaling of immune cell receptors and affecting the function and survival of both innate and adaptive immune cells. The host immune response to HCV includes both innate and adaptive immunity. Innate immunity involves interferons and natural killer cells, while adaptive immunity involves CD4+ and CD8+ T cells [4]. The immune response mechanism of HCV infection activates an innate immune response, including the production of interferons, which can inhibit viral replication [5,6]. Additionally, host cytotoxic T lymphocytes and natural killer cells play important roles in the process of viral clearance. However, the HCV uses various strategies to evade the host immune response, including avoiding the effects of neutralizing antibodies and spreading through cells [7,8]. Although anti-HCV antibodies are not protective, they represent the immune response of the host immune system to HCV infection. Therefore, exploring HCV antibody titers is valuable [9]. While anti-HCV is not a protective antibody against HCV, its development appears to be an immune response in which T cells are responsible for attacking the virus and signaling natural killer cells and interferons to protect the host. Acute cellular rejection (ACR) remains a significant problem after liver transplantation (LT). It represents a genetic pathological response of the immune system to the potential effects of immunosuppressants, particularly in HCV-associated LT [10]. Liver grafts may face an immune response from the host and a residual viral biological attack. The IL-28B single-nucleotide polymorphism (SNP) rs8099917 is known to help fight and spontaneously clear HCV infection due to its genetic characteristics [11]. Additionally, the host's own small non-coding RNAs, including miRNA-122, miRNA-301, miRNA-133a, and miRNA-21, may be expressed in the liver with varying degrees of genetic upregulation and downregulation under different pathogenic conditions [12]. In particular, miRNA-122 is indeed a significant player in the context of ACR after LT [13]. Meanwhile, the host factors, such as genetic polymorphisms, can influence the outcome of HCV infection and the response to treatment. Studies have shown that the rs8099917 genotype may influence the outcomes of liver transplantation, including the risk of ACR. However, the exact mechanisms and the extent of this influence are still under investigation [14,15]. This study aims to explore the relationship between the IL-28B SNP rs8099917 genotype, miRNA-122 expression, and the pathological mechanism of anti-HCV levels on ACR after LT.

2. Materials and Methods

2.1. Patients

A total of 45 anti-HCV-positive patients receiving LDLT were included in this study, comprising 29 males and 16 females, with a mean age of 58.16 $\pm$ 7.02 years, of whom 42.2% had hepatocellular carcinoma. The inclusion criteria required anti-HCV-positive patients receiving LDLT who experienced a sudden onset of alanine aminotransferase (ALT) flare-up greater than twice the normal limit (U/L), who experienced acute jaundice with total serum bilirubin greater than 1.4 mg/dL, and who underwent percutaneous liver biopsy for histopathologic diagnosis. The exclusion criteria included patients with negative serum anti-HCV antibodies, positive serum HBsAg levels, underlying psychiatric conditions, primary biliary cirrhosis, alcohol-related liver disease, or a history of pediatric LT. The complete clinical patient profile is shown in Table 1. According to IL-28B SNP rs8099917 genotyping, the patients were divided into two groups: 29 cases with the TT genotype and 16 cases with the GT genotype. All patients received interferon plus ribavirin or direct-acting antiviral agents, including Harvoni (i.e., 400 mg of sofosbuvir plus 90 mg of ledipasvir), or a combination of 400 mg of sofosbuvir/60 mg of daclatasvir/800 mg of ribavirin for HCV infection whenever possible. All recipients received the same immunosuppression protocol, being administered mycophenolate mofetil at 250 mg of Q12H PO and tacrolimus at 1 mg of QD PO, with doses adjusted according to mycophenolic acid plasma levels > 3.0 µg/mL and tacrolimus blood concentrations of 5–10 ng/mL. Our exclusion criteria included pediatric transplantation, deceased donor liver transplantation, HBsAg-positive patients, primary biliary cholangitis, biliary atresia, Wilson's disease, polycystic disease, and re-transplantation.

Table 1. The clinical profiles of anti-HCV-positive recipients with different IL-28B SNP rs8099917 TT and GT genotypes underwent living donor liver transplantation.

Category	TT Group No. (%)	GT Group No. (%)	*p* Value
Cases	29 (64.4)	16 (35.6)	
Gender, M/F	17/12	12/4	NS
Age (mean ± SD)	57.41 ± 6.65	59.94 ± 7.84	NS
Combined with HCC	25 (86.2)	10 (62.5)	NS
HCV genotype (1a/1b/2/undetected)	2/12/5/10	0/4/5/7	
Pre-LT			
AST	39.69 ± 16.77	63.81 ± 57.88	<0.05
ALT	33.83 ± 21.17	67.94 ± 108.88	NS
Bil. T	1.40 ± 0.72	2.03 ± 1.15	<0.05
HCV RNA	11 (37.9)	6 (37.5)	NS
Anti-HCV titer	52.73 ± 26.52	29.17 ± 14.29	<0.005
DAA treatment	6 (20.7)	5 (31.3)	NS
Post-LT			
AST	35.41 ± 22.55	64.13 ± 41.17	<0.005
ALT	42.62 ± 39.00	73.13 ± 45.98	<0.05
Bil. T	0.64 ± 0.28	0.69 ± 0.37	NS
HCV RNA	6 (20.7)	2 (12.5)	NS
Anti-HCV titer	79.39 ± 45.56	50.30 ± 40.21	<0.05
DAA treatment	10 (34.5)	6 (37.5)	NS

Alb: albumin; ALT: alanine aminotransferase; AST: aspartate aminotransferase; Bil. T: total bilirubin; HCC: hepatocellular carcinoma; LT: liver transplantation.

2.2. Anti-HCV Antibody

Serum samples for anti-HCV were collected on the day before LT and at the same time as liver biopsy when clinically necessary after LT. Anti-HCV was detected using the Cobas e411 analyzer through an automated electrochemiluminescence immunoassay, executed with the Elecsys Anti-HCV II assay kit (Roche Diagnostics GmbH, Mannheim, Germany) [16].

2.3. IL-28B Single-Nucleotide Polymorphism

Genomic DNA was extracted from the peripheral blood mononuclear cells of recipients on the pre-operative day using the QIAamp DNA Blood Mini Kit (Qiagen, Hilden, Germany). IL-28B SNP rs8099917 genotypes were studied using a 7500 Fast Real-Time PCR System (Applied Biosystems, Foster City, CA, USA) with Custom TaqMan SNP Genotyping Assays (Applied Biosystems) for allele discrimination. The real-time PCR reactions were performed in 96-well microplates, using the ABI 7500 Fast Real-Time Polymerase Chain Reaction System (Applied Biosystems International, Framingham, MA, USA) in accordance with the manufacturer's instructions. The IL-28B SNP rs8099917 was defined as TT, GT, or GG genotype, as recommended by the manufacturer. All genotypes of IL-28B SNP rs8099917 were assayed in duplicate to assess inter-assay precision [17,18]. Only two genotypes, TT and GT, were detected in this study.

2.4. MicroRNA-122

Samples of miRNA-122 were collected at the same time as liver biopsy when clinically necessary after LT. miRNA-122 levels were extracted from the plasma using the miRNeasy Mini Kit (Qiagen 217004, Beijing, China) according to the manufacturer's protocol. Reverse transcription (RT) was performed with 1 μg of RNA using the First-Strand cDNA Synthesis Kit (Promega, Madison, WI, USA) or miScript RT Kit (Qiagen, Hilden, Germany) for the transcription of miRNA according to the manufacturer's instructions. Using the ABI TaqMan Fast Universal PCR master mix or TaqMan Universal PCR master mix for miRNA (Applied Biosystems, Foster City, CA, USA), we performed RT-PCR on an ABI 7500 Fast Real-Time PCR System with the SDS 1.4 program. The cycle threshold (Ct value), which was inversely correlated with miRNA level, was defined as the number of cycles required for the fluorescent signal to cross the threshold in quantitative PCR. Comparative RT-PCR data, including non-template controls, were obtained in triplicate. Due to the very low concentration of miRNA-122 in blood, it was crucial to select the appropriate reference gene and use sensitive detection methods. One of these methods was to use U6 snRNA as the reference gene, which was corrected and relatively stable in this study. The fold increase in cytokine mRNA expression was calculated using the comparative $2^{-\Delta\Delta Ct}$ method, where Ct represents the threshold cycle for each transcript. The expression of the miRNA was defined based on Ct, and relative expression levels were calculated as 2ˆ[(Ct of miR−122) − (Ct of U6)] after normalization with reference to the expression of small nuclear RNA U6 [19].

2.5. Liver Biopsy Interpretation

All suspected clinical organ rejections after LT were assessed by the first author of this study. The histopathological diagnostic criteria for all pathological graft rejections were defined based on the 1995 Banff classification, and the severity grades were characterized by the rejection activity index [20].

2.6. Institutional Review Board Statement

The study protocol was approved and authorized by the ethics committee of our hospital (approval number: 202300159B0) and performed in accordance with the ethical principles of the Declaration of Helsinki.

2.7. Informed Consent Statement

Each participant provided written informed consent before participating in this study.

2.8. Statistics

Statistical analyses were performed using SPSS (version 22.0; SPSS Inc., Chicago, IL, USA). Descriptive values were expressed as mean ± standard deviation and percentages. Categorical variables were compared using the Mann–Whitney U test, the chi-squared test, or Fisher's exact test, and continuous variables were compared using Student's *t*-test. All tests were two-tailed, and a *p*-value of <0.05 was considered statistically significant. Log Rank (Mantel-Cox) and Kaplan–Meier curves were used for outcome analysis. Multivariate analysis was used to correlate the relationship between the co-factors.

3. Results

Although the IL-28B SNP rs8099917 can be defined as three genotypes (TT, GT, or GG), only two genotypes (TT and GT) were detected in this study. Therefore, the subjects were divided into TT and GT groups for further discussion.

3.1. The Relationship Between miRNA-122 Expression and IL-28B rs8099917

In the rs8099917 genotype, the TT group was significantly associated with higher serum miRNA-122 levels compared to the GT group (0.0170 ± 0.0283 vs. 0.0016 ± 0.0038, $p < 0.001$) (Figure 1).

3.2. The Relationship Between Acute Cellular Rejection and IL-28B rs8099917

The ACR rates were 27.6% (8/29) for the TT genotype and 62.5% (10/16) for the GT genotype, indicating a significantly higher rate in the GT group compared to the TT group ($p = 0.024$) (Table 2).

The time to rejection ranged from 1 to 8 months, with a mean of 2.8 months in the GT group and 3.9 months in the TT group. The TT group had significantly better outcomes compared to the GT group ($p = 0.005$), as determined by Log Rank (Mantel–Cox) overall comparisons, which are illustrated in the Kaplan–Meier curve (Figure 2).

Table 2. The relationship between acute cellular rejection and IL-28B SNP rs8099917 in patients with HCV relative living donor liver transplantation.

IL-28B SNP rs8099917	Acute Cellular Rejection N (%)	*p* Value
Genotype TT, $n = 29$	8 (27.6)	0.024
Genotype GT, $n = 16$	10 (62.5)	

HCV: hepatitis C virus; IL: interleukin; SNP: single-nucleotide polymorphism. Mann–Whitney U test (2-tailed); <0.05 indicates a significant difference.

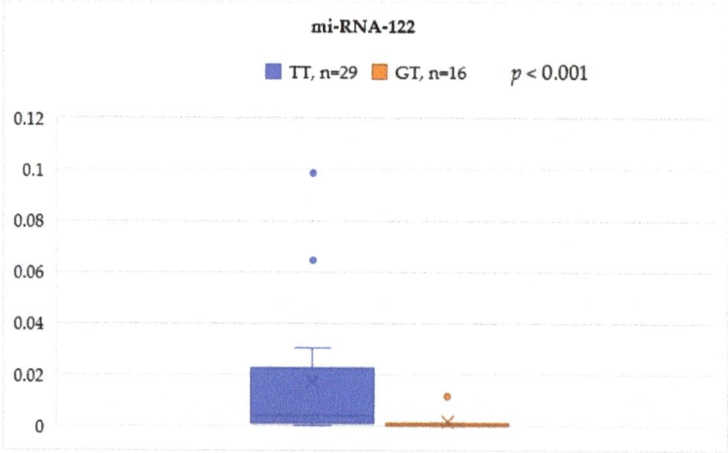

Figure 1. Mann–Whitney test analysis for the miRNA-122 expression comparisons between the IL-28B SNP rs8099917 genotypes TT and GT.

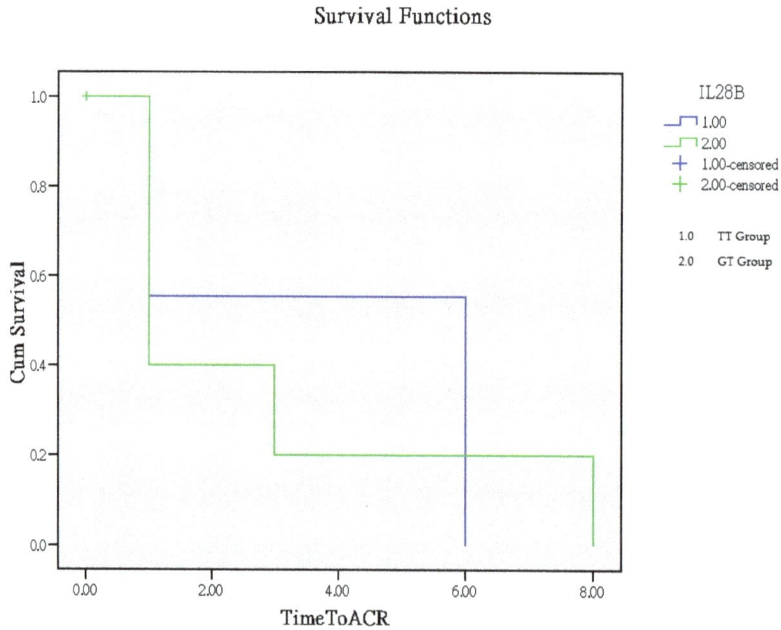

Figure 2. Analysis of clinical outcomes using the Kaplan–Meier curve for time to rejection.

3.3. The Relationship Between Anti-HCV Titer and IL-28B rs8099917

The mean rank and rank sum of pre-LT anti-HCV and post-LT anti-HCV showed that the IL-28B genotype TT group had higher values for both variables. The Mann–Whitney U test indicated a significant difference in pre-LT anti-HCV and post-LT anti-HCV between the different IL-28B genotypes (TT and GT) (p values of 0.006 and 0.027, respectively), suggesting that the IL-28B genotype significantly affects these variables. A negative Z-score indicates a higher rank sum in the group with the IL-28B genotype TT. These results suggest that the IL-28B genotype has a significant effect on pre-LT and post-LT anti-HCV titers, with the TT group showing higher values for these variables (Table 3A,B).

Table 3. (**A**) The Mann–Whitney U test for analyzing the relationship between pre-/post-LT anti-HCV titer and IL-28B TT and GT genotypes. (**B**) Significant difference between the pre-/post-LT anti-HCV titer and IL-28B TT and GT genotypes.

(A)				
Ranks	**IL28B**	**N**	**Mean Rank**	**Sum of Ranks**
Pre-LT anti-HCV	TT	29	26.97	782.00
	GT	16	15.81	253.00
	Total	45		
Post-LT anti-HCV	TT	29	26.21	760.00
	GT	16	17.19	275.00
	Total	45		

(B)		
Test Statistics (a)	**Pre-LT Anti-HCV**	**Post-LT Anti-HCV**
Mann–Whitney U	117.000	139.000
Wilcoxon W	253.000	275.000
Z	−2.729	−2.207
Asymp. Sig. (two-tailed)	0.006	0.027

(**a**), grouping variable: IL28B.

Multivariate analysis with 95% confidence intervals showed the correlation between IL-28B rs8099917 genotyping, pre-LT/post-LT anti-HCV titers, and ACR (Tables 4 and 5). Table 4 shows the effects of different variables on multiple dependent variables (miRNA-122, pre-LT anti-HCV, post-LT anti-HCV, ACR).

Table 4. Multivariate tests [c] for IL-28B SNP rs8099917 and co-factors.

	Effect	**Value**	**F**	**Hypothesis df**	**Error df**	**Sig.**	**Observed Power [a]**
Intercept	Pillai's Trace	0.811	42.847 [b]	4.000	40.000	0.000	1.000
	Wilks' Lambda	0.189	42.847 [b]	4.000	40.000	0.000	1.000
	Hotelling's Trace	4.285	42.847 [b]	4.000	40.000	0.000	1.000
	Roy's Largest Root	4.285	42.847 [b]	4.000	40.000	0.000	1.000
rs8099917	Pillai's Trace	0.037	5.074 [b]	4.000	40.000	0.002	0.943
	Wilks' Lambda	0.663	5.074 [b]	4.000	40.000	0.002	0.943
	Hotelling's Trace	0.507	5.074 [b]	4.000	40.000	0.002	0.943
	Roy's Largest Root	0.507	5.074 [b]	4.000	40.000	0.002	0.943

a: Computed using alpha = 0.05; b: Exact statistic; c: Design = Intercept+IL28B.

Table 5. Parameter estimates for IL-28B SNP rs8099917 and co-factors.

Dependent Variable	Parameter	B	Std. Error	t	Sig.	95% Confidence Interval		Observed Power [a]
						Lower Bound	Upper Bound	
Pre-LT anti-HCV	Intercept	29.173	5.751	5.037	0.000	17.575	40.770	0.999
	[IL-28B TT]	23.559	7.164	3.209	0.002	9.111	30.006	0.095
	[IL-28B GT]	0 [b]	.	.	.	.	.	.
Post-LT anti-HCV	Intercept	50.304	10.943	4.597	0.000	28.236	72.372	0.994
	[IL-28B TT]	29.082	13.631	2.134	0.039	1.593	56.572	0.550
	[IL-28B GT]	0 [b]	.	.	.	.	.	0.
miRNA-122	Intercept	0.002	0.006	0.284	0.778	−0.010	0.013	0.059
	[IL-28B TT]	0.015	0.007	2.143	0.038	0.001	0.030	0.554
	[IL-28B GT]	0 [b]	.	.	.	.	.	.
ACR	Intercept	0.625	0.118	5.307	0.000	387	0.863	0.999
	[IL-28B TT]	−0.349	0.147	−2.380	0.022	−0.645	−0.053	0.643
	[IL-28B GT]	0 [b]	.	.	.	.	.	.

a: Computed using alpha = 0.05; b: This parameter is set to zero because it is redundant. The observed power indicates the performance of the test, and the closer the value is to 1, the higher the performance of the test.

3.3.1. Pre-LT Anti-HCV Titer

- Intercept: B = 29.173, indicating a predicted value of 29.173 for pre-LT anti-HCV when all other variables are zero, which is a highly significant result ($p < 0.001$).
- IL-28B TT: B = 23.559 ($p = 0.002$, 95% CI: 9.111 to 38.006, observed power = 0.895), indicating that the predicted value of pre-LT anti-HCV increased by 23.559 when IL-28B was the TT genotype, which is significant ($p = 0.002$).

3.3.2. Post-LT Anti-HCV Titer

- Intercept: B = 50.304, indicating a predicted value of 50.304 for post-LT anti-HCV when all other variables are zero, which is a highly significant result ($p < 0.001$).
- IL-28B TT: B = 29.082 ($p = 0.039$, 95% CI: 1.593 to 56.572, observed power = 0.550), indicating that the predicted value of post-LT anti-HCV increased by 29.082 when IL-28B was the TT genotype, which is significant ($p = 0.039$).

3.3.3. miRNA-122

- Intercept: B = 0.002, indicating a predicted value of 0.002 for miRNA-122 when all other variables are zero. This result is not significant ($p = 0.778$).
- IL-28B TT: B = 0.015 ($p = 0.038$, 95% CI: 0.001 to 0.030, observed power = 0.554), indicating that the predicted value of miRNA-122 increases by 0.015 when IL-28B is the TT genotype, which is significant ($p = 0.038$).

3.3.4. ACR

- Intercept: B = 0.625, indicating a predicted value of 0.625 for ACR when all other variables are zero, which is a highly significant result ($p < 0.001$).
- IL-28B TT: B = −0.349 ($p = 0.022$, 95% CI: −0.is to −0.053, observed power = 0.643), indicating that the predicted value of ACR was reduced by 0.349 when IL-28B is the TT genotype, which is significant ($p = 0.022$).

For the IL-28B TT genotype, the parameter estimate (B) is −0.349, with a standard error of 0.147. The t-value is −2.380, and the *p*-value (Sig.) is 0.022, which is less than 0.05, indicating that this result is statistically significant. The 95% confidence interval for this parameter ranges from −0.645 to −0.053. This means that individuals with the IL-28B TT genotype have an ACR that is, on average, 0.349 units lower than those with the IL-28B GT genotype. The negative B value suggests a negative association between the IL-28B TT genotype and ACR. The observed power for this parameter is 0.643, indicating a moderate level of statistical power. In summary, the IL-28B TT genotype is significantly associated with a lower ACR compared to the IL-28B GT genotype.

In conclusion, IL-28B has a significant effect on all dependent variables. Specifically, when the SNP rs8099917 is the TT genotype, the values of miRNA-122, pre-LT anti-HCV, and post-LT anti-HCV increase, while the value of ACR decreases. The intercept term is significant in all models, suggesting that the base level of the dependent variables is important even after controlling for other variables. These results suggest that the IL-28B rs8099917 genotype has a significant effect on these biomarkers, which may be of great significance for related medical research and clinical applications.

4. Discussion

In the context of LT, the immune mechanism of HCV infection is extremely complex, involving interactions between the host, the new liver graft, and viral agents. This study aims to explore the roles of IL-28B and microRNA, which have shown positive effects on HCV, in addressing the challenge posed by the lack of an effective vaccine. Indeed, serum HCV antibodies are not protective and cannot eliminate the virus from the body. Based on our current results, anti-HCV titers may signal an association with acute cellular rejection (ACR) after LT. Specifically, lower pre-LT anti-HCV titers and higher post-LT anti-HCV titers may enhance the immune response to liver grafts. The immune response mechanism of HCV infection activates an innate immune response, including the production of interferons. Host cytotoxic T lymphocytes and natural killer cells play important roles in the process of viral clearance. Meanwhile, the host's immune system also faces the new liver graft and expresses miRNA-122 for upregulation, leading to the avoidance of ACR after LT. Although anti-HCV antibodies are not protective, they represent the immune response of the host immune system to HCV infection. Therefore, exploring HCV antibody titers is valuable [9]. While anti-HCV is not a protective antibody against HCV, its development appears to be an immune response in which T cells are responsible for attacking the virus and signaling natural killer cells and interferons to protect the host. Regarding the outcome of LT, the anti-HCV response to host immunity seems to not only improve HCV treatment control but also protect against ACR in LT situations, as evidenced by our current results. Since anti-HCV is not a protective antibody, a transplanted organ is perceived as a foreign organ. Even in patients with the TT genotype and sustained virologic response (SVR), there is an immune response that enhances the body's immunoglobulin, leading to a relative increase in anti-HCV titers after LT.

It is well known that the IL-28B SNP rs8099917 genotype benefits HCV clearance, whether spontaneously or through PEG interferon administration [21,22]. In our current study, it is suggested that the rs8099917 GT genotype may be significantly associated with ACR after LT compared to the TT genotype. The IL-28B genotype has been studied in the context of LT, and research indicates that this genotype can influence the immune response and outcomes post-LT. The TT genotype is associated with a different immune response compared to other genotypes [21]. The rs8099917 SNP of the IL-28B gene significantly affects the regulation of the immune system. IL-28B belongs to the interferon-lambda (IFN-λ) family, and these polymorphisms affect its expression levels, which in turn modulate the immune response. Studies have shown that individuals carrying the rs8099917 TG/GG genotype have a significant increase in seroconversion after influenza vaccination. Individuals with these genotypes had increased IL-4 production and HLA-DR expression in T cells and B cells after vaccination. In addition, IL-28B can promote the production of Th1 cytokines and IFN-γ while inhibiting the production of Th2 cytokines, such as IL-4, IL-5, and IL-13, thereby regulating the balance of immune responses. Overall, IL-28B rs8099917 SNPs play an important role in the regulation of the immune system by influencing Th1/Th2 homeostasis and B-cell responses [21,22]. The frequency of IL28B rs8099917 genotypes can vary significantly depending on the population and ethnicity. Studies from Uruguay show that the distribution of IL28B rs8099917 genotypes in the general population was 60.9% TT, 33.7% TG, and 5.4% GG [23]. This distribution is similar to findings from other populations. In our previous study, the genotyping of IL28B rs8099917 showed that 41 (82%) were TT, 9 (18%) were GT, and 0 were GG on post-transplant day 30 (KW Chiu, et al. PLoS One, 2016, 11, e0156846). These results are consistent with the findings of our current study.

In our recent report, different clinicopathological situations express different miRNAs with varying upregulation and downregulation effects in the context of LT [24,25]. Specifically, miRNA-122 has a close correlation with ACR in LT patients. The expression of liver miRNAs is significantly higher compared to serum miRNAs. When organ tissues are not available, serum miRNAs can be a simple method to aid in the diagnosis of acute jaundice after LT. miRNA-122 plays an important role in organ immunity and rejection. In terms of immunomodulation, miRNA-122 is predominantly expressed in the liver and plays a crucial role in liver disease and LT. It can modulate the immune response and affect the liver's immune tolerance. Studies have shown that miRNA-122 can influence the activity and function of immune cells by regulating various immune-related genes. Regarding the rejection of transplanted organs, miRNA-122 is considered a potential biomarker for predicting and diagnosing liver transplant rejection in LT. Studies have found that the expression level of miRNA-122 is associated with rejection after LT, with low expression levels potentially predicting the occurrence of rejection. Overall, miRNA-122 significantly impacts organ immunity and rejection by modulating immune responses and serving as a biomarker for rejection. Recent reports have shown that lower levels of miRNA-122 are associated with an increased risk of rejection. Hence, monitoring miRNA-122 levels could potentially serve as a biomarker to predict or diagnose ACR, allowing for earlier intervention and management [25]. Based on our previous studies, the IL-28B SNP genotype and miR-122 levels may differ between donor liver tissue and recipient blood. This discrepancy can be an interesting topic to explore. We propose a so-called homogeneous phenomenon when there are different genotypes between the donor and the recipient. When the genotype does not match the recipient's immune response and histopathological rejection conditions, it may affect the stability of the liver graft [9]. Regarding the outcome of LT, the anti-HCV response to host immunity appears to not only improve HCV treatment control but also protect against ACR in LT situations. Based on our current results, the IL-28B TT genotype is associated with higher anti-HCV titers both before and after LT compared to the GT genotype. This could indicate a stronger immune response in individuals with the TT genotype. Higher miRNA-122 levels might be linked to better regulation of immune responses, potentially reducing the risk of acute cellular rejection. Our study suggests that the IL-28B TT genotype may play a significant role in modulating immune responses, both in terms of anti-HCV titers and the risk of acute cellular rejection, possibly mediated through miRNA-122 levels. The IL28B genotype, particularly the rs8099917 TT genotype, is associated with a better immune response and better treatment outcomes in chronic HCV infection. It may provide a positive response on the immune system against HCV. However, its direct impact on specifically enhancing the immune response to ACR after LT is still under investigation. Our current results may provide new scientific insights into the genetic regulation of IL28B and miRNA-122 and their roles in the mechanism of ACR.

Our study has some limitations. First, the sample size is relatively small, and the study is single-centered. Nevertheless, we consider the risk of bias to be very small due to the standardization of pre-LT clinical and laboratory assessments, sophisticated LT surgical techniques, and post-LT protocols for immunosuppression regimens and graft outcome surveillance. Second, we could determine the association between changes in anti-HCV antibody titers before and after LT procedures and allograft injury only through a retrospective review of medical records. There were a lack of data regarding the immunosuppressive treatment schedule and the serum levels of immunosuppressive drugs to enable comparisons between anti-HCV antibody titers. Future prospective studies may be warranted to both identify the role of molecular expression levels in allograft rejection and clarify the mechanisms underlying the association between DAA-induced anti-HCV antibody titer fluctuations and immune reactions to post-LT allograft injury. The findings of such studies may shed light on measures for optimizing pre-LT plans and donor selection, as well as serve as a reference for modifying anti-HCV treatment protocols for potential liver allograft recipients to reduce the incidence and severity of ACR. In particular, a simple Student *t*-test interpretation was statistically significant in the analysis of IL-28B TT and miRNA-122. Nonetheless, we judged the risk of bias to be very small, as the standardization of pre-LT clinical and laboratory assessments, the monitoring of immunosuppressive regimens and graft outcomes after LT, the use of multivariate analyses to examine the strength of statistical data to confirm their reliable associations, and increased study funding with a view to collecting more cases may be more convincing. Moreover, immune response to an implanted graft may not be same as that of the HCV virus. Besides the small sample size, the major drawback of this study is not taking into consideration the genetic backgrounds of donors, as the outcome of liver transplant critically depends on both recipient and donor genetic makeups.

In conclusion, IL-28B has a significant effect, especially when the SNP rs8099917 is the TT genotype. The current results suggest that the IL-28B rs8099917 genotype has a significant effect on these biomarkers, which may be of great significance for related medical research and clinical applications.

Author Contributions: K.-W.C. and Y.-C.L. were responsible for the study concept and design; W.-F.L., K.-T.H. and L.-W.H. performed data analysis and interpretation; K.-W.C. and C.-C.W. drafted the manuscript and provided critical revisions. All authors have read and agreed to the published version of the manuscript.

Funding: This work was supported by a grant from the Chang Gung Memorial Hospital of Taiwan (grant number: CMRPG8N0761).

Institutional Review Board Statement: The study protocol was approved and authorized by the Ethics Committee of our hospital (approval number: 202300159B0), which were performed in accordance with the ethical principles of the Declaration of Helsinki.

Informed Consent Statement: Each participant gave written informed consent before participating in the study.

Data Availability Statement: The original contributions presented in this study are included in the article; further inquiries can be directed to the corresponding author.

Conflicts of Interest: The authors declare no conflicts of interest.

Abbreviations

ALF	Acute liver failure
ALT	Alanine aminotransferase
AST	Aspartate aminotransferase
HCV	Hepatitis C virus
HCC	Hepatocellular carcinoma
INR	International normalized ratio
LT	Liver transplantation
LDLT	Living-donor liver transplantation
miRNA	MicroRNA
SNP	Single-nucleotide polymorphism

References

1. Chigbu, D.I.; Loonawat, R.; Sehgal, M.; Patel, D.; Jain, P. Hepatitis C virus infection: Host-Virus interaction and mechanisms of viral persistence. *Cells* **2019**, *8*, 376. [CrossRef] [PubMed]
2. Tang, W.; Chen, W.; Amini, A.; Boeras, D.; Falconer, J.; Kelly, H.; Peeling, R.; Varsaneux, O.; Tucker, J.D.; Easterbrook, P. Diagnostic accuracy of tests to detect Hepatitis C antibody: A meta-analysis and review of the literature. *BMC Infect. Dis.* **2017**, *17*, 695. [CrossRef]
3. Ghany, M.G.; Morgan, T.R. AASLD-IDSA Hepatitis C Guidance Panel Hepatitis C guidance 2019 update: American association for the study of liver diseases-infectious diseases society of America recommendations for testing, managing, and treating hepatitis C virus infection. *Hepatology* **2020**, *71*, 686–721. [CrossRef]
4. Olivera, S.; Perez, A.; Falcon, V.; Urquiza, D.; Pichardo, D.; Martinez-Donato, G. Protective cellular immune response against hepatitis C virus elicited by chimeric protein formulations in BALB/c mice. *Arch. Virol.* **2020**, *165*, 593–607. [CrossRef] [PubMed]
5. Liu, H.Y.; Lin, Y.H.; Lin, P.J.; Tsai, P.C.; Liu, S.F.; Huang, Y.C.; Tsai, J.J.; Huang, C.I.; Yeh, M.L.; Liang, P.C.; et al. Anti-HCV antibody titer highly predicts HCV viremia in patients with hepatitis B virus dual-infection. *PLoS ONE* **2021**, *16*, e0254028. [CrossRef]
6. Law, M. Antibody Responses in Hepatitis C Infection. *Cold Spring Harb. Perspect. Med.* **2021**, *11*, a036962. [CrossRef] [PubMed]
7. Luxenburger, H.; Neumann-Haefelin, C.; Thimme, R.; Boettler, T. HCV specific T cell responses during and after chronic HCV infection. *Viruses* **2018**, *10*, 645. [CrossRef] [PubMed]
8. Osburn, W.O.; Snider, A.E.; Wells, B.L.; Latanich, R.; Bailey, J.R.; Thomas, D.L.; Cox, A.L.; Ray, S.C. Clearance of hepatitis C infection is associated with the early appearance of broad neutralizing antibody responses. *Hepatology* **2014**, *59*, 2140–2151. [CrossRef]
9. Jing, J.S.; Wang, Z.Q.; Jiang, Y.K.; Zhang, X.Y.; Jiang, W.M. Association of cytokine gene polymorphisms with chronic hepatitis C virus genotype 1b infection in Chinese Han population: An observational study. *Medicine* **2020**, *99*, e22362. [CrossRef]
10. Miri, H.H.; Fazeli, P.; Ali-Hassanzadeh, M.; Bemani, P.; Kabelitz, D.; Kalantar, K. Correlation between IL-28 polymorphism and spontaneous clearance in HCV patients: Systematic review and meta-analysis. *Arch. Virol.* **2021**, *166*, 2469–2478. [CrossRef]
11. Hung, C.H.; Chang, K.C.; Lu, S.N.; Wang, J.H.; Chen, C.H.; Lee, C.M.; Hu, T.H. Spontaneous clearance of hepatitis C virus in an interleukin 28B favorable genotype highly prevalent area. *Hepatology* **2013**, *57*, 2089–2090. [CrossRef]
12. Mikulski, D.; Kościelny, K.; Dróżdż, I.; Mirocha, G.; Nowicki, M.; Misiewicz, M.; Perdas, E.; Strzałka, P.; Wierzbowska, A.; Fendler, W. Serum levels of miR-122-5p and miR-125a-5p predict hepatotoxicity occurrence in patients undergoing autologous hematopoietic stem cell transplantation. *Int. J. Mol. Sci.* **2024**, *25*, 4355. [CrossRef]
13. Schmuck, R.B.; Reutzel-Selke, A.; Raschzok, N.; Morgul, H.M.; Struecker, B.; Lippert, S.; de Carvalho Fischer, C.; Schmelzle, M.; Boas-Knoop, S.; Bahra, M.; et al. Bile: miRNA pattern and protein-based biomarkers may predict acute cellular rejection after liver transplantation. *Biomarkers* **2017**, *22*, 19–27. [CrossRef]
14. Koch, P.F.; Ludwig, K.; Krenzien, F.; Hillebrandt, K.H.; Schöning, W.; Pratschke, J.; Raschzok, N.; Sauer, I.M.; Moosburner, S. miRNA as potential biomarkers after liver transplantation: A systematic review. *Transplant Rev (Orlando).* **2024**, *38*, 100831. [CrossRef] [PubMed]
15. Biggins, S.W.; Trotter, J.; Gralla, J.; Burton, J.R.; Bambha, K.M.; Dodge, J.; Brocato, M.; Cheng, L.; McQueen, M.; Forman, L.; et al. Differential effects of donor and recipient IL28B and DDX58 SNPs on severity of HCV after liver transplantation. *J. Hepatol.* **2013**, *58*, 969–976. [CrossRef] [PubMed]
16. Chang, T.S.; Chang, K.C.; Chen, W.M.; Hsu, N.T.; Lee, C.Y.; Lin, Y.C.; Huang, W.C.; Chiu, W.N.; Hu, J.H.; Huang, T.J.; et al. Village-to-village screening for hepatitis B and C using quantitative HBsAg and anti-HCV testing with reflex HCV core antigen tests in the remote communities of a resource-rich setting: A population-based prospective cohort study. *BMJ Open* **2021**, *11*, e046115. [CrossRef] [PubMed]
17. Augestad, E.H.; Holmboe Olesen, C.; Grønberg, C.; Soerensen, A.; Velázquez-Moctezuma, R.; Fanalista, M.; Bukh, J.; Wang, K.; Gourdon, P.; Prentoe, J. The hepatitis C virus envelope protein complex is a dimer of heterodimers. *Nature* **2024**, *633*, 704–709. [CrossRef] [PubMed]
18. Shaheen, M.; Afzal, S.; Khubaib, B.; Amin, I.; Shahid, M.; Idrees, M. Association of Interleukin 28B Polymorphism with Clearance of Hepatitis C Virus: A Mini Review. *Crit. Rev. Eukaryot. Gene Expr.* **2020**, *30*, 223–229. [CrossRef]
19. Kamo, Y.; Ichikawa, T.; Miyaaki, H.; Uchida, S.; Yamaguchi, T.; Shibata, H.; Takuya Honda, T.; Taura, N.; Isomoto, H.; Takeshima, F.; et al. Significance of miRNA-122 in chronic hepatitis C patients with serotype 1 on interferon therapy. *Hepatol. Res.* **2015**, *45*, 88–96. [CrossRef]
20. Torbenson, M.; Washington, K. Pathology of liver disease: Advances in the last 50 years. *Hum. Pathol.* **2020**, *95*, 78–98. [CrossRef]
21. Kawaoka, T.; Takahashi, S.; Takaki, S.; Hiramatsu, A.; Waki, K.; Hiraga, N.; Miki, D.; Tsuge, M.; Imamura, M.; Kawakami, Y.; et al. Interleukin-28B single nucleotide polymorphism of donors and recipients can predict viral response to pegylated interferon/ribavirin therapy in patients with recurrent hepatitis C after living donor liver transplantation. *J. Gastroenterol. Hepatol.* **2012**, *27*, 1467–1472. [CrossRef] [PubMed]
22. Yan, B.; Gao, J.; Guo, J.; Yang, D.; Li, D. Interleukin-28B dampens protease-induced lung inflammation via IL-25 and TSLP inhibition in epithelial cells. *Sci. Rep.* **2020**, *10*, 20973. [CrossRef] [PubMed]
23. Echeverría, N.; Chiodi, D.; López, P.; Ciceron, A.S.; Angulo, J.; Marcelo López-Lastra, M.; Silvera, P.; Canavesi, A.; Bianchi, C.; Colistro, V.; et al. IL28B gene polymorphism rs12979860, but not rs8099917, contributes to the occurrence of chronic HCV infection in Uruguayan patients. *Virol. J.* **2018**, *15*, 40. [CrossRef] [PubMed]

24. Joshi, D.; Salehi, S.; Brereton, H.; Arno, M.; Quaglia, A.; Heaton, N.; O'Grady, J.; Agarwal, K.; Aluvihare, V. Distinct microRNA profiles are associated with the severity of hepatitis C virus recurrence and acute cellular rejection after liver transplantation. *Liver Transpl.* **2013**, *19*, 383–394. [CrossRef]
25. Zhang, P.; Guo, Z.; Zhong, K.; Li, Q.; Ouyang, J.; Chen, M.; Hu, A.; Jiao, X.; Zhu, X.; He, X. Evaluation of Immune Profiles and MicroRNA Expression Profiles in Peripheral Blood Mononuclear Cells of Long-Term Stable Liver Transplant Recipients and Recipients with Acute Rejection Episodes. *Transplant. Proc.* **2015**, *47*, 2907–2915. [CrossRef]

current issues in molecular biology

MDPI

Review

New Insights into the Interplay Between Simple Sugars and Liver Diseases

Simona Parisse *, Erika Coltorti, Monica Mischitelli, Flaminia Ferri and Stefano Ginanni Corradini *

Department of Translational and Precision Medicine, Sapienza University of Rome, Viale dell'Università 37, 00185 Rome, Italy; erika.coltorti@uniroma1.it (E.C.); monica.mischitelli@uniroma1.it (M.M.); flaminia.ferri@uniroma1.it (F.F.)

* Correspondence: simona.parisse@uniroma1.it (S.P.); stefano.corradini@uniroma1.it (S.G.C.)

Abstract: In hepatology, there is growing interest in identifying the mechanisms and risk factors underlying liver diseases with increasing incidence, with particular focus on metabolic dysfunction-associated steatotic liver disease (MASLD) and its complications. Simple sugars have been recognized as key contributors to liver injury and disease progression, not only in the context of MASLD but also beyond. As a result, numerous studies have aimed to elucidate their role in liver pathophysiology. Specifically, simple sugars have been associated with pivotal mechanisms involved in the onset of liver diseases, including inflammation, de novo lipogenesis, oxidative stress, insulin resistance, and dysbiosis with increased intestinal permeability. These mechanisms collectively contribute to a significant association between simple sugar intake and liver diseases of varying stages and severity. The scientific evidence available to date has not only clarified potential pathogenic mechanisms and clinical correlations but also led to the identification of potential therapeutic targets, encompassing both lifestyle interventions and molecular approaches. This review aims to provide a comprehensive analysis of the associations between simple sugar intake, liver injury, and liver diseases. To this end, we conducted an extensive review of the literature, selecting the most relevant and up-to-date studies on the topic.

Keywords: fructose; sugar; liver; MASLD; diet; cirrhosis; steatosis

1. Introduction

The significant epidemiological changes observed in the field of hepatology have prompted a discussion on the role of environmental factors in the onset and progression of chronic liver diseases. Over the past few decades, there has been a notable shift in the prevalence of liver disease etiologies. Viral etiologies have been declining, largely due to the introduction of effective antiviral therapies for HCV and widespread vaccination programs for HBV [1,2]. In contrast, metabolic etiologies have risen. Epidemiological projections indicate that, within the next 25 years, metabolic dysfunction-associated steatotic liver disease (MASLD—formerly NAFLD) will become the leading cause of liver cirrhosis and, similarly, the predominant etiology in patients with hepatocellular carcinoma (HCC) [1,3–6]. Consequently, there has been a growing need to investigate the factors driving this shift. Studies have increasingly focused on the lifestyle and dietary habits of patients, identifying simple sugars as key contributors to this emerging liver disease epidemic.

Through complex mechanisms, simple sugars can contribute both directly and indirectly to liver and systemic injury. Specifically, the cellular absorption of simple sugars leads to the formation of acetyl-CoA, which is used for ATP synthesis; when present in excess, acetyl-CoA is redirected toward de novo lipogenesis, resulting in lipid accumulation in the liver and adipose tissue. Hepatotoxic effects also arise from the generation of reactive oxygen species (ROS) and through the Maillard reaction. Moreover, it is worth highlighting that the intake of simple sugars, particularly fructose, significantly affects the composition and function of the intestinal microbiota, thereby promoting hepatic injury and fostering both hepatic and systemic inflammation. Several of these mechanisms, especially ROS generation, inflammation, and microbiota alterations, also play a key role in the development of insulin resistance, which further amplifies liver and systemic damage. This review will explore in detail these pathogenic mechanisms and their manifestation in various liver diseases [7–11]. The mechanisms underlying these pathological processes are intricate, involving not only the liver but also extrahepatic organs and complex systemic regulatory networks. When persistent, these alterations drive disease progression from hepatic steatosis to steatohepatitis and may ultimately lead to cirrhosis and/or the development of primary liver malignancies, particularly hepatocellular carcinoma (HCC) [1,12]. Among the interconnected pathogenetic mechanisms, the intestinal microbiota warrants special attention, not merely as a passive bystander, but as a critical player in the pathogenesis of simple sugar-induced liver damage [13]. This review aims to synthesize the most recent and relevant scientific literature, sourced from MEDLINE and PubMed, to provide an in-depth and up-to-date analysis of the role of simple sugars in liver diseases.

2. Definition and Classification

From a biochemical perspective, simple sugars refer to a category of molecules belonging to the carbohydrate group. Carbohydrates represent a broad biochemical class, encompassing molecules of varying complexity, all of which are fundamentally composed of carbon and water. These molecules can be simplistically represented by the biochemical formula $C_n(H_2O)_n$ [14].

Carbohydrates are macronutrients and their primary classification is based on degree of polymerization (DP) and type of linkage (α or β), as agreed at the Food and Agriculture Organization/World Health Organization Expert Consultation in 1997. It is possible to distinguish three main groups: simple sugars (DP1–2), oligosaccharides (short-chain carbohydrates) (DP3–9) and polysaccharides (DPX10) [14].

Simple sugars comprise monosaccharides and disaccharides.

There are several definitions of monosaccharides; in general, a monosaccharide is a molecule that is not decomposable by hydrolysis and that represents a building block from which all carbohydrates are built. Furthermore, monosaccharides can be classified by the number X of carbon atoms they contain: triose [3], tetrose [4], pentose [5], hexose [6], heptose [7], and so on [14].

The most commonly considered in human health and frequently found in the diet include glucose, fructose, and galactose—they are all hexoses. Glucose is classified as an aldose due to the presence of an aldehyde group (–CHO) at the terminal position of its carbon chain, whereas fructose is a ketose, characterized by a ketone group (–CO) located within the carbon backbone. Galactose, on the other hand, is a structural epimer of glucose, differing in the configuration around a single carbon atom [15–17].

Disaccharides are molecules composed of two carbohydrate units linked by a glycosidic bond. The most common disaccharides include sucrose, which consists of fructose and glucose; lactose, formed by the bond between galactose and glucose; and maltose, composed of two glucose molecules [16,17]. When ingested through the diet, specific enzymes present in the intestinal lumen hydrolyze the glycosidic bond in disaccharides, converting them into monosaccharides to facilitate absorption.

Polysaccharides are long-chain molecules composed of multiple monosaccharide units linked together. They can be classified as homopolysaccharides when they are composed of the same building blocks or heteropolysaccharides when they are composed of different building blocks [16].

Among the most common polysaccharides are cellulose, which is composed of β-1,4 glycosidic bonds, and starch. Starch exists in two forms: amylose, the unbranched form consisting solely of α-1,4 glycosidic linkages, and amylopectin, the branched form that contains both α-1,4 and α-1,6 glycosidic bonds. Glycogen, the primary glucose storage polysaccharide in animals, is also composed of α-1,4 and α-1,6 glycosidic linkages. In both amylopectin and glycogen, the α-1,4 bonds form the linear segments of the molecule, while the α-1,6 bonds are responsible for the branching points in their structures [16,18].

From a general perspective, carbohydrates are essential for fundamental cellular functions and the regulation of the energetical homeostasis [9,18]. From a hepatological clinical perspective, simple sugars, as previously mentioned, have gained increasing attention due to their growing recognition as direct contributors to the development and progression of liver diseases [7,8,19].

3. Diet

Naturally occurring sugars are simple sugars abundant in commonly consumed foods like fruits and vegetables, and they are also used to produce natural sweeteners such as honey and table sugar. These natural sweeteners are attractive to consumers, who identify natural products as healthier options, so they are widely used in the sweetened food market to receive consumers who indicate them as healthier choices. Among natural sweeteners, fructose is the sweetest natural sugar, and it is abundant in fruit, agave nectar, honey, and some vegetables; however, the most widely used in several commercial beverages and baked products is glucose [16,20,21].

In the modern era, processed and industrially manufactured foods are being consumed more frequently, particularly in Western countries, where excessive sugar intake is increasingly recognized as a key factor in the rising prevalence of various lifestyle-related diseases [22].

Processed and industrially manufactured foods contain significantly higher concentrations of simple sugars compared to non-industrial foods [23]. Consequently, in Western societies, the primary dietary source of simple sugars now derives from industrial food products rather than natural food sources, with a consequent negative impact on health [10,23–25]. In industrially processed foods, the sweet component is utilized to improve the pleasantness of the flavor and might have an impact on products' texture, color, preservation and caloric value [26].

The most commonly used sweeteners in the food industry are sucrose and high-fructose corn syrup (HFCS) [27]. HFCS represents a relatively inexpensive sugar source and is, therefore, widely utilized in the Western food industry due to its ease of production and low cost. It is industrially produced by enzymatically converting corn starch into mono- and disaccharides, including maltose, fructose, and glucose. In HFCS, fructose constitutes approximately 42–55% of the total sugar content [27,28].

HFCS is commonly found in a variety of food products, including beverages (both carbonated and other sugary drinks such as fruit juices), confectionery, jams, jellies, baked goods, and dairy products [10].

With increasing industrialization, the consumption of HFCS-containing products has steadily risen over the years. Large-scale studies conducted in the United States provide significant insights into this trend. Indeed, from the 1970s to the early 2000s, there was a continuous and significant increase in the consumption of sugary beverages and HFCS in the U.S. [29]. More specifically, between 1970 and 1997, the annual per capita consumption of HFCS increased from 64 g/day to 81 g/day [30]. While sucrose intake has declined over time, overall fructose consumption has simultaneously increased by approximately 25%. This indicates that the majority of dietary fructose is now derived from industrial sweeteners, including HFCS, rather than from natural sources [27,30,31].

The exponential increase in HFCS consumption correlates with the increased incidence of obesity and type 2 diabetes mellitus (T2DM), and steatosis and steatohepatitis, certain cancers like those of the liver or pancreas, and finally cardiovascular and kidney diseases [22].

Over the past few decades, due to increased public health awareness and policy initiatives, HFCS consumption has begun to decline, although it remains widespread in the population. In particular, per capita HFCS consumption in the United States began to decrease in the early 2000s, reaching 27 g/day in 2018 [27].

Another interesting category of sugars commonly found in the general population's diet, originating either from natural sources or through artificial production, is represented by sugar alcohols. Sugar alcohols, or polyols, are compounds commonly used as sweeteners and bulking agents [32]. Naturally present in various plant-derived foods, they are considered sugar substitutes due to their lower caloric content compared to simple sugars and their minimal or absent requirement for insulin during metabolism. For this reason, they are particularly popular among individuals with diabetes, although they are often consumed unknowingly by the general population [33]. Among the main sugar alcohols is sorbitol, which is produced by the reduction of glucose, whereby the aldehyde group (–CHO) is converted into a primary alcohol group (CH_2OH) [34]. Mannitol, another polyol, occurs naturally in fruits, vegetables, and seaweeds but can also be synthesized by hydrogenating mannose, similarly converting its aldehyde group into a primary alcohol. Xylitol, often referred to as "wood sugar" due to its natural occurrence in birch trees, can also be produced via biomass hydrolysis to xylose or through catalytic conversion from xylose [35]. It is widely used in sugar-free chewing gums and oral care products due to its dental health benefits [35,36]. Lastly, erythritol is a linear polyol composed of four carbon atoms, each bearing a hydroxyl group. It is virtually noncaloric (0.2 kcal/g, less than 5% of the energy value of sucrose), and it occurs naturally in fruits such as melons, grapes, and pears, as well as in fermented foods like wine and cheese. Uniquely among polyols, erythritol is not produced industrially through chemical reduction but rather via microbial fermentation. However, despite their potential benefits, emerging evidence indicates that excessive intake of foods containing sugar alcohols may exert detrimental effects. Specifically, the intracellular metabolism of polyols can activate pathways involved in cellular damage, promote de novo lipogenesis, and contribute to the formation of advanced glycation end-products (AGEs), several of which have been significantly associated with various pathological conditions and their complications, particularly in individuals with diabetes [37,38]. Moreover, recent studies have suggested that high consumption of these compounds may also be linked to poorer outcomes in patients with cirrhosis [39]. Therefore, while sugar alcohols may represent viable alternatives to traditional sweeteners, it is important to recognize the potential risks associated with their overconsumption that need to be further investigated and clarified.

Given the published data on the adverse effects of excessive sugar consumption, primarily linked to sugary beverages, including metabolic syndrome, visceral adiposity, hypertriglyceridemia, insulin resistance, and liver damage, the World Health Organization (WHO) has strongly recommended reducing sugar intake in both adults and children. Specifically, WHO guidelines advise limiting sugar intake to less than 10% of total daily energy consumption [27,29,40,41].

4. Metabolism

The metabolism of simple sugars varies among different types, and their distinct metabolic pathways explain their differing contributions to liver diseases. Simple sugars that reach the intestinal lumen through the diet, either in their free form or as a result of specific endoluminal enzymatic activity, are absorbed across the intestinal barrier and transported to the liver via the portal circulation [42,43].

Regarding glucose, it is essential to consider that it represents the main source of cellular energy. A significant oscillation of its concentration can be recorded between the fasting and postprandial states, ranging from 5 mmol/L to 8–10 mmol/L, respectively. An increase in blood glucose levels leads to a stimulation of insulin secretion. Insulin, in turn, facilitates the absorption of glucose in peripheral tissues, particularly in muscle and adipose tissue. In the intestinal lumen, glucose is transported mainly into cells via the membrane transporter GLUT2. In adipocytes, only a little part of the glucose taken up is stored as glycogen (<5%) of glucose taken up, since most of sugar uptake, stimulated by insulin, contributes to triacylglyceride (TG) synthesis by providing glycerol-phosphate as the backbone for free fatty acid esterification as well as through the de novo synthesis of free fatty acids. Therefore, at most, 50% of glucose is stored as TG [44].

It is estimated that only 15–25% of glucose metabolism occurs in the intestine and liver [45].

The intracellular metabolism of glucose begins with its phosphorylation to glucose-6-phosphate (G6P) by glucokinase. Subsequently, G6P can follow different pathways depending on cellular needs: (1) storage as glycogen for energy reserves, (2) glycolysis to produce ATP and pyruvate, or (3) the pentose phosphate pathway for nucleotide synthesis [46,47]. Glucose entering the glycolytic pathway is converted into fructose-6-phosphate and subsequently directed into glycolysis, regulated by phosphofructokinase, through its intermediate stages of fructose-1,6-bisphosphate, triose phosphate, pyruvate, and ultimately ATP + CO_2 (Figure 1). This pathway is regulated by negative feedback, mainly controlled by high ATP and citrate concentrations [45,47]. In the liver, the primary goal of glucose metabolism is to generate glycogen reserves that can be utilized during fasting states [18,47]. When hepatic glucose levels exceed energy demands, the de novo lipogenesis pathway is activated using mainly glycolysis-derived Acetyl CoA. In this scenario, excess glucose is converted into fatty acids, which can be stored in lipid droplets within hepatocytes, incorporated into membrane structures, or assembled into very-low-density lipoproteins (VLDL) for extrahepatic transport to adipose tissue [46].

Notably, intracellular glucose may enter the polyol pathway, wherein aldose reductase catalyzes its reduction to sorbitol, which can be subsequently converted to fructose. Interestingly, this metabolic pathway, through the generation of endogenous fructose, enables the activation of multiple intracellular signaling cascades characteristic of fructose metabolism [22,37,38].

A notable distinction exists in the metabolic handling of fructose. Under basal conditions, its concentration in peripheral plasma is markedly lower than that of glucose, averaging approximately 0.04 mM. Nonetheless, following food intake, a rapid and transient elevation, up to tenfold, can be observed, with circulating fructose levels typically returning to baseline within two hours. This regulation of plasma fructose concentrations is primarily attributed to extraction process exerted by the liver, which is capable of removing up to 70% of an orally administered fructose load postprandially [48].

The main sites of fructose metabolism are the liver, intestine, and other organs and cells, such as the kidney and immune system cells [22,45,49,50].

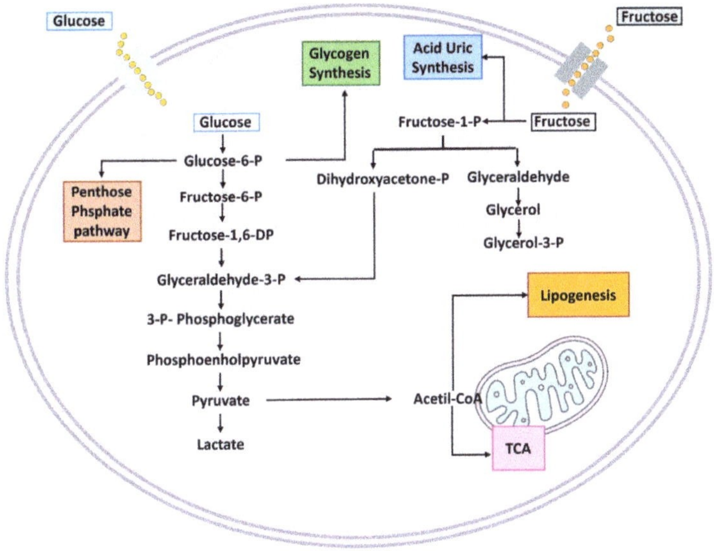

Figure 1. Schematic figure representing glucose end fructose intracellular metabolism. Abbreviations: TCA: tricarboxylic acids; P: phosphate; DP; diphosphate.

In the intestinal lumen the sugar is transported into enterocytes via the GLUT5 membrane transporter and subsequently reaches the liver through the portal circulation [45]. A particularly interesting aspect is that, unlike glucose, fructose metabolism is unrestricted; in fact, it lacks negative feedback and is not subject to regulation by circulating hormones such as insulin [45,51].

Therefore, fructose loads can lead to large, rapid expansions in the hexose- and triose-phosphate pools, potentially providing increased substrate for all central carbon metabolic pathways, including glycolysis, glycogenesis, gluconeogenesis, lipogenesis, and oxidative phosphorylation (Figure 1) [48].

Another possible metabolic fate of fructose is its conversion into lactate, which is then released into systemic circulation and transported to other tissues [31,52].

These significant differences in the metabolism of the two primary simple sugars and the epidemiological data explain the reason why fructose is considered a key factor in the development of hepatic and cardiometabolic diseases, supporting recent dietary recommendations to limit sugar consumption published by several public health agencies, including the WHO, and the Dietary Guidelines Advisory Committee [48].

5. Mechanism of Liver Damage

Simple sugars participate in several molecular mechanisms responsible for liver damage. Most of these mechanisms are dose-dependent and vary among individuals based on a complex interplay of genetic, epigenetic, and environmental factors. From a general perspective, sugar-mediated liver damage should be considered as a multifactorial process consisting of multiple interconnected hits, with highly complex regulatory mechanisms. In the following paragraphs, we outline some of the main mechanisms to consider when addressing liver pathologies (Figure 2).

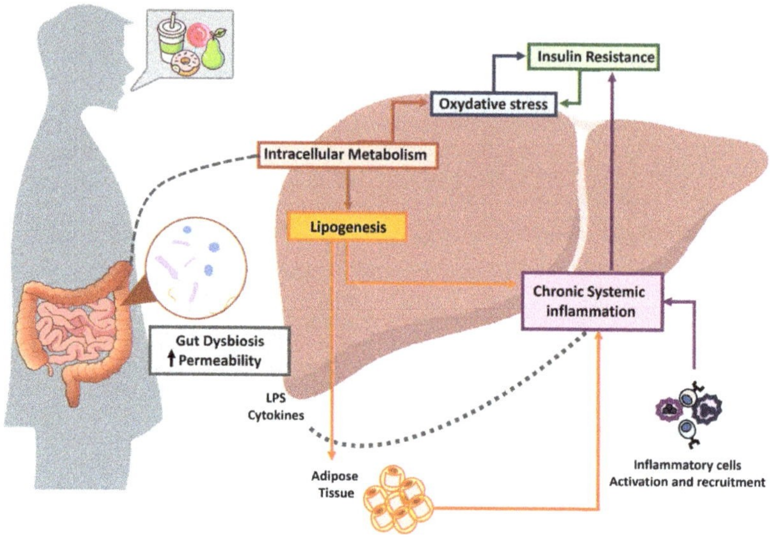

Figure 2. Schematic representation illustrating the interplay of damage mechanisms induced by simple sugars, with a specific focus on hepatic involvement the consumption of foods high in simple sugars can induce gut dysbiosis and increase intestinal permeability, leading to systemic inflammation and chronic hepatic inflammation. Hepatic metabolism of simple sugars may activate de novo lipogenesis, resulting in increased fatty acid synthesis and their subsequent accumulation in both hepatic and adipose tissues. This lipid overload can trigger immune activation, contributing to immune-mediated tissue injury and further promoting chronic systemic inflammation. Intrahepatic metabolism of simple sugars is also associated with the induction of oxidative stress, a key mediator of cellular damage. In this context, insulin resistance may be driven by both oxidative stress, creating a self-perpetuating cycle, and the persistent systemic inflammatory state induced by excessive simple sugar intake.

5.1. Oxidative Stress

One of the key mechanisms of liver damage mediated by simple sugars is oxidative stress. A common feature among different simple sugars is that part of their metabolism is directed toward ATP production through direct or indirect activation of the glycolytic pathway. As a consequence, excessive simple sugar intake stimulates mitochondria to produce increased amounts of reactive oxygen species (ROS) [53,54].

Another mechanism that can be easily triggered by glucose or fructose overload is hyperinsulinemia and insulin resistance (see Section 5.4). When present in excess, insulin can lead to the activation of NADPH oxidase, resulting in the formation of large amounts of ROS [53–55].

Fructose plays an even more significant role than glucose in triggering oxidative stress mechanisms. Indeed, in addition to the aforementioned pathways, it has been described that protein fructosylation, which occurs in the presence of excess fructose and is a faster process than glucose-induced glycation, generates ROS [56]. Moreover, excessive dietary fructose intake leads to a reduction in antioxidant enzymes such as superoxide dismutase, glutathione peroxidase, and glutathione with consequent increase in ROS [57–60]. It is interesting to consider that even the consumption of HFCS has been associated with a significant increase in ROS, leading to clinical negative effects [61].

An additional interesting fructose-specific oxidative stress mechanism is linked to its metabolism via fructokinase. This enzyme is ATP-dependent and, as previously described, lacks a negative feedback regulation system. Therefore, in cases of fructose excess, the excessive ATP consumption by this enzyme leads to an increase in uric acid concentration as a byproduct of nucleotide metabolism. Interestingly, while uric acid is widely recognized as a potent extracellular antioxidant, it can exert pro-oxidant effects within cells, leading to oxidative stress. Specifically, it has been shown to activate mitochondrial NADPH oxidase, which utilizes NADPH leading to the production of ROS. In addition, uric acid can promote citrate accumulation by inhibiting aconitase, an oxidation-sensitive enzyme of the Krebs cycle, thereby enhancing de novo lipogenesis [62].

5.2. Inflammation

A central aspect of the mechanisms through which simple sugars exert their damaging effects at the hepatic level is inflammation. These sugars have the distinctive ability to induce a persistent systemic inflammatory state. Sugar-mediated inflammation is not limited to the liver but also involves key sites such as adipose tissue, the intestine, and, through a distinct pathway, the central nervous system [63].

This complex chronic inflammatory state involves an intricate network of cytokines, which in turn are responsible for multiple intracellular and extracellular signaling cascades. Among the main mediators of intra- and extrahepatic inflammation are C-reactive protein (CRP), peroxisome proliferator-activated receptors (PPARs), Toll-like receptor 4 (TLR-4), and most of the most common cytokines like interleukin-6 (IL-6), interleukin-4 (IL-4), and tumor necrosis factor-alpha (TNF-α) [7,63].

Studies conducted on both human subjects and animal models have reinforced existing knowledge regarding the association between simple sugar intake and elevated CRP levels [64–68]. Some studies indicate that, among simple sugars, fructose is linked to a notable early post-prandial increase in CRP compared to glucose [64]. However, the evidence on this topic remains debated. Notably, a compelling randomized controlled trial conducted on human found no significant differences in CRP levels after a short-term (8-day) diet high in fructose, glucose, or HFCS [65]. One possible reason for this seemingly conflicting result could be the limited duration of the high simple sugar diet. In fact, more recent studies that have analyzed patients' habitual diets using food diaries, despite the methodological limitations, have found noteworthy associations between CRP levels and simple sugar intake [67,68].

Among simple sugars, fructose consumption has been specifically described as capable of inducing inflammation by altering intestinal permeability both through its intestinal metabolism and through specific pathways [22,45,49]. Specifically, fructose may trigger the activation of TLR4, which is naturally present in innate immune response cells and enterocytes. The activation of this receptor initiates a signaling cascade that plays a crucial role in triggering the innate immune response [69]. This process leads to the production of pro-inflammatory cytokines, including TNF-α, IL-6, and IL-1β, as well as the activation of the NLRP3 inflammasome [7,56].

Within the framework of sugar-mediated inflammation, it is essential to consider that adipose tissue, which increases due to de novo lipogenesis induced by excess simple sugars, plays a crucial role [46]. It is now well established that adipose tissue can produce and release pro-inflammatory cytokines that contribute to the systemic metabolic disorder induced by simple sugars and to liver damage [70,71].

Another noteworthy mechanism for triggering the inflammatory cascade is the direct interaction between fructose and immune cells [49]. It has been shown that a fraction of circulating fructose is taken up and directly metabolized by myeloid cells, both circulating and in bone marrow, resulting in modulation of lymphopoiesis, inflammatory cell activation, and release of pro-inflammatory cytokines [49,72,73].

A final mechanism through which simple sugars can promote chronic inflammation involves their interaction with the central nervous system. Excessive consumption of simple sugars or artificial sweeteners, particularly sucralose and fructose, has been shown to stimulate the release of corticotropin-releasing hormone (CRH) and alter the regulation of the hypothalamic–pituitary–adrenal axis [74]. CRH, in turn, stimulates the production of adrenocorticotropic hormone (ACTH), which leads to cortisol release from the adrenal glands [75]. In this way, excessive sugars establish a central modulation of stress and inflammation.

In summary, excessive simple sugar consumption unequivocally induces a systemic inflammatory state that is strongly dependent on the quantity and duration of exposure, with activation pathways that are intricate and involve multiple organs.

5.3. Intestinal Permeability and Gut Microbiota

As previously mentioned, another potential mechanism through which sugars can cause liver damage is the intestine.

Excessive simple sugars intake and its metabolism, which, as previously mentioned, partially occurs in enterocytes, can compromise the integrity of the intestinal barrier [22,49].

Notably, fructose appears to be the primary simple sugar responsible for these alterations, largely due to the action of fructokinase. It has been described that, in the absence of fructokinase, the alterations in intercellular junctions would not be observed [76].

These alterations are specifically characterized by the reduction in crucial enterocytes membrane proteins, including adherens junction proteins (β-catenin and E-cadherin) and tight junction proteins (claudin-1, claudin-4, ZO-1 and occludin) [58].

These changes, along with the aforementioned inflammatory mechanisms and sugar-induced shifts in gut microbiota composition, can also lead to enterocyte apoptosis [58,76].

From a general perspective, it is crucial to consider that the loss of intercellular junctions and enterocytes inevitably results in increased intestinal permeability, consequently promoting greater translocation of bacterial products into the liver through the portal circulation, thereby contributing to hepatic damage and inflammation [22,58,76,77].

Regarding changes in gut microbiota composition, although extensively studied, they are not yet fully understood. In general, the Western diet, characterized by high saturated fat and sugar intake, reduces gut microbiota diversity, promoting the growth of Firmicutes and Proteobacteria while decreasing Bacteroidetes and Fusobacteria. The reduction in Bacteroides, in particular, appears to be the alteration most strongly associated with the development of MASLD and LPS translocation [78].

In recent years, significant efforts have been directed toward understanding the relationship between gut microbiota and sugar-mediated liver damage, primarily to identify potential therapeutic targets. These may include, for example, the administration of prebiotics or probiotics, alone or in combination with omega-3 fatty acids (FA), to mitigate or prevent liver damage [79].

5.4. Insulin Resistance

One of the systemic mechanisms underlying metabolic syndrome and its associated liver damage is insulin resistance (IR) [80]. IR is a condition characterized by an inadequate cellular response to insulin, leading to a low sensitivity to insulin stimulation.

The damaging mechanisms induced by simple sugars, as described in previous sections, contribute to the onset of IR. In turn, IR promotes metabolic dysregulation, which exacerbates simple sugar-induced damage [53].

Specifically, oxidative stress and the subsequent production of reactive oxygen species (ROS) stimulate insulin production, and certain oxidative stress biomarkers have been associated with the presence of IR [81].

The chronic systemic inflammatory state caused by excessive simple sugar intake, characterized by elevated circulating levels of LPS, IL-1β, IL-6, and TNF-α, can inhibit insulin receptor signaling. This results in reduced insulin sensitivity in key organs such as the liver, leading to hepatic steatosis and fibrosis [53,80].

PPARα inhibition is also involved in the development of IR. This transcription factor plays a crucial role in insulin sensitivity, as demonstrated by the fact that its activation is a therapeutic target for IR treatment [82].

Endoplasmic reticulum (ER) stress, typically observed in conditions of excessive sugar intake, contributes to the IR process [80,83]. ER stress triggers the activation of JNK and NF-κB signaling pathways. Specifically, JNK induces phosphorylation of the insulin receptor substrate 1 (IRS1), negatively affecting its function [83]. Furthermore, ER stress can inhibit the Akt/PKB signaling pathway, leading to reduced insulin-stimulated glucose uptake [84].

In the setting of IR, peripheral adipose tissue plays a key role in metabolic dysregulation. IR enhances lipolytic activity, leading to elevated plasma concentrations of free fatty acids (FFAs) [85]. These excess FFAs are predominantly taken up by the liver, where they are stored as triglycerides, resulting in hepatic steatosis [86]. Once inside hepatocytes, high levels of FFAs can induce lipotoxicity by activating intracellular stress responses, including ER stress, autophagy, and the production of pro-inflammatory cytokines [87]. This lipotoxic environment, driven by sustained FFA overload, also stimulates the activation of Kupffer cells and hepatic stellate cells, fostering a pro-inflammatory and profibrotic milieu within the liver [87].

Given its high prevalence and significant impact on human health, understanding the mechanisms responsible for impaired insulin action is crucial to improving IR and elucidating the beneficial effects of insulin-sensitizing agents. Among the drugs currently widely used in this context, metformin is certainly one of the most commonly prescribed. Although its mechanism of action remains partially unclear, metformin is still considered a first-line agent for the treatment of IR. Its use enhances peripheral glucose utilization, likely through the upregulation of GLUT4 expression, and it is thought to act via both direct and indirect pathways that modulate insulin signaling [88].

6. Liver Disease Development and Worsening

There is increasing evidence linking excessive dietary intake of simple sugars to liver diseases. In particular, excessive consumption of simple sugars can, either alone or in combination with other stimuli, play a significant role in the onset and progression of specific liver diseases. Notably, it can lead to the accumulation of hepatic fat, resulting in MASLD. If persistent inflammatory damage occurs, it can progress to metabolic dysfunction-associated steatohepatitis (MASH). Furthermore, excessive sugar intake has also been associated with more severe clinical conditions such as cirrhosis and HCC.

6.1. Metabolic Dysfunction-Associated Steatotic Liver Disease and Metabolic Dysfunction-Associated Steatohepatits: Preclinical Data

The dietary consumption of simple sugars, including fructose, glucose, sucrose, and high-fructose corn syrup, has been widely associated with the development of hepatic steatosis and steatohepatitis and metabolic syndrome in experimental models [62,89–93].

Regarding the relationship between dietary habits and the onset of hepatic steatosis, prolonged exposure (8–24 weeks) to a high-fructose diet has been described as a factor contributing to its development [89,94].

An interesting observation to consider is that hepatic steatosis can also arise in the presence of a hypocaloric diet that is high in sugar (40%) and in the absence of weight gain [95,96].

From a qualitative perspective, comparative analyses regarding the type of sugars consumed in excess in the diet have revealed that, compared to isocaloric diets containing sucrose (glucose-fructose), a mixture of 50% free glucose and fructose is slightly more associated with fat accumulation in the liver [91].

The primary mechanism underlying fructose-induced hepatic steatosis is the stimulation of de novo lipogenesis. Key hepatic lipid metabolism enzymes affected by fructose stimulation include sterol regulatory element-binding protein 1c (SREBP-1c) and carbohydrate response element-binding protein (ChREBP). These two elements play a pivotal regulatory role in activating de novo lipogenesis and induce the expression of enzymes necessary for lipid synthesis. This metabolic regulation leads to the simultaneous activation of hepatic lipogenesis and inhibition of fatty acid β-oxidation [77,97–100].

Another mechanism through which fructose promotes lipogenesis is gene expression modulation, favoring a lipogenic profile through the negative regulation of PPARα [101,102]. Specifically, fructose enhances protein synthesis via the MTORC-RPS6KB1 pathway, promoting the accumulation of misfolded proteins and consequently inducing ER stress, which increases the expression of pro-lipogenic proteins [99]. An intriguing aspect of the effects of a fructose-enriched diet emerges from studies demonstrating that only a fructose-supplemented diet leads to a significant increase not only in the expression of enzymes involved in de novo lipogenesis metabolism but also in the upregulation of patatin-like phospholipase domain-containing protein 3 (PNPLA3) [101]. This lipase mobilizes stored triglycerides for VLDL secretion, and its polymorphism is known to be associated with lipid accumulation in hepatocytes [103].

Another critical mechanism involves systemic fatty acid transport imbalances. This metabolic shift is characterized by increased CD36 expression on hepatocyte membranes, facilitating the uptake of long-chain fatty acids from circulation, thereby enhancing hepatic lipid accumulation [77,97].

The progression of fructose-induced liver damage can lead to the transition from hepatic steatosis to steatohepatitis. This transition is mediated by oxidative stress activation and the upregulation of pro-inflammatory cytokines (see Sections 5.1 and 5.2).

6.2. Metabolic Dysfunction-Associated Steatohepatitis: Data on Humans

Based on evidence obtained from animal and in vitro models, several human studies have supported the negative impact of simple sugars in the development of MASLD and metabolic syndrome.

An RCT involving four groups of 24 subjects showed that beverages containing fructose or sucrose (80 g/day for 7 weeks) increased hepatic lipogenesis compared to glucose or sugar abstention. No significant differences were observed in glycemic or insulinemic parameters, suggesting that a longer exposure time may be necessary to develop metabolic complications [104]. Similarly, another recent double-blind study reported a linear increase in hepatic steatosis, evaluated by MRI, and in insulin resistance, assessed through OGTT, in healthy individuals following dietary supplementation with HFCS beverages [105]. Mirror data were derived from the FRUITLESS study, an RCT in which the authors demonstrated that a low-fructose diet (<10 g/day for 6 weeks) significantly reduced hepatic fat content in individuals with hepatic steatosis [106]. In accordance with this finding, an RCT conducted on adolescents showed a significant reduction in hepatic fat (6.23%) and improvements in ALT and fasting insulin concentration after 8 weeks of a low-sugar diet (<3% of total daily caloric intake) [107].

An important aspect to consider is the potential relationship between the harmful effects associated with the intake of simple sugars and the dietary sources from which they are derived. Although current evidence is promising, additional research is necessary to thoroughly evaluate the impact of various food sources, further research is required to comprehensively assess the impact of different food types [108]. On this purpose, in the Maastricht Study, conducted on a population of 3981 adults (50% women), greater intrahepatic fat accumulation, evaluated via MRI, was observed when fructose was consumed in liquid form, through fruit juices and sugar-sweetened beverages, with a particularly pronounced effect in subjects with type 2 diabetes [109]. Data from a large study involving over 15,000 confirmed this association: the risk of MASLD increased by 18% for total added sugars and 20% for liquid sugars, while no significant association was found for solid sugars [110]. It has been hypothesized that fructose from whole fruits does not have the same harmful impact, thanks to the presence of fiber, antioxidants, vitamins, and flavonoids, which may counteract or modulate the steatogenic effect of fructose [111].

Lastly, Global Burden of Disease (GBD) 2017 data showed an association between the high consumption of fructose-containing sugar-sweetened beverages, (>3 g/day from carbonated beverages, sodas, energy drinks, fruit drinks) and MASLD-related mortality, further exacerbated by a diet low in Mediterranean foods (fruit, whole grains, fish, nuts) [112]. Excessive sugar consumption was particularly harmful in MASLD patients, contributing significantly to cardiovascular mortality (aHR = 2.83; 95% CI: 1.01–7.91) [113].

Overall, several human studies have demonstrated a significant association between the consumption of simple sugars and hepatic steatosis, also showing that a reduction in simple sugar intake can lead to a reversal of the condition. However, from a quantitative perspective, the studies conducted so far differ in terms of observation periods lengths, target populations, and the amounts of simple sugars investigated. As a result, it remains challenging to define the precise threshold of simple sugar intake that can be considered safe in the context of hepatology.

6.3. Metabolic Dysfunction and Alcohol-Related Liver Disease: Preclinical Data

Alongside the rising prevalence of MASLD, alcohol-induced liver injury is also on the rise, particularly in Western countries. Moreover, the coexistence of these two conditions is becoming increasingly frequent, to the extent that a distinct nosological entity has recently been proposed: metabolic dysfunction-associated alcohol-related liver disease (MetALD) [1,6]. MetALD is defined as a condition in which excessive alcohol intake—ranging from 140 to 350 g/week for women and 210 to 420 g/week for men—coexists with the metabolic criteria required for an MASLD diagnosis [6].

Beyond the identification of this distinct clinical entity, it has been demonstrated over time that the combined intake of alcohol and simple sugars has a synergistically deleterious effect on liver health. Supporting this concept, a preliminary in vitro study from 2012 showed that concurrent exposure of hepatocytes to glucose and alcohol exacerbates liver cell damage, mainly through oxidative stress, and significantly increases hepatocyte apoptosis [114].

Further evidence from murine models has revealed that the combination of fructose and alcohol, compared to diets rich in either alcohol or fructose alone, markedly intensifies pathogenic mechanisms. Specifically, it enhances lipogenesis, insulin resistance, inflammation, oxidative stress, and hepatic fibrosis [115,116].

An intriguing preclinical mouse model further demonstrated that short-term alcohol consumption combined with a diet high in fats and simple sugars leads to a significant increase in hepatic inflammation, characterized by the recruitment of neutrophils and macrophages, along with elevated circulating neutrophil levels. These findings suggest that the consumption of unbalanced diets rich in sugars and fats, common in Western countries, may predispose individuals to more severe liver injury when alcohol is also consumed [117].

Recently, these findings have been further substantiated using a murine model representative of MetALD. In this study, mice were subjected to a high-fat, high-sugar diet for three months, in combination with daily alcohol administration and weekly binge alcohol exposure. This regimen led to a markedly accelerated progression of liver disease. Specifically, the experimental group exhibited significantly elevated serum levels of aminotransferases and bilirubin, along with (1) enhanced cytokine secretion and increased hepatic infiltration of inflammatory cells (neutrophils and macrophages), (2) increased collagen deposition and exacerbated steatosis, and (3) impaired hepatic protein synthesis and hepatocellular dedifferentiation [118].

6.4. Metabolic Dysfunction and Alcohol-Related Liver Disease: Data on Humans

Current human research on the potential hepatic damage resulting from the combined intake of alcohol and simple sugars remains scarce. This is due in part to ethical limitations in experimentally replicating and analyzing the effects of these two elements when consumed together, and in part to the relatively recent recognition of the nosological category MetALD. Consequently, the existing literature predominantly focuses on the clinical outcomes of alcohol consumption in individuals with underlying metabolic risk [119–123].

A pivotal finding from recent large-scale cohort studies derived from the NHANES database is that individuals with MetALD exhibit a significantly higher risk of all-cause and cancer-related mortality compared to those with MASLD. These findings indicate a synergistically harmful effect of alcohol when combined with a lifestyle typical of metabolic syndrome [119,120].

An additional consideration is that alcohol consumption—especially among younger individuals—often occurs in the form of binge drinking, characterized by the rapid ingestion of large quantities of alcohol. This behavior, known to be more detrimental than chronic intake, frequently involves mixed drinks containing alcohol and sugar-sweetened beverages. Such combinations may amplify hepatotoxic effects by increasing oxidative stress, inflammation, and lipogenesis. Nonetheless, evidence exploring the interaction between binge drinking and high-sugar diets is still limited [124].

Data from Northern European population studies have shown that even low levels of alcohol intake significantly elevate the risk of liver disease and cancer in individuals with pre-existing hepatic steatosis. Moreover, binge drinking has been associated with approximately a 25% increase in the risk of developing hepatic steatosis [123].

A noteworthy study employing MRI to quantitatively assess hepatic fat content found that each additional daily unit of alcohol intake was associated with a significant increase in liver fat. Interestingly, an identical increase in hepatic fat was observed when the alcohol unit was substituted with an isocaloric sugar-sweetened beverage [122]. This supports the hypothesis that simultaneous consumption of alcohol and sugary beverages could lead to an even greater accumulation of hepatic fat.

More recently, a study assessed the risk of liver fibrosis in individuals with MASLD and low-to-moderate alcohol consumption using liver elastography. The results showed a dose-dependent increase in the prevalence of significant fibrosis with higher weekly alcohol intake [121].

In conclusion, although human data on this topic are still limited, it is reasonable to hypothesize that the combination of alcohol and a diet high in simple sugars may potentiate liver damage and have detrimental effects on overall health.

6.5. Cirrhosis: Preclinical Data

As previously highlighted, chronic hepatic injury mechanisms triggered by excessive simple sugar intake may ultimately culminate in liver cirrhosis. Although existing evidence on this topic remains somewhat limited and not always consistent, it is generally accepted that in the setting of chronic high sugar consumption, cirrhosis can develop through the synergistic effects of multiple hepatotoxic processes.

Among these, oxidative stress and chronic inflammation are central and mutually reinforcing contributors to liver fibrogenesis. Elevated production of ROS activates Toll-like receptors, particularly TLR4, driving Kupffer cells toward a pro-inflammatory state. This activation enhances the secretion of pro-inflammatory cytokines (TNF-α, IL-6, IL-1β), amplifying both local and systemic inflammatory responses in a self-sustaining feedback loop [125,126].

Chronic inflammation, which may also be exacerbated by increased intestinal permeability, plays a key role in hepatic stellate cell (HSC) activation. Once activated, HSCs can transdifferentiate into myofibroblasts and drive the production and deposition of type I and III collagen. The TGF-β1/SMAD3 signaling pathway is the primary driver of this process, as it (1) promotes extracellular matrix (ECM) accumulation, (2) suppresses matrix metalloproteinase activity, thereby impairing ECM degradation, and (3) enhances HSC proliferation, further perpetuating fibrosis [58,127].

Additionally, simple sugars contribute to fibrogenesis by downregulating PPAR-γ, a nuclear receptor involved in maintaining HSC quiescence, promoting fatty acid oxidation, and suppressing fibrotic signaling. Impaired PPAR-γ activity removes these regulatory constraints, thereby facilitating the progression of hepatic fibrosis [58,128].

6.6. Cirrhosis: Data on Humans

Current human data exploring the link between simple sugar consumption and liver cirrhosis are sparse and largely preliminary. Despite significant efforts by the scientific community to establish appropriate dietary guidelines for patients with cirrhosis, adherence to such recommendations remains low, and there is a lack of conclusive evidence regarding the role of simple sugars in the progression of the disease [129,130].

A recent observational study, which analyzed food records from patients with liver cirrhosis, highlighted a significant independent association between higher simple sugar intake and increased disease severity over time [131]. Notably, this worsening was more pronounced in individuals with elevated visceral fat, indicating that simple sugars may not only be implicated in the development of cirrhosis, as outlined previously, but may also play a role in accelerating its clinical progression, particularly in the presence of underlying metabolic dysfunction [131].

Further studies are essential to deepen our understanding of this relationship, with the aim of enhancing the evidence base and informing more effective clinical management strategies for patients with liver cirrhosis.

6.7. Hepatocellular Carcinoma: Preclinical Data

Due to their role in cellular metabolism, inflammation, and insulin resistance, simple sugars have frequently been investigated for their potential contribution to carcinogenesis, although results have been variable [132].

Studies specifically investigating HCC and the potential contribution of sugars to carcinogenesis remain limited and sometimes contradictory. For instance, experimental studies on HCC cell lines have shown that cancer cells may downregulate specific enzymes involved in fructose metabolism, thereby shifting their metabolic preference from fructose to glucose [133–135].

It is well known that cancer cells often adopt the so-called "Warburg effect", relying predominantly on anaerobic glycolysis for ATP and lactate production even in the presence of oxygen [136].

Moreover, as described above, fructose metabolism leads to rapid ATP depletion and ROS production, which are typically associated with cellular damage and apoptosis, thereby potentially limiting tumor growth. However, under metabolic stress, tumor cells may compensate by redirecting metabolism towards the pentose phosphate pathway and mitochondrial oxidative phosphorylation. This adaptation increases antioxidant production (e.g., NADPH and glutathione) to mitigate ROS excess. This metabolic reprogramming, referred to as the "reverse Warburg effect", may be induced by fructose [137].

Interestingly, while fructose does not seem to promote tumor growth in vitro, studies have reported that high-dose fructose administration in vivo does not inhibit tumor cell growth either [138]. This suggests that HCC cells may adapt their metabolism to utilize alternative energy sources.

In addition, tumor cells may promote their proliferation by hyperactivating the insulin receptor while simultaneously downregulating fructose-metabolizing enzymes such as aldolase B (ALDOB). Under physiological conditions, ALDOB can negatively regulate insulin signaling and de novo lipogenesis [139–141].

Concurrently, an upregulation of aldose reductase (AKR1B1) has been observed. When overexpressed, this enzyme diverts approximately 30% of intracellular glucose through the polyol pathway, thereby increasing endogenous fructose production [142].

This combination of events establishes a pro-tumorigenic vicious cycle characterized by enhanced glycolysis (Warburg effect), lipid accumulation, and oxidative stress.

Lastly, dietary consumption of simple sugars may indirectly contribute to hepatic carcinogenesis via intestinal sugar metabolism [143,144].

6.8. Hepatocellular Carcinoma: Data on Humans

With regard to the evidence linking simple sugars and HCC, it is important to first acknowledge that MASLD with or without advanced fibrosis represents a significant risk factor for this malignancy, with approximately 15% of HCC cases associated with this condition [145]. Moreover, current projections estimate that MASLD-associated HCC will become increasingly prevalent over the coming years [4]. This epidemiological trend has naturally led to increased efforts within the scientific community to identify prognostic indicators and risk factors for HCC development, particularly in the context of MASLD [12,146].

To date, there are only a limited number of human studies that have directly investigated the association between simple sugar consumption and the risk of HCC. As previously discussed, experimental studies have provided a molecular basis suggesting a potential link between simple sugars and tumorigenesis. Nevertheless, it is essential to consider that cancer development is inherently multifactorial, affected by demographics, genetic background, lifestyle, and comorbidities. Therefore, asserting a direct and unequivocal relationship between simple sugar intake and HCC onset would be overly reductive [147,148].

Some studies have reported that diets high in total sugars, glycemic load, or sugary beverage consumption are associated with an increased incidence of HCC [149–151].

However, a high glycemic index or load does not always correlate with a statistically significant increase in HCC risk [149,152].

Nevertheless, the findings from these various studies are undoubtedly compelling and highlight a highly relevant topic that warrants further in-depth investigation.

A recent meta-analysis aiming to synthesize the current evidence, comprising seven studies (four cohort and three case-control), found that high dietary glycemic load was associated with an increased risk of HCC among individuals with cirrhosis of viral etiology (HBV and/or HCV). In contrast, no significant association was observed in virus-negative groups [153].

In conclusion, further studies in human populations are necessary to establish whether a true causal relationship exists between simple sugar intake and HCC onset—ideally integrating clinical and experimental data within a translational research approach.

7. Dietary Alternatives to Simple Sugars

Sweeteners are substances of natural or artificial origin used in the food industry to replace dietary sugars, with the aim of producing low-calorie foods while maintaining a sweet and pleasant taste. Among natural sweeteners, examples include stevia, agave syrup, honey, and polyols (such as xylitol and erythritol), whereas artificial sweeteners include aspartame, sucralose, and saccharin. However, despite their reduced caloric intake, the consumption of sweeteners may negatively affect glucose and lipid metabolism, potentially leading to effects similar to those of sugars.

While studies on natural sweeteners have reported positive effects on glucose and lipid metabolism, greater concern has arisen from the use of artificial sweeteners [93].

From a general perspective, one of the primary detrimental effects of artificial sweeteners appears to be their role in inducing insulin resistance. These sweeteners can promote both de novo lipogenesis and oxidative stress, in addition to altering the gut microbiota [154]. Over time, this complex interplay of pathogenic mechanisms may lead to insulin system dysfunction, facilitating the development of insulin resistance [155].

A recent study conducted on a population of 3739 U.S. adults highlighted a significant association between the consumption of artificially sweetened beverages (ASBs) and the development of hepatic steatosis, independent of caloric intake and BMI [92]. Although the exact mechanisms underlying this association have not yet been fully elucidated, the most widely accepted hypothesis suggests that intestinal dysbiosis induced by these sweeteners acts as the primary trigger. Additionally, ASBs have been shown to play a role in reducing T-cell-mediated immune responses, which theoretically should exert a protective effect against MASLD. Given the limited number of available studies and the controversial association between artificial sweetener consumption and liver injury, further research is needed to better understand this relationship. In this regard, it is worth noting that a recent meta-analysis reported a negative association between artificial sweeteners and transaminase elevation, thereby opening the debate on the actual threshold beyond which their consumption may become harmful to humans [156].

A detailed analysis of the major artificial sweeteners has documented that sucralose, one of the most commonly used artificial sweeteners in the food industry, is poorly absorbed at the intestinal level and directly interacts with gut microbiota, promoting the proliferation of bacteria belonging to the genera Bacteroides and Clostridioides [157]. Simultaneously, this sweetener reduces bacterial metabolism of bile acids, leading to decreased activation of the farnesoid X receptor (FXR) in hepatocytes and contributing to the establishment of an inflammatory state [157].

Studies on murine models have further demonstrated that sucralose, through activation of the TIR3 receptor, induces endoplasmic reticulum stress in the liver, an increase in ROS, and upregulation of genes involved in lipogenesis (such as SREBP1, ChREBP, and others) [158].

Another well-known artificial sweetener is aspartame. The primary effects associated with this sweetener are inflammatory, including activation of the NLRP3 inflammasome, reduction in antioxidant molecule availability (such as glutathione), and increased extracellular matrix deposition in the liver with the increase in plasma ALT activity as [159,160].

It remains unclear whether long-term artificial sweetener consumption plays a role in carcinogenesis [161].

Overall, while artificial sweeteners are commonly used as an alternative to counteract the negative effects of excessive simple sugar consumption, the current literature suggests that these products may themselves act through similar pathogenic mechanisms, contributing to significant hepatic and systemic manifestations.

8. Conclusions

In conclusion, significant progress has been made in recent years in understanding the preclinical and clinical relationships between simple sugars and liver diseases. The available evidence supports a robust causal link between the intake of simple sugars, their metabolic pathways, and the activation of mechanisms contributing to liver injury and, consequently, to the development of chronic liver diseases of varying severity.

An important aspect to consider is that the effects of simple sugars on the liver result not only from direct mechanisms, but also from complex intracellular processes involving multiple organ systems. The impact of simple sugars should, therefore, be regarded as systemic, with a central role played by adipose tissue, the immune system, the central nervous system, the gut, and the intestinal microbiota.

With regard to liver diseases, certain associations between simple sugar intake and disease onset still require further investigation and characterization through translational studies, in order to elucidate the precise mechanisms underlying clinical observations (e.g., in the context of liver cirrhosis and HCC).

Among the objectives of future research, beyond the identification of novel molecular and microbial therapeutic targets (such as probiotics and prebiotics), it would be valuable to determine, with reasonable accuracy, the "safe" intake threshold for each type of simple sugar in relation to specific clinical conditions, in alignment with the principles of personalized and precision medicine.

Author Contributions: Conceptualization, S.P. and S.G.C.; methodology, E.C., M.M. and F.F.; investigation, E.C., M.M. and F.F.; data curation, S.P.; writing—original draft preparation, S.P. and E.C.; writing—review and editing, S.G.C. and M.M. All authors have read and agreed to the published version of the manuscript.

Funding: This research received no external funding.

Institutional Review Board Statement: Not applicable.

Informed Consent Statement: Not applicable.

Data Availability Statement: No new data were created or analyzed in this study.

Acknowledgments: The authors thank Fabio Parisse for his contribution in the graphic support of the production of the figures in this paper.

Conflicts of Interest: The authors declare no conflicts of interest.

Abbreviations

ACTH	Adrenocorticotropic Hormone
ALDOB	Aldolase B
ASB	Artificial Sweetened Beverage
ATP	Adenosine Triphosphate
ChREBP	Carbohydrate Response Element Binding Protein

CRH	Corticotropin-Releasing Hormone
CRP	C Reactive Protein
DP	Degree of Polymerization
ECM	Extracellular Matrix
ER	Endoplasmatic Reticulum
HBV	Hepatitis B Virus
HCC	Hepatocellular Carcinoma
HCV	Hepatitis C Virus
HFCS	High Fructose Corn Syrup
IL	Interleukin
IR	Insulin Resistance
IRS1	Insulin Receptor Substrate 1
G6P	Glucose-6-Posphate
LPS	Lipopolysaccharide
MASLD	Metabolic Dysfunction-Associated Steatotic Liver Disease
MASH	Metabolic Dysfunction-Associated Steatohepatitis
METAld	Metabolic Dysfunction and Alcohol-Related Liver Disease
MRI	Magnetic Resonance Imaging
NAFLD	Nonalcoholic fatty liver disease
PNPLA3	patatin-like phospholipase domain-containing protein 3
ROS	Reactive oxygen species
SREBP-1c	Sterol regulatory element binding transcription factor 1
T2DM	Type 2 Diabetes Mellitus
WHO	World Health Organization

References

1. Huang, D.Q.; Terrault, N.A.; Tacke, F.; Gluud, L.L.; Arrese, M.; Bugianesi, E.; Loomba, R. Global epidemiology of cirrhosis—Aetiology, trends and predictions. *Nat. Rev. Gastroenterol. Hepatol.* **2023**, *20*, 388–398. [CrossRef] [PubMed]
2. Devarbhavi, H.; Asrani, S.K.; Arab, J.P.; Nartey, Y.A.; Pose, E.; Kamath, P.S. Global burden of liver disease: 2023 update. *J. Hepatol.* **2023**, *79*, 516–537. [CrossRef] [PubMed]
3. Huang, D.Q.; Singal, A.G.; Kono, Y.; Tan, D.J.H.; El-Serag, H.B.; Loomba, R. Changing global epidemiology of liver cancer from 2010 to 2019: NASH is the fastest growing cause of liver cancer. *Cell Metab.* **2022**, *34*, 969–977.e2. [CrossRef] [PubMed]
4. Vitale, A.; Svegliati-Baroni, G.; Ortolani, A.; Cucco, M.; Dalla Riva, G.V.; Giannini, E.G.; Piscaglia, F.; Rapaccini, G.; Di Marco, M.; Caturelli, E.; et al. Epidemiological trends and trajectories of MAFLD-associated hepatocellular carcinoma 2002–2033: The ITA.LI.CA database. *Gut* **2023**, *72*, 141–152. [CrossRef]
5. Younossi, Z.M.; Golabi, P.; Paik, J.M.; Henry, A.; Van Dongen, C.; Henry, L. The global epidemiology of nonalcoholic fatty liver disease (NAFLD) and nonalcoholic steatohepatitis (NASH): A systematic review. *Hepatology* **2023**, *77*, 1335–1347. [CrossRef]
6. Rinella, M.E.; Lazarus, J.V.; Ratziu, V.; Francque, S.M.; Sanyal, A.J.; Kanwal, F.; Romero, D.; Abdelmalek, M.F.; Anstee, Q.M.; Arab, J.P.; et al. A multisociety Delphi consensus statement on new fatty liver disease nomenclature. *J. Hepatol.* **2023**, *79*, 1542–1556. [CrossRef]
7. Federico, A.; Rosato, V.; Masarone, M.; Torre, P.; Dallio, M.; Romeo, M.; Persico, M. The Role of Fructose in Non-Alcoholic Steatohepatitis: Old Relationship and New Insights. *Nutrients* **2021**, *13*, 1314. [CrossRef]
8. Jung, S.; Bae, H.; Song, W.S.; Jang, C. Dietary Fructose and Fructose-Induced Pathologies. *Annu. Rev. Nutr.* **2022**, *42*, 45–66. [CrossRef]
9. Qi, X.; Tester, R.F. Fructose, galactose and glucose—In health and disease. *Clin. Nutr. ESPEN* **2019**, *33*, 18–28. [CrossRef]
10. Yki-Järvinen, H.; Luukkonen, P.K.; Hodson, L.; Moore, J.B. Dietary carbohydrates and fats in nonalcoholic fatty liver disease. *Nat. Rev. Gastroenterol. Hepatol.* **2021**, *18*, 770–786. [CrossRef]
11. Drożdż, K.; Nabrdalik, K.; Hajzler, W.; Kwiendacz, H.; Gumprecht, J.; Lip, G.Y.H. Metabolic-Associated Fatty Liver Disease (MAFLD), Diabetes, and Cardiovascular Disease: Associations with Fructose Metabolism and Gut Microbiota. *Nutrients* **2021**, *14*, 103. [CrossRef] [PubMed]
12. Huang, D.Q.; El-Serag, H.B.; Loomba, R. Global epidemiology of NAFLD-related HCC: Trends, predictions, risk factors and prevention. *Nat. Rev. Gastroenterol. Hepatol.* **2021**, *18*, 223–238. [CrossRef] [PubMed]

13. Park, G.; Jung, S.; Wellen, K.E.; Jang, C. The interaction between the gut microbiota and dietary carbohydrates in nonalcoholic fatty liver disease. *Exp. Mol. Med.* **2021**, *53*, 809–822. [CrossRef] [PubMed]
14. Cummings, R.D. A periodic table of monosaccharides. *Glycobiology* **2024**, *34*, cwad088. [CrossRef]
15. Chandel, N.S. Carbohydrate Metabolism. *Cold Spring Harb. Perspect. Biol.* **2021**, *13*, a040568. [CrossRef]
16. Cummings, J.H.; Stephen, A.M. Carbohydrate terminology and classification. *Eur. J. Clin. Nutr.* **2007**, *61* (Suppl. S1), S5–S18. [CrossRef]
17. Clemente-Suárez, V.J.; Mielgo-Ayuso, J.; Martín-Rodríguez, A.; Ramos-Campo, D.J.; Redondo-Flórez, L.; Tornero-Aguilera, J.F. The Burden of Carbohydrates in Health and Disease. *Nutrients* **2022**, *14*, 3809. [CrossRef]
18. Judge, A.; Dodd, M.S. Metabolism. *Essays Biochem.* **2020**, *64*, 607–647. [CrossRef]
19. Basaranoglu, M.; Basaranoglu, G.; Bugianesi, E. Carbohydrate intake and nonalcoholic fatty liver disease: Fructose as a weapon of mass destruction. *Hepatobiliary Surg. Nutr.* **2015**, *4*, 109–116.
20. Azaïs-Braesco, V.; Sluik, D.; Maillot, M.; Kok, F.; Moreno, L.A. A review of total & added sugar intakes and dietary sources in Europe. *Nutr. J.* **2017**, *16*, 6. [CrossRef]
21. Castro-Muñoz, R.; Correa-Delgado, M.; Córdova-Almeida, R.; Lara-Nava, D.; Chávez-Muñoz, M.; Velásquez-Chávez, V.F.; Hernández-Torres, C.E.; Gontarek-Castro, E.; Ahmad, M.Z. Natural sweeteners: Sources, extraction and current uses in foods and food industries. *Food Chem.* **2022**, *370*, 130991. [CrossRef] [PubMed]
22. Febbraio, M.A.; Karin, M. "Sweet death": Fructose as a metabolic toxin that targets the gut-liver axis. *Cell Metab.* **2021**, *33*, 2316–2328. [CrossRef] [PubMed]
23. Martini, D.; Godos, J.; Bonaccio, M.; Vitaglione, P.; Grosso, G. Ultra-Processed Foods and Nutritional Dietary Profile: A Meta-Analysis of Nationally Representative Samples. *Nutrients* **2021**, *13*, 3390. [CrossRef] [PubMed]
24. Lane, M.M.; Gamage, E.; Du, S.; Ashtree, D.N.; McGuinness, A.J.; Gauci, S.; Baker, P.; Lawrence, M.; Rebholz, C.M.; Srour, B.; et al. Ultra-processed food exposure and adverse health outcomes: Umbrella review of epidemiological meta-analyses. *BMJ* **2024**, *384*, e077310. [CrossRef]
25. Zhang, Y.; Giovannucci, E.L. Ultra-processed foods and health: A comprehensive review. *Crit. Rev. Food Sci. Nutr.* **2023**, *63*, 10836–10848. [CrossRef]
26. Grembecka, M. Natural sweeteners in a human diet. *Rocz. Panstw. Zakl. Hig.* **2015**, *66*, 195–202.
27. Khorshidian, N.; Shadnoush, M.; Zabihzadeh Khajavi, M.; Sohrabvandi, S.; Yousefi, M.; Mortazavian, A.M. Fructose and high fructose corn syrup: Are they a two-edged sword? *Int. J. Food Sci. Nutr.* **2021**, *72*, 592–614. [CrossRef]
28. White, J.S.; Hobbs, L.J.; Fernandez, S. Fructose content and composition of commercial HFCS-sweetened carbonated beverages. *Int. J. Obes.* **2015**, *39*, 176–182. [CrossRef]
29. Bray, G.A.; Nielsen, S.J.; Popkin, B.M. Consumption of high-fructose corn syrup in beverages may play a role in the epidemic of obesity. *Am. J. Clin. Nutr.* **2004**, *79*, 537–543. [CrossRef]
30. Elliott, S.S.; Keim, N.L.; Stern, J.S.; Teff, K.; Havel, P.J. Fructose, weight gain, and the insulin resistance syndrome. *Am. J. Clin. Nutr.* **2002**, *76*, 911–922. [CrossRef]
31. Tappy, L.; Lê, K.A. Metabolic Effects of Fructose and the Worldwide Increase in Obesity. *Physiol. Rev.* **2010**, *90*, 23–46. [CrossRef] [PubMed]
32. Dills, W.L. Sugar Alcohols as Bulk Sweeteners. *Annu. Rev. Nutr.* **1989**, *9*, 161–186. [CrossRef] [PubMed]
33. Msomi, N.Z.; Erukainure, O.L.; Islam, M.d.S. Suitability of Sugar Alcohols as Antidiabetic Supplements: A Review. *J. Food Drug Anal.* **2021**, *29*, 1–14. [CrossRef] [PubMed]
34. Zhang, W.; Chen, J.; Chen, Q.; Wu, H.; Mu, W. Sugar alcohols derived from lactose: Lactitol, galactitol, and sorbitol. *Appl. Microbiol. Biotechnol.* **2020**, *104*, 9487–9495. [CrossRef]
35. Hong, S.J.; Ahn, M.H.; Sangshetti, J.; Arote, R.B. Sugar alcohol-based polymeric gene carriers: Synthesis, properties and gene therapy applications. *Acta Biomater.* **2019**, *97*, 105–115. [CrossRef]
36. Gasmi Benahmed, A.; Gasmi, A.; Arshad, M.; Shanaida, M.; Lysiuk, R.; Peana, M.; Pshyk-Titko, I.; Adamiv, S.; Shanaida, Y.; Bjørklund, G. Health benefits of xylitol. *Appl. Microbiol. Biotechnol.* **2020**, *104*, 7225–7237. [CrossRef]
37. Yamaguchi, H.; Nagai, R. Insights from the fructose-derived product glucoselysine: Revisiting the polyol pathway in diabetic complications. *J. Diabetes. Investig.* **2025**, *16*, 569–577. [CrossRef]
38. Delannoy, P.; Tolan, D.R.; Lanaspa, M.A.; San Millán, I.; Bae, S.Y.; Johnson, R.J. Aldose reductase, fructose and fat production in the liver. *Biochem. J.* **2025**, *482*, 295–307. [CrossRef]
39. Fotros, D.; Hekmatdoost, A.; Pashayee-Khamene, F.; Karimi, S.; Ahmadzadeh, S.; Saberifiroozi, M.; Hatami, B.; Yari, Z. Fermentable oligosaccharides, disaccharides, monosaccharides, and polyols (FODMAPs) and mortality among survivors of liver cirrhosis: A prospective cohort study. *Nutr. J.* **2025**, *24*, 15. [CrossRef]
40. Word Health Organization. *Guideline: Sugars Intake for Adults and Children*; World Health Organization: Geneva, Switzerland, 2015.
41. Young, J.; Scott, S.; Clark, L.; Lodge, J.K. Associations between free sugar intake and markers of health in the UK population: An analysis of the National Diet and Nutrition Survey rolling programme. *Br. J. Nutr.* **2022**, *128*, 225–236. [CrossRef]

42. Wright, E.M.; Martín, M.G.; Turk, E. Intestinal absorption in health and disease—Sugars. *Best Pr. Res. Clin. Gastroenterol.* **2003**, *17*, 943–956. [CrossRef] [PubMed]
43. Ferraris, R.P.; Choe, J.Y.; Patel, C.R. Intestinal Absorption of Fructose. *Annu. Rev. Nutr.* **2018**, *38*, 41–67. [CrossRef] [PubMed]
44. Santoro, A.; McGraw, T.E.; Kahn, B.B. Insulin action in adipocytes, adipose remodeling, and systemic effects. *Cell Metab.* **2021**, *33*, 748–757. [CrossRef] [PubMed]
45. Tappy, L. Metabolism of sugars: A window to the regulation of glucose and lipid homeostasis by splanchnic organs. *Clin. Nutr.* **2021**, *40*, 1691–1698. [CrossRef]
46. Rui, L. Energy Metabolism in the Liver. In *Comprehensive Physiology*; Wiley: Hoboken, NJ, USA, 2014; pp. 177–197.
47. Han, H.S.; Kang, G.; Kim, J.S.; Choi, B.H.; Koo, S.H. Regulation of glucose metabolism from a liver-centric perspective. *Exp. Mol. Med.* **2016**, *48*, e218. [CrossRef]
48. Hannou, S.A.; Haslam, D.E.; McKeown, N.M.; Herman, M.A. Fructose metabolism and metabolic disease. *J. Clin. Investig.* **2018**, *128*, 545–555. [CrossRef]
49. Lodge, M.; Dykes, R.; Kennedy, A. Regulation of Fructose Metabolism in Nonalcoholic Fatty Liver Disease. *Biomolecules* **2024**, *14*, 845. [CrossRef]
50. Westerbeke, F.H.M.; Rios-Morales, M.; Attaye, I.; Nieuwdorp, M. Fructose catabolism and its metabolic effects: Exploring host–microbiota interactions and the impact of ethnicity. *J. Physiol.* **2025**. [CrossRef]
51. Diggle, C.P.; Shires, M.; Leitch, D.; Brooke, D.; Carr, I.M.; Markham, A.F.; Hayward, B.E.; Asipu, A.; Bonthron, D.T. Ketohexokinase: Expression and Localization of the Principal Fructose-metabolizing Enzyme. *J. Histochem. Cytochem.* **2009**, *57*, 763–774. [CrossRef]
52. Geidl-Flueck, B.; Gerber, P. Insights into the Hexose Liver Metabolism—Glucose versus Fructose. *Nutrients* **2017**, *9*, 1026. [CrossRef]
53. Ziolkowska, S.; Binienda, A.; Jabłkowski, M.; Szemraj, J.; Czarny, P. The Interplay between Insulin Resistance, Inflammation, Oxidative Stress, Base Excision Repair and Metabolic Syndrome in Nonalcoholic Fatty Liver Disease. *Int. J. Mol. Sci.* **2021**, *22*, 11128. [CrossRef] [PubMed]
54. Papachristoforou, E.; Lambadiari, V.; Maratou, E.; Makrilakis, K. Association of Glycemic Indices (Hyperglycemia, Glucose Variability, and Hypoglycemia) with Oxidative Stress and Diabetic Complications. *J. Diabetes Res.* **2020**, *2020*, 7489795. [CrossRef] [PubMed]
55. Sarmiento-Ortega, V.E.; Moroni-González, D.; Diaz, A.; García-González, M.Á.; Brambila, E.; Treviño, S. Hepatic Insulin Resistance Model in the Male Wistar Rat Using Exogenous Insulin Glargine Administration. *Metabolites* **2023**, *13*, 572. [CrossRef] [PubMed]
56. Jegatheesan, P.; De Bandt, J. Fructose and NAFLD The Multifaceted Aspects of Fructose Metabolism. *Nutrients* **2017**, *9*, 230. [CrossRef]
57. Feillet-Coudray, C.; Fouret, G.; Vigor, C.; Bonafos, B.; Jover, B.; Blachnio-Zabielska, A.; Rieusset, J.; Casas, F.; Gaillet, S.; Landrier, J.F.; et al. Long-Term Measures of Dyslipidemia, Inflammation, and Oxidative Stress in Rats Fed a High-Fat/High-Fructose Diet. *Lipids* **2019**, *54*, 81–97. [CrossRef]
58. Cho, Y.; Kim, D.; Seo, W.; Gao, B.; Yoo, S.; Song, B. Fructose Promotes Leaky Gut, Endotoxemia, and Liver Fibrosis Through Ethanol-Inducible Cytochrome P450-2E1–Mediated Oxidative and Nitrative Stress. *Hepatology* **2021**, *73*, 2180–2195. [CrossRef]
59. Kanazawa, J.; Kakisaka, K.; Suzuki, Y.; Yonezawa, T.; Abe, H.; Wang, T.; Takikawa, Y. Excess fructose enhances oleatic cytotoxicity via reactive oxygen species production and causes necroptosis in hepatocytes. *J. Nutr. Biochem.* **2022**, *107*, 109052. [CrossRef]
60. Campos-Maldonado, F.; González-Dávalos, M.L.; Piña, E.; Anaya-Loyola, M.A.; Shimada, A.; Varela-Echavarria, A.; Mora, O. Fructose promotes more than glucose the adipocytic differentiation of pig mesenchymal stem cells. *J. Food Biochem.* **2022**, *46*, e14429. [CrossRef]
61. Wang, L.; Ji, T.; Yuan, Y.; Fu, H.; Wang, Y.; Tian, S.; Hu, J.; Wang, L.; Wang, Z. High-fructose corn syrup promotes proinflammatory Macrophage activation via ROS-mediated NF-κB signaling and exacerbates colitis in mice. *Int. Immunopharmacol.* **2022**, *109*, 108814. [CrossRef]
62. Lanaspa, M.A.; Sanchez-Lozada, L.G.; Choi, Y.J.; Cicerchi, C.; Kanbay, M.; Roncal-Jimenez, C.A.; Ishimoto, T.; Li, N.; Marek, G.; Duranay, M.; et al. Uric Acid Induces Hepatic Steatosis by Generation of Mitochondrial Oxidative Stress. *J. Biol. Chem.* **2012**, *287*, 40732–40744. [CrossRef]
63. Ma, X.; Nan, F.; Liang, H.; Shu, P.; Fan, X.; Song, X.; Hou, Y.; Zhang, D. Excessive intake of sugar: An accomplice of inflammation. *Front. Immunol.* **2022**, *13*, 988481. [CrossRef] [PubMed]
64. Jameel, F.; Phang, M.; Wood, L.G.; Garg, M.L. Acute effects of feeding fructose, glucose and sucrose on blood lipid levels and systemic inflammation. *Lipids Health Dis.* **2014**, *13*, 195. [CrossRef] [PubMed]
65. Kuzma, J.N.; Cromer, G.; Hagman, D.K.; Breymeyer, K.L.; Roth, C.L.; Foster-Schubert, K.E.; Holte, S.E.; Weigle, D.S.; Kratz, M. No differential effect of beverages sweetened with fructose, high-fructose corn syrup, or glucose on systemic or adipose tissue inflammation in normal-weight to obese adults: A randomized controlled trial. *Am. J. Clin. Nutr.* **2016**, *104*, 306–314. [CrossRef] [PubMed]

66. Tamer, F.; Ulug, E.; Akyol, A.; Nergiz-Unal, R. The potential efficacy of dietary fatty acids and fructose induced inflammation and oxidative stress on the insulin signaling and fat accumulation in mice. *Food Chem. Toxicol.* **2020**, *135*, 110914. [CrossRef]
67. Lin, W.T.; Kao, Y.H.; Li, M.S.; Luo, T.; Lin, H.Y.; Lee, C.H.; Seal, D.W.; Hu, C.Y.; Chen, L.S.; Tseng, T.S. Sugar-Sweetened Beverages Intake, Abdominal Obesity, and Inflammation among US Adults without and with Prediabetes—An NHANES Study. *Int. J. Environ. Res. Public. Health* **2022**, *20*, 681. [CrossRef]
68. Hou, J.; Cui, Y.; Gao, J.; Rong, M. Dietary simple sugar intake, metabolic indicators, markers of inflammation, and injury among semi-professional football players. *Food Nutr. Res.* **2025**, *69*, 11036. [CrossRef]
69. Todoric, J.; Di Caro, G.; Reibe, S.; Henstridge, D.C.; Green, C.R.; Vrbanac, A.; Ceteci, F.; Conche, C.; McNulty, R.; Shalapour, S.; et al. Fructose stimulated de novo lipogenesis is promoted by inflammation. *Nat. Metab.* **2020**, *2*, 1034–1045. [CrossRef]
70. Chavakis, T.; Alexaki, V.I.; Ferrante, A.W. Macrophage function in adipose tissue homeostasis and metabolic inflammation. *Nat. Immunol.* **2023**, *24*, 757–766. [CrossRef]
71. Rodrigues, R.M.; He, Y.; Hwang, S.; Bertola, A.; Mackowiak, B.; Ahmed, Y.A.; Seo, W.; Ma, J.; Wang, X.; Park, S.H.; et al. E-Selectin-Dependent Inflammation and Lipolysis in Adipose Tissue Exacerbate Steatosis-to-NASH Progression via S100A8/9. *Cell Mol. Gastroenterol. Hepatol.* **2022**, *13*, 151–171. [CrossRef]
72. Jaiswal, N.; Agrawal, S.; Agrawal, A. High fructose-induced metabolic changes enhance inflammation in human dendritic cells. *Clin. Exp. Immunol.* **2019**, *197*, 237–249. [CrossRef]
73. Tan, J.; Ni, D.; Wali, J.A.; Cox, D.A.; Pinget, G.V.; Taitz, J.; Daïen, C.I.; Senior, A.; Read, M.N.; Simpson, S.J.; et al. Dietary carbohydrate, particularly glucose, drives B cell lymphopoiesis and function. *iScience* **2021**, *24*, 102835. [CrossRef] [PubMed]
74. Zhang, Y.; Luo, C.; Huang, P.; Chen, L.; Ma, Y.; Ding, H. Effects of chronic exposure to a high fat diet, nutritive or non-nutritive sweeteners on hypothalamic-pituitary-adrenal (HPA) and -gonadal (HPG) axes of male Sprague-Dawley rats. *Eur. J. Nutr.* **2024**, *63*, 2209–2220. [CrossRef] [PubMed]
75. Cawley, N.X. Sugar Making Sugar: Gluconeogenesis Triggered by Fructose via a Hypothalamic-Adrenal-Corticosterone Circuit. *Endocrinology* **2012**, *153*, 3561–3563. [CrossRef] [PubMed]
76. Johnson, R.J.; Rivard, C.; Lanaspa, M.A.; Otabachian-Smith, S.; Ishimoto, T.; Cicerchi, C.; Cheeke, P.R.; Macintosh, B.; Hess, T. Fructokinase, Fructans, Intestinal Permeability, and Metabolic Syndrome: An Equine Connection? *J. Equine. Vet. Sci.* **2013**, *33*, 120–126. [CrossRef]
77. Zhou, X.; Zhang, X.; Niu, D.; Zhang, S.; Wang, H.; Zhang, X.; Nan, F.; Jiang, S.; Wang, B. Gut microbiota induces hepatic steatosis by modulating the T cells balance in high fructose diet mice. *Sci. Rep.* **2023**, *13*, 6701. [CrossRef]
78. Romualdo, G.R.; Valente, L.C.; Sprocatti, A.C.; Bacil, G.P.; de Souza, I.P.; Rodrigues, J.; Rodrigues, M.A.M.; Vinken, M.; Cogliati, B.; Barbisan, L.F. Western diet–induced mouse model of non-alcoholic fatty liver disease associated with metabolic outcomes: Features of gut microbiome-liver-adipose tissue axis. *Nutrition* **2022**, *103*, 111836. [CrossRef]
79. Kizilaslan, N.; Zekiye Erdem, N.; Katar, M.; Gevrek, F. The Effects of Probiotics and Omega-3 Fatty Acids in Liver Steatosis Induced in Rats by High-Fructose Corn Syrup. *Int. J. Clin. Pract.* **2022**, *2022*, 1–14. [CrossRef]
80. Lee, S.H.; Park, S.Y.; Choi, C.S. Insulin Resistance: From Mechanisms to Therapeutic Strategies. *Diabetes Metab. J.* **2022**, *46*, 15–37. [CrossRef]
81. Henriksen, E.J.; Diamond-Stanic, M.K.; Marchionne, E.M. Oxidative stress and the etiology of insulin resistance and type 2 diabetes. *Free Radic. Biol. Med.* **2011**, *51*, 993–999. [CrossRef]
82. Quarta, C.; Stemmer, K.; Novikoff, A.; Yang, B.; Klingelhuber, F.; Harger, A.; Bakhti, M.; Bastidas-Ponce, A.; Baugé, E.; Campbell, J.E.; et al. GLP-1-mediated delivery of tesaglitazar improves obesity and glucose metabolism in male mice. *Nat. Metab.* **2022**, *4*, 1071–1083. [CrossRef]
83. Brown, M.; Dainty, S.; Strudwick, N.; Mihai, A.D.; Watson, J.N.; Dendooven, R.; Paton, A.W.; Paton, J.C.; Schröder, M. Endoplasmic reticulum stress causes insulin resistance by inhibiting delivery of newly synthesized insulin receptors to the cell surface. *Mol. Biol. Cell* **2020**, *31*, 2597–2629. [CrossRef] [PubMed]
84. Du, K.; Herzig, S.; Kulkarni, R.N.; Montminy, M. TRB3: A tribbles homolog that inhibits Akt/PKB activation by insulin in liver. *Science* **2003**, *300*, 1574–1577. [CrossRef] [PubMed]
85. Sears, B.; Perry, M. The role of fatty acids in insulin resistance. *Lipids Health Dis.* **2015**, *14*, 121. [CrossRef]
86. Malhi, H.; Gores, G. Molecular Mechanisms of Lipotoxicity in Nonalcoholic Fatty Liver Disease. *Semin. Liver Dis.* **2008**, *28*, 360–369. [CrossRef] [PubMed]
87. Rada, P.; González-Rodríguez, Á.; García-Monzón, C.; Valverde, Á.M. Understanding lipotoxicity in NAFLD pathogenesis: Is CD36 a key driver? *Cell Death Dis.* **2020**, *11*, 802. [CrossRef]
88. Herman, R.; Kravos, N.A.; Jensterle, M.; Janež, A.; Dolžan, V. Metformin and Insulin Resistance: A Review of the Underlying Mechanisms behind Changes in GLUT4-Mediated Glucose Transport. *Int. J. Mol. Sci.* **2022**, *23*, 1264. [CrossRef]
89. Jensen, T.; Abdelmalek, M.F.; Sullivan, S.; Nadeau, K.J.; Green, M.; Roncal, C.; Nakagawa, T.; Kuwabara, M.; Sato, Y.; Kang, D.H.; et al. Fructose and sugar: A major mediator of non-alcoholic fatty liver disease. *J. Hepatol.* **2018**, *68*, 1063–1075. [CrossRef]

90. Ackerman, Z.; Oron-Herman, M.; Grozovski, M.; Rosenthal, T.; Pappo, O.; Link, G.; Sela, B.A. Fructose-Induced Fatty Liver Disease. *Hypertension* **2005**, *45*, 1012–1018. [CrossRef]
91. Sánchez-Lozada, L.G.; Mu, W.; Roncal, C.; Sautin, Y.Y.; Abdelmalek, M.; Reungjui, S.; Le, M.; Nakagawa, T.; Lan, H.Y.; Yu, X.; et al. Comparison of free fructose and glucose to sucrose in the ability to cause fatty liver. *Eur. J. Nutr.* **2010**, *49*, 1–9. [CrossRef]
92. Tseng, T.S.; Lin, W.T.; Ting, P.S.; Huang, C.K.; Chen, P.H.; Gonzalez, G.V.; Lin, H.Y. Sugar-Sweetened Beverages and Artificially Sweetened Beverages Consumption and the Risk of Nonalcoholic Fatty Liver (NAFLD) and Nonalcoholic Steatohepatitis (NASH). *Nutrients* **2023**, *15*, 3997. [CrossRef]
93. Valle, M.; St-Pierre, P.; Pilon, G.; Marette, A. Differential Effects of Chronic Ingestion of Refined Sugars versus Natural Sweeteners on Insulin Resistance and Hepatic Steatosis in a Rat Model of Diet-Induced Obesity. *Nutrients* **2020**, *12*, 2292. [CrossRef] [PubMed]
94. Nakagawa, T.; Hu, H.; Zharikov, S.; Tuttle, K.R.; Short, R.A.; Glushakova, O.; Ouyang, X.; Feig, D.I.; Block, E.R.; Herrera-Acosta, J.; et al. A causal role for uric acid in fructose-induced metabolic syndrome. *Am. J. Physiol.-Ren. Physiol.* **2006**, *290*, F625–F631. [CrossRef] [PubMed]
95. Schultz, A.; Neil, D.; Aguila, M.; Mandarim-de-Lacerda, C. Hepatic Adverse Effects of Fructose Consumption Independent of Overweight/Obesity. *Int. J. Mol. Sci.* **2013**, *14*, 21873–21886. [CrossRef] [PubMed]
96. Roncal-Jimenez, C.A.; Lanaspa, M.A.; Rivard, C.J.; Nakagawa, T.; Sanchez-Lozada, L.G.; Jalal, D.; Andres-Hernando, A.; Tanabe, K.; Madero, M.; Li, N.; et al. Sucrose induces fatty liver and pancreatic inflammation in male breeder rats independent of excess energy intake. *Metabolism* **2011**, *60*, 1259–1270. [CrossRef]
97. Deng, S.; Ge, Y.; Zhai, Z.; Liu, H.; Zhang, X.; Chen, Y.; Yang, Y.; Wu, Z. Fructose induces hepatic steatosis in adolescent mice linked to the disorders of lipid metabolism, bile acid metabolism, and autophagy. *J. Nutr. Biochem.* **2024**, *129*, 109635. [CrossRef]
98. Singh, S.; Sharma, A.; Ahmad, S.; Guru, B.; Gulzar, F.; Kumar, P.; Ahmad, I.; Tamrakar, A.K. Convergence of Fructose-Induced NLRP3 Activation with Oxidative Stress and ER Stress Leading to Hepatic Steatosis. *Inflammation* **2023**, *46*, 217–233. [CrossRef]
99. Raza, S.; Shahi, A.; Medhe, P.; Tewari, A.; Gupta, P.; Rajak, S.; Chakravarti, B.; Sinha, R.A. Fructose-induced perturbation in cellular proteostasis via RPS6KB1 promotes hepatic steatosis. *Biochim. Et Biophys. Acta BBA Mol. Cell Res.* **2024**, *1871*, 119597. [CrossRef]
100. Bhat, S.F.; Pinney, S.E.; Kennedy, K.M.; McCourt, C.R.; Mundy, M.A.; Surette, M.G.; Sloboda, D.M.; Simmons, R.A. Exposure to high fructose corn syrup during adolescence in the mouse alters hepatic metabolism and the microbiome in a sex-specific manner. *J. Physiol.* **2021**, *599*, 1487–1511. [CrossRef]
101. Velázquez, A.M.; Bentanachs, R.; Sala-Vila, A.; Lázaro, I.; Rodríguez-Morató, J.; Sánchez, R.M.; Alegret, M.; Roglans, N.; Laguna, J.C. ChREBP-driven DNL and PNPLA3 Expression Induced by Liquid Fructose are Essential in the Production of Fatty Liver and Hypertriglyceridemia in a High-Fat Diet-Fed Rat Model. *Mol. Nutr. Food Res.* **2022**, *66*, 2101115. [CrossRef]
102. Papadopoulos, G.; Legaki, A.I.; Georgila, K.; Vorkas, P.; Giannousi, E.; Stamatakis, G.; Moustakas, I.I.; Petrocheilou, M.; Pyrina, I.; Gercken, B.; et al. Integrated omics analysis for characterization of the contribution of high fructose corn syrup to non-alcoholic fatty liver disease in obesity. *Metabolism* **2023**, *144*, 155552. [CrossRef]
103. Ferri, F.; Carotti, S.; Carpino, G.; Mischitelli, M.; Cantafora, A.; Molinaro, A.; Argenziano, M.E.; Parisse, S.; Corsi, A.; Riminucci, M.; et al. The Propensity of the Human Liver to Form Large Lipid Droplets Is Associated with PNPLA3 Polymorphism, Reduced INSIG1 and NPC1L1 Expression and Increased Fibrogenetic Capacity. *Int. J. Mol. Sci.* **2021**, *22*, 6100. [CrossRef] [PubMed]
104. Geidl-Flueck, B.; Hochuli, M.; Németh, Á.; Eberl, A.; Derron, N.; Köfeler, H.C.; Tappy, L.; Berneis, K.; Spinas, G.A.; Gerber, P.A. Fructose- and sucrose- but not glucose-sweetened beverages promote hepatic de novo lipogenesis: A randomized controlled trial. *J. Hepatol.* **2021**, *75*, 46–54. [CrossRef] [PubMed]
105. Sigala, D.M.; Hieronimus, B.; Medici, V.; Lee, V.; Nunez, M.V.; Bremer, A.A.; Cox, C.L.; Price, C.A.; Benyam, Y.; Abdelhafez, Y.; et al. The Dose-Response Effects of Consuming High Fructose Corn Syrup-Sweetened Beverages on Hepatic Lipid Content and Insulin Sensitivity in Young Adults. *Nutrients* **2022**, *14*, 1648. [CrossRef] [PubMed]
106. Simons, N.; Veeraiah, P.; Simons, P.I.H.G.; Schaper, N.C.; Kooi, M.E.; Schrauwen-Hinderling, V.B.; Feskens, E.J.M.; van der Ploeg, E.M.C.L.; Van den Eynde, M.D.G.; Schalkwijk, C.G.; et al. Effects of fructose restriction on liver steatosis (FRUITLESS); A double-blind randomized controlled trial. *Am. J. Clin. Nutr.* **2021**, *113*, 391–400. [CrossRef]
107. Schwimmer, J.B.; Ugalde-Nicalo, P.; Welsh, J.A.; Angeles, J.E.; Cordero, M.; Harlow, K.E.; Alazraki, A.; Durelle, J.; Knight-Scott, J.; Newton, K.P.; et al. Revised Dietary Intake Estimates for Protein and Fat. *JAMA* **2019**, *322*, 469.
108. Lee, D.; Chiavaroli, L.; Ayoub-Charette, S.; Khan, T.A.; Zurbau, A.; Au-Yeung, F.; Cheung, A.; Liu, Q.; Qi, X.; Ahmed, A.; et al. Important Food Sources of Fructose-Containing Sugars and Non-Alcoholic Fatty Liver Disease: A Systematic Review and Meta-Analysis of Controlled Trials. *Nutrients* **2022**, *14*, 2846. [CrossRef]
109. Buziau, A.M.; Eussen, S.J.P.M.; Kooi, M.E.; van der Kallen, C.J.H.; van Dongen, M.C.J.M.; Schaper, N.C.; Henry, R.M.A.; Schram, M.T.; Dagnelie, P.C.; van Greevenbroek, M.M.J.; et al. Fructose Intake from Fruit Juice and Sugar-Sweetened Beverages Is Associated With Higher Intrahepatic Lipid Content: The Maastricht Study. *Diabetes Care* **2022**, *45*, 1116–1123. [CrossRef]

110. Zhang, S.; Li, H.; Meng, G.; Zhang, Q.; Liu, L.; Wu, H.; Gu, Y.; Zhang, T.; Wang, X.; Zhang, J.; et al. Added sugar intake and its forms and sources in relation to risk of non-alcoholic fatty liver disease: Results from the Tianjin Chronic Low-grade Systemic Inflammation and Health cohort study. *Br. J. Nutr.* **2023**, *129*, 2094–2101. [CrossRef]
111. Pixner, T.; Stummer, N.; Schneider, A.M.; Lukas, A.; Gramlinger, K.; Julian, V.; Thivel, D.; Mörwald, K.; Maruszczak, K.; Mangge, H.; et al. The Role of Macronutrients in the Pathogenesis, Prevention and Treatment of Non-Alcoholic Fatty Liver Disease (NAFLD) in the Paediatric Population—A Review. *Life* **2022**, *12*, 839. [CrossRef]
112. Paik, J.M.; Mir, S.; Alqahtani, S.A.; Younossi, Y.; Ong, J.P.; Younossi, Z.M. Dietary Risks for Liver Mortality in NAFLD: Global Burden of Disease Data. *Hepatol. Commun.* **2022**, *6*, 90–100. [CrossRef]
113. Zhou, J.; Liu, C.; Liu, L.; Li, L. Sugar-sweetened beverage intake and long-term mortality in individuals with metabolic dysfunction-associated steatotic liver disease: A longitudinal analysis of the National Health and Nutrition Examination Survey database. *Eur. J. Gastroenterol. Hepatol.* **2024**, *36*, 608–614. [CrossRef] [PubMed]
114. Chandrasekaran, K.; Swaminathan, K.; Mathan Kumar, S.; Clemens, D.L.; Dey, A. In Vitro Evidence for Chronic Alcohol and High Glucose Mediated Increased Oxidative Stress and Hepatotoxicity. *Alcohol. Clin. Exp. Res.* **2012**, *36*, 1004–1012. [CrossRef] [PubMed]
115. Thomes, P.G.; Benbow, J.H.; Brandon-Warner, E.; Thompson, K.J.; Jacobs, C.; Donohue, T.M., Jr.; Schrum, L.W. Dietary fructose augments ethanol-induced liver pathology. *J. Nutr. Biochem.* **2017**, *43*, 141–150. [CrossRef] [PubMed]
116. Alwahsh, S.M.; Xu, M.; Schultze, F.C.; Wilting, J.; Mihm, S.; Raddatz, D.; Ramadori, G. Combination of Alcohol and Fructose Exacerbates Metabolic Imbalance in Terms of Hepatic Damage, Dyslipidemia, and Insulin Resistance in Rats. *PLoS ONE* **2014**, *9*, e104220. [CrossRef]
117. Gonçalves, J.L.; Lacerda-Queiroz, N.; Sabino, J.F.L.; Marques, P.E.; Galvão, I.; Gamba, C.O.; Cassali, G.D.; de Carvalho, L.M.; da Silva, D.A.; Versiani, A.; et al. Evaluating the effects of refined carbohydrate and fat diets with acute ethanol consumption using a mouse model of alcoholic liver injury. *J. Nutr. Biochem.* **2017**, *39*, 93–100. [CrossRef]
118. Babuta, M.; Morel, C.; de Carvalho Ribeiro, M.; Datta, A.A.; Calenda, C.; Copeland, C.; Nasser, I.; Szabo, G. A novel experimental model of MetALD in male mice recapitulates key features of severe alcohol-associated hepatitis. *Hepatol. Commun.* **2024**, *8*, e0450. [CrossRef]
119. Li, M.; Chen, W.; Deng, Y.; Xie, W. Impacts of cardiometabolic risk factors and alcohol consumption on all-cause mortality among MASLD and its subgroups. *Nutr. Metab. Cardiovasc. Dis.* **2024**, *34*, 2085–2094. [CrossRef]
120. Aboona, M.B.; Danpanichkul, P.; Chen, V.L.; Rangan, P.; Kim, D.; Alkhouri, N.; Fallon, M.B.; Noureddin, M.; Arab, J.P.; Wijarnpreecha, K. Mortality outcomes in individuals with MASLD *versus* MASLD and increased alcohol intake. *J. Gastroenterol. Hepatol.* **2024**, *39*, 2456–2463. [CrossRef]
121. Marti-Aguado, D.; Calleja, J.L.; Vilar-Gomez, E.; Iruzubieta, P.; Rodríguez-Duque, J.C.; Del Barrio, M.; Puchades, L.; Rivera-Esteban, J.; Perelló, C.; Puente, A.; et al. Low-to-moderate alcohol consumption is associated with increased fibrosis in individuals with metabolic dysfunction-associated steatotic liver disease. *J. Hepatol.* **2024**, *81*, 930–940. [CrossRef]
122. van Eekelen, E.; Beulens, J.W.J.; Geelen, A.; Schrauwen-Hinderling, V.B.; Lamb, H.; de Roos, A.; Rosendaal, F.; de Mutsert, R. Consumption of Alcoholic and Sugar-Sweetened Beverages is Associated with Increased Liver Fat Content in Middle-Aged Men and Women. *J. Nutr.* **2019**, *149*, 649–658. [CrossRef]
123. Åberg, F.; Puukka, P.; Salomaa, V.; Männistö, S.; Lundqvist, A.; Valsta, L.; Perola, M.; Färkkilä, M.; Jula, A. Risks of Light and Moderate Alcohol Use in Fatty Liver Disease: Follow-Up of Population Cohorts. *Hepatology* **2020**, *71*, 835–848. [CrossRef] [PubMed]
124. Skinner, R.C.; Hagaman, J.A. The interplay of Western diet and binge drinking on the onset, progression, and outlook of liver disease. *Nutr. Rev.* **2022**, *80*, 503–512. [CrossRef] [PubMed]
125. Lodge, M.; Scheidemantle, G.; Adams, V.R.; Cottam, M.A.; Richard, D.; Breuer, D.; Thompson, P.; Shrestha, K.; Liu, X.; Kennedy, A. Fructose regulates the pentose phosphate pathway and induces an inflammatory and resolution phenotype in Kupffer cells. *Sci. Rep.* **2024**, *14*, 4020. [CrossRef] [PubMed]
126. Kumar, S.; Duan, Q.; Wu, R.; Harris, E.N.; Su, Q. Pathophysiological communication between hepatocytes and non-parenchymal cells in liver injury from NAFLD to liver fibrosis. *Adv. Drug Deliv. Rev.* **2021**, *176*, 113869. [CrossRef]
127. Oliveira-Cordeiro, B.; Fernandes-DA-Silva, A.; Silva-Veiga, F.M.; Miranda, C.S.; Martins, F.F.; Souza-Mello, V. Long-term hepatic damage in high-fructose-fed C57BL/6 mice: Hepatic fibrogenesis, endoplasmic reticulum stress markers, and fibrosis. *An. Acad. Bras. Cienc.* **2023**, *95* (Suppl. S2), e20220784.
128. Xin, X.; Cai, B.Y.; Chen, C.; Tian, H.J.; Wang, X.; Hu, Y.Y.; Feng, Q. High-trans fatty acid and high-sugar diets can cause mice with non-alcoholic steatohepatitis with liver fibrosis and potential pathogenesis. *Nutr. Metab.* **2020**, *17*, 40. [CrossRef]
129. Georgiou, A.; Yannakoulia, M.; Papatheodoridis, G.V.; Deutsch, M.; Alexopoulou, A.; Vlachogiannakos, J.; Ioannidou, P.; Papageorgiou, M.V.; Voulgaris, T.; Papadopoulos, N.; et al. Assessment of dietary habits and the adequacy of dietary intake of patients with cirrhosis-the KIRRHOS study. *Clin. Nutr.* **2021**, *40*, 3992–3998. [CrossRef]

130. Bischoff, S.C.; Bernal, W.; Dasarathy, S.; Merli, M.; Plank, L.D.; Schütz, T.; Plauth, M. ESPEN practical guideline: Clinical nutrition in liver disease. *Clin. Nutr.* **2020**, *39*, 3533–3562. [CrossRef]
131. Parisse, S.; Carnevale, S.; Di Bartolomeo, F.; Poli, E.; Miceli, F.; Ferri, F.; Mischitelli, M.; Rocco, B.; Lai, Q.; Lucatelli, P.; et al. A Low Daily Intake of Simple Sugars in the Diet Is Associated with Improved Liver Function in Cirrhotic Liver Transplant Candidates. *Nutrients* **2023**, *15*, 1575. [CrossRef]
132. Hasan, N.; Yazdanpanah, O.; Khaleghi, B.; Benjamin, D.J.; Kalebasty, A.R. The role of dietary sugars in cancer risk: A comprehensive review of current evidence. *Cancer Treat. Res. Commun.* **2024**, *43*, 100876. [CrossRef]
133. Li, X.; Qian, X.; Peng, L.X.; Jiang, Y.; Hawke, D.H.; Zheng, Y.; Xia, Y.; Lee, J.H.; Cote, G.; Wang, H.; et al. A splicing switch from ketohexokinase-C to ketohexokinase-A drives hepatocellular carcinoma formation. *Nat. Cell Biol.* **2016**, *18*, 561–571. [CrossRef] [PubMed]
134. Liu, G.M.; Li, Q.; Zhang, P.F.; Shen, S.L.; Xie, W.X.; Chen, B.; Wu, J.; Hu, W.J.; Huang, X.Y.; Peng, B.G. Restoration of FBP1 suppressed Snail-induced epithelial to mesenchymal transition in hepatocellular carcinoma. *Cell Death Dis.* **2018**, *9*, 1132. [CrossRef] [PubMed]
135. Tee, S.S.; Kim, N.; Cullen, Q.; Eskandari, R.; Mamakhanyan, A.; Srouji, R.M.; Chirayil, R.; Jeong, S.; Shakiba, M.; Kastenhuber, E.R.; et al. Ketohexokinase-mediated fructose metabolism is lost in hepatocellular carcinoma and can be leveraged for metabolic imaging. *Sci. Adv.* **2022**, *8*, eabm7985. [CrossRef] [PubMed]
136. Parks, S.K.; Mueller-Klieser, W.; Pouysségur, J. Lactate and Acidity in the Cancer Microenvironment. *Annu. Rev. Cancer Biol.* **2020**, *4*, 141–158. [CrossRef]
137. Dewdney, B.; Roberts, A.; Qiao, L.; George, J.; Hebbard, L. A Sweet Connection? Fructose's Role in Hepatocellular Carcinoma. *Biomolecules* **2020**, *10*, 496. [CrossRef]
138. Dewdney, B.; Alanazy, M.; Gillman, R.; Walker, S.; Wankell, M.; Qiao, L.; George, J.; Roberts, A.; Hebbard, L. The effects of fructose and metabolic inhibition on hepatocellular carcinoma. *Sci. Rep.* **2020**, *10*, 16769. [CrossRef]
139. Li, M.; He, X.; Guo, W.; Yu, H.; Zhang, S.; Wang, N.; Liu, G.; Sa, R.; Shen, X.; Jiang, Y.; et al. Aldolase B suppresses hepatocellular carcinogenesis by inhibiting G6PD and pentose phosphate pathways. *Nat. Cancer* **2020**, *1*, 735–747. [CrossRef]
140. Liu, G.; Wang, N.; Zhang, C.; Li, M.; He, X.; Yin, C.; Tu, Q.; Shen, X.; Zhang, L.; Lv, J.; et al. Fructose-1, 6-Bisphosphate Aldolase B Depletion Promotes Hepatocellular Carcinogenesis Through Activating Insulin Receptor Signaling and Lipogenesis. *Hepatology* **2021**, *74*, 3037–3055. [CrossRef]
141. Tao, Q.F.; Yuan, S.X.; Yang, F.; Yang, S.; Yang, Y.; Yuan, J.H.; Wang, Z.G.; Xu, Q.G.; Lin, K.Y.; Cai, J.; et al. Aldolase B inhibits metastasis through Ten–Eleven Translocation 1 and serves as a prognostic biomarker in hepatocellular carcinoma. *Mol. Cancer* **2015**, *14*, 170. [CrossRef]
142. Syamprasad, N.P.; Jain, S.; Rajdev, B.; Panda, S.R.; Kumar, G.J.; Shaik, K.M.; Shantanu, P.A.; Challa, V.S.; Jorvekar, S.B.; Borkar, R.M.; et al. AKR1B1 drives hyperglycemia-induced metabolic reprogramming in MASLD-associated hepatocellular carcinoma. *JHEP Rep.* **2024**, *6*, 100974. [CrossRef]
143. Zhou, P.; Chang, W.Y.; Gong, D.A.; Xia, J.; Chen, W.; Huang, L.Y.; Liu, R.; Liu, Y.; Chen, C.; Wang, K.; et al. High dietary fructose promotes hepatocellular carcinoma progression by enhancing O-GlcNAcylation via microbiota-derived acetate. *Cell Metab.* **2023**, *35*, 1961–1975.e6. [CrossRef] [PubMed]
144. Esquea, E.M.; Young, R.G.; Reginato, M.J. Fructose promotes liver cancer via microbial acetate-induced O-GlcNAcylation. *Trends Endocrinol. Metab.* **2024**, *35*, 88–90. [CrossRef] [PubMed]
145. Tan, D.J.H.; Ng, C.H.; Lin, S.Y.; Pan, X.H.; Tay, P.; Lim, W.H.; Teng, M.; Syn, N.; Lim, G.; Yong, J.N.; et al. Clinical characteristics, surveillance, treatment allocation, and outcomes of non-alcoholic fatty liver disease-related hepatocellular carcinoma: A systematic review and meta-analysis. *Lancet Oncol.* **2022**, *23*, 521–530. [CrossRef]
146. Toh, M.R.; Wong, E.Y.T.; Wong, S.H.; Ng, A.W.T.; Loo, L.H.; Chow, P.K.; Ngeow, J. Global Epidemiology and Genetics of Hepatocellular Carcinoma. *Gastroenterology* **2023**, *164*, 766–782. [CrossRef]
147. Ha, S.; Wong, V.W.S.; Zhang, X.; Yu, J. Interplay between gut microbiome, host genetic and epigenetic modifications in MASLD and MASLD-related hepatocellular carcinoma. *Gut* **2025**, *74*, 141–152. [CrossRef]
148. Mullish, B.H. Risk Factors for Liver Cancer and Chronic Liver Disease-related Death: Are Sugar Substitutes Better Than the Real Thing? *Gastroenterology* **2024**, *166*, 213–214. [CrossRef]
149. Fedirko, V.; Lukanova, A.; Bamia, C.; Trichopolou, A.; Trepo, E.; Nöthlings, U.; Schlesinger, S.; Aleksandrova, K.; Boffetta, P.; Tjønneland, A.; et al. Glycemic index, glycemic load, dietary carbohydrate, and dietary fiber intake and risk of liver and biliary tract cancers in Western Europeans. *Ann. Oncol.* **2013**, *24*, 543–553. [CrossRef]
150. Zhao, L.; Zhang, X.; Coday, M.; Garcia, D.O.; Li, X.; Mossavar-Rahmani, Y.; Naughton, M.J.; Lopez-Pentecost, M.; Saquib, N.; Shadyab, A.H.; et al. Sugar-Sweetened and Artificially Sweetened Beverages and Risk of Liver Cancer and Chronic Liver Disease Mortality. *JAMA* **2023**, *330*, 537. [CrossRef]

151. Rossi, M.; Lipworth, L.; Maso, L.D.; Talamini, R.; Montella, M.; Polesel, J.; McLaughlin, J.K.; Parpinel, M.; Franceschi, S.; Lagiou, P.; et al. Dietary glycemic load and hepatocellular carcinoma with or without chronic hepatitis infection. *Ann. Oncol.* **2009**, *20*, 1736–1740. [CrossRef]
152. Lagiou, P.; Rossi, M.; Tzonou, A.; Georgila, C.; Trichopoulos, D.; La Vecchia, C. Glycemic load in relation to hepatocellular carcinoma among patients with chronic hepatitis infection. *Ann. Oncol.* **2009**, *20*, 1741–1745. [CrossRef]
153. Yang, L.S.; Yan, L.J.; Meng, G.X.; Ding, Z.N.; Yao, S.Y.; Li, H.C.; Dong, Z.R.; Chen, Z.Q.; Hong, J.G.; Li, T. The Association of Glycemic Index, Glycemic Load, and Daily Carbohydrates Intake with the Risk of Hepatocellular Carcinoma: A Systematic Review and Meta-Analysis. *Nutr. Cancer* **2023**, *75*, 461–469. [CrossRef] [PubMed]
154. Suez, J.; Cohen, Y.; Valdés-Mas, R.; Mor, U.; Dori-Bachash, M.; Federici, S.; Zmora, N.; Leshem, A.; Heinemann, M.; Linevsky, R.; et al. Personalized microbiome-driven effects of non-nutritive sweeteners on human glucose tolerance. *Cell* **2022**, *185*, 3307–3328.e19. [CrossRef]
155. Tsai, M.J.; Li, C.H.; Wu, H.T.; Kuo, H.Y.; Wang, C.T.; Pai, H.L.; Chang, C.J.; Ou, H.Y. Long-Term Consumption of Sucralose Induces Hepatic Insulin Resistance through an Extracellular Signal-Regulated Kinase 1/2-Dependent Pathway. *Nutrients* **2023**, *15*, 2814. [CrossRef] [PubMed]
156. Golzan, S.A.; Movahedian, M.; Haghighat, N.; Asbaghi, O.; Hekmatdoost, A. Association between non-nutritive sweetener consumption and liver enzyme levels in adults: A systematic review and meta-analysis of randomized clinical trials. *Nutr. Rev.* **2023**, *81*, 1105–1117. [CrossRef]
157. Chi, L.; Yifei, Y.; Bian, X.; Gao, B.; Tu, P.; Ru, H.; Lu, K. Chronic sucralose consumption inhibits farnesoid X receptor signaling and perturbs lipid and cholesterol homeostasis in the mouse livers, potentially by altering gut microbiota functions. *Sci. Total Environ.* **2024**, *919*, 169603. [CrossRef]
158. Wu, H.T.; Lin, C.H.; Pai, H.L.; Chen, Y.C.; Cheng, K.P.; Kuo, H.Y.; Li, C.H.; Ou, H.Y. Sucralose, a Non-nutritive Artificial Sweetener Exacerbates High Fat Diet-Induced Hepatic Steatosis Through Taste Receptor Type 1 Member 3. *Front Nutr.* **2022**, *9*, 823723. [CrossRef]
159. Finamor, I.; Pérez, S.; Bressan, C.A.; Brenner, C.E.; Rius-Pérez, S.; Brittes, P.C.; Cheiran, G.; Rocha, M.I.; da Veiga, M.; Sastre, J.; et al. Chronic aspartame intake causes changes in the trans-sulphuration pathway, glutathione depletion and liver damage in mice. *Redox Biol.* **2017**, *11*, 701–707. [CrossRef]
160. Finamor, I.A.; Bressan, C.A.; Torres-Cuevas, I.; Rius-Pérez, S.; da Veiga, M.; Rocha, M.I.; Pavanato, M.A.; Pérez, S. Long-Term Aspartame Administration Leads to Fibrosis, Inflammasome Activation, and Gluconeogenesis Impairment in the Liver of Mice. *Biology* **2021**, *10*, 82. [CrossRef]
161. Sergi, C.M. MASLD and aspartame: Are new studies in the horizon? *Front. Med.* **2023**, *10*, 1266918. [CrossRef]

current issues in molecular biology

MDPI

Review

Advances in Understanding Lipopolysaccharide-Mediated Hepatitis: Mechanisms and Pathological Features

Kazuhiko Nakadate *, Hayate Saitoh, Miina Sakaguchi, Fumito Miruno, Naoto Muramatsu, Nozomi Ito, Kanako Tadokoro and Kiyoharu Kawakami

Department of Functional Morphology, Meiji Pharmaceutical University, 2-522-1 Noshio, Kiyose, Tokyo 204-8588, Japan; y201111@std.my-pharm.ac.jp (H.S.); y201117@std.my-pharm.ac.jp (M.S.); y201290@std.my-pharm.ac.jp (F.M.); y201295@std.my-pharm.ac.jp (N.M.); s212009@std.my-pharm.ac.jp (N.I.); s212052@std.my-pharm.ac.jp (K.T.); k-kawakami@my-pharm.ac.jp (K.K.)

* Correspondence: nakadate@my-pharm.ac.jp

Abstract: Lipopolysaccharide (LPS), a key component of Gram-negative bacterial membranes, plays a central role in the pathogenesis of inflammatory liver diseases. In this review, we aimed to explore the role of LPS in hepatic injury. Upon hepatic infiltration, LPS activates Kupffer cells via toll-like receptor 4 (TLR4) signaling, inducing proinflammatory cytokines such as tumor necrosis factor-α and interleukin-1β. These mediators amplify hepatocyte apoptosis, endothelial damage, and platelet aggregation, thereby contributing to sinusoidal thrombosis and tissue ischemia. Pathological features, such as hepatocyte shrinkage, sinusoidal expansion, and fibrin deposition, are hallmark indicators of LPS-induced hepatic inflammation. Therapeutically, aspirin shows promise for attenuating cytokine release, protecting endothelial integrity, and reducing thrombogenesis. Emerging strategies include targeting TLR4 pathways, modulating the gut–liver axis, and utilizing biomolecular approaches such as RNA interference for LPS suppression. The integration of public health interventions, such as dietary optimization and microbiome regulation, offers additional preventive measures. In this review, the dual roles of LPS in inflammation and thrombosis have been emphasized. Advancing our understanding of LPS-driven mechanisms and enhancing treatment strategies are pivotal for managing hepatic inflammation and its systemic implications. Future research should focus on refining biomarkers, optimizing therapeutic efficacy, and addressing safety concerns for clinical applications.

Keywords: lipopolysaccharide; TLR4; hepatitis; pathological features; histopathological diagnosis; electron microscopic observation; CLEM

1. Introduction

Hepatitis occurs in acute and chronic forms and has significant personal and public health consequences worldwide [1,2]. Acute hepatitis is often caused by short-term exposure to viral infections, toxins, alcohol, or drugs and can lead to acute liver failure and life-threatening conditions [1,3,4]. In contrast, chronic hepatitis progresses through long-term inflammation and can lead to serious organ damage, including cirrhosis and hepatocellular carcinoma [5]. Such liver diseases not only significantly reduce the quality of life of patients but also result in increased medical costs, loss of work capacity, and a growing social and economic burden [6,7]. Therefore, understanding the mechanisms of hepatitis pathogenesis and progression and developing effective treatment strategies based on this understanding have become key challenges in medicine and public health.

The causes of hepatitis are diverse and include viral infections, alcohol ingestion, drug-induced disorders, metabolic abnormalities, and autoimmune diseases [8–10]. Recently, lipopolysaccharide (LPS), a cell wall component of Gram-negative bacteria produced by intestinal microbiota, has been shown to play an important role in the development and progression of hepatitis. It acts as a powerful trigger for the immune system, reaching the liver through the portal vein when intestinal permeability and dysbiosis are increased [11–13]. The liver is a major organ responsible for detoxification and metabolism and is an immunological "front line" that receives blood directly from the intestinal tract [14]. Therefore, LPS is considered a major factor in the activation of immune cells in the liver and causes inflammatory liver injury [15].

Central to the inflammatory response induced by LPS in the liver is the activation of Kupffer cells (liver-specific macrophages), which produce large amounts of inflammatory cytokines such as tumor necrosis factor (TNF)-α, interleukin (IL)-1β, and IL-6 that trigger an inflammatory cascade in the liver [16–19]. This process leads to apoptosis and necrosis of hepatocytes, vascular endothelial cell damage, platelet activation, and thrombus formation [20–22]. In particular, LPS acts directly on the vascular endothelium of the liver, increasing vascular permeability and causing disrupted blood flow and inadequate oxygen supply, which accelerate the decline in liver function [23].

Furthermore, the LPS-induced inflammatory response has systemic effects that are not limited to the liver. The presence of LPS in the blood promotes the production of C-reactive protein (CRP), an acute phase response protein. CRP is a major biomarker of systemic inflammation, and its elevation suggests effects on other organs in addition to liver inflammation [24,25]. Moreover, platelet activation upon LPS stimulation can lead to progressive thrombus formation, which can cause further tissue damage [26]. Systemic inflammation and vascular damage may also contribute to the progression of chronic and cardiovascular diseases, indicating that LPS has a broad spectrum of effects. Therapeutic strategies are being developed in relation to the inflammatory mechanisms of LPS, and the use of anti-inflammatory drugs, antioxidants, and probiotics has been highlighted as a way to reduce LPS-induced inflammation [27–29]. Antiplatelet and anticoagulant agents have also been shown to be useful for preventing thrombus formation [30,31]. In particular, aspirin is considered a promising agent for the treatment of hepatitis and systemic inflammation caused by LPS because of its anti-inflammatory and antithrombotic effects [32–34].

Although much research has been conducted on the impact of LPS on liver disease as described above, there are several important research gaps that have yet to be resolved in order to establish a comprehensive treatment and prevention strategy for LPS-induced hepatitis. Filling these gaps is essential to advance the field of research and improve patient outcomes. Among them, it is known that systemic inflammation caused by LPS exacerbates cardiovascular and chronic diseases, but there is a lack of detailed research on specific biomarkers and pathways linking hepatitis to systemic effects. Identification of these could help predict complications and guide appropriate treatment strategies for each patient. To this end, it is first necessary to clarify in detail the mechanisms by which LPS induces hepatitis and thrombosis in the liver and its pathogenesis. Then, the usefulness of currently available drugs and new therapeutic approaches, including public health strategies, will be required.

In this review, we aimed to explore the effects of LPS on the liver and to describe its detailed molecular mechanisms. Moreover, we summarized the pathological diagnostic methods that can lead to therapeutic strategies to control LPS-induced inflammation. These findings are intended to contribute to improving treatments and optimizing public health strategies for patients with liver diseases.

2. Materials and Methods

For this review, we used the database contained in PubMed to extract several empirical articles from the relevant literature. Keywords used in the search included "thrombus", "LPS", "hepatitis", "aspirin", "condition", and "diagnosis". In addition, a comprehensive search of the relevant research literature was conducted using PubMed and drug databases for relevance to pathologies used in general practice. We did not use the systematic review and meta-analysis approach in this study.

As for the histological analysis in the present study, we also used the results of the analysis using the technique previously reported by us [33]. That study in animal study was reviewed and approved by the relevant committees of Meiji Pharmaceutical University (No. 2704, 1 April 2023–2024).

3. Action of LPS in the Liver

3.1. Structure and Immunological Significance of LPS

LPS is the major lipid complex that constitutes the outer membrane of Gram-negative bacteria, and its structure and biological functions are closely related to bacterial pathogenicity (Figure 1).

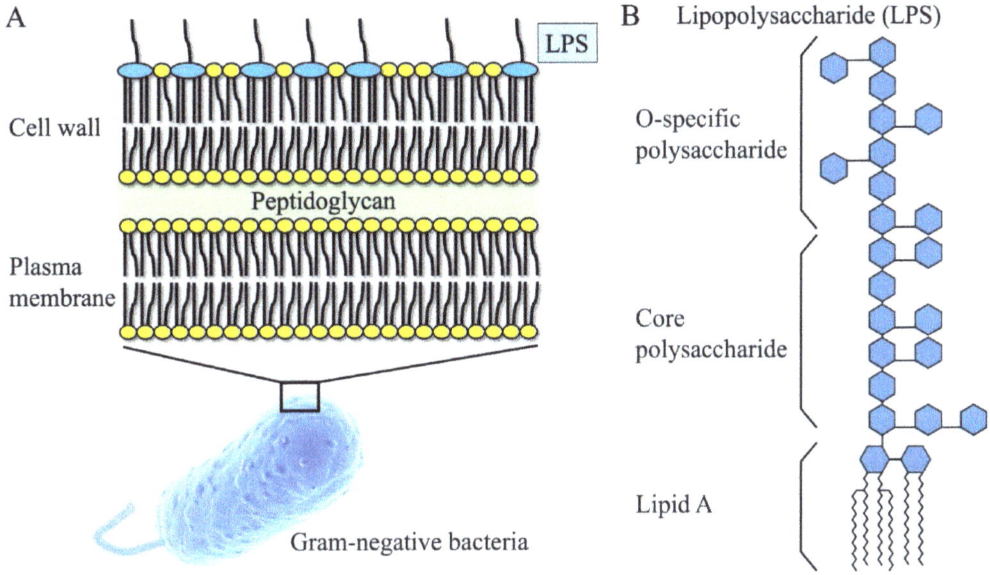

Figure 1. (**A**) Schematic representation of the cell surface structure of Gram-negative bacteria; (**B**) structure of lipopolysaccharide (LPS).

1. Lipid A: Lipid A is the hydrophobic portion of LPS and is at the core of its toxicity and biological activity. Lipid A consists of a disaccharide of glucosamine phosphate attached to a fatty acid that binds directly to toll-like receptor 4 (TLR4) on the cell membrane and triggers an immune response. The composition of fatty acids and the number of phosphate groups vary among bacterial species, contributing to differences in pathogenicity and immune response.
2. Lipooligosaccharide (core polysaccharide): A hydrophilic lipooligosaccharide is the linkage between lipid A and O antigen. This region contains sugar molecules such as keto-oxononanoic acid and heptose, which provide structural stability to LPS.

3. Hydrophilic O-specific polysaccharide (O antigen): This is a long-chain polysaccharide structure located in the outermost layer and is a characteristic region that determines bacterial serotypes. Mutation or loss of the O antigen affects the immune evasion ability of bacteria and is associated with infectivity.

The main mechanism by which LPS induces an immune response is through the activation of signaling pathways via TLR4, a pattern recognition receptor on the cell surface that forms a complex with CD14 and MD-2 to recognize LPS with high affinity [35,36]. This binding triggers intracellular signaling and leads to the activation of nuclear factor kappa-light-chain-enhancer of activated B cells, which promotes the transcription of inflammatory genes [37]. This results in the production of inflammatory cytokines such as TNF-α, IL-1β, and IL-6, leading to acute and chronic inflammatory responses [38–40].

In addition, TLR4 signaling functions through MyD88-dependent and TRIF-dependent pathways, which activate distinct pathways. The MyD88-dependent pathway is primarily responsible for early inflammatory responses, whereas the TRIF-dependent pathway induces delayed responses, including interferon-β production. These pathways complement each other and provide the basis for the rapid and sustained response of the immune system to bacterial infections [41,42].

3.2. LPS and Inflammatory Response in the Liver

The liver receives blood from the intestinal tract via the portal vein and is at the forefront of the LPS derived from intestinal bacteria. As shown in Figure 2, within the liver, cells of the immune system, such as Kupffer cells (liver-specific macrophages) and monocytes, serve as the major LPS response cells [43,44]. Kupffer cells express high levels of TLR4 and CD14 and are capable of rapid LPS recognition and response [45,46].

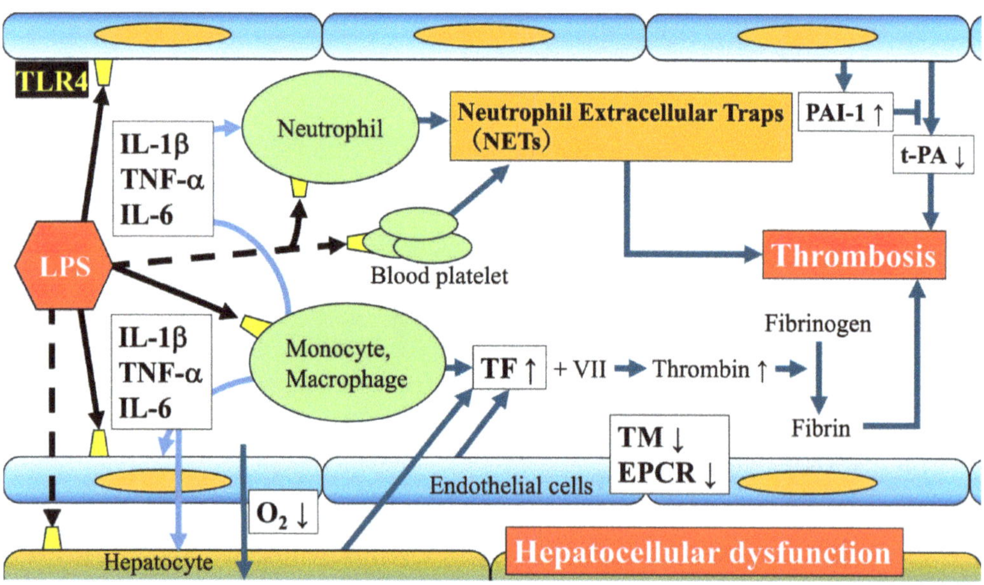

Figure 2. Schematic diagram of intrahepatic sinusoids, various factors induced by LPS, and the mechanism of thrombogenesis. LPS—lipopolysaccharide; TLR4—toll-like receptor 4; IL-1β—interleukin-1β; TNF-α—tumor necrosis factor-α; IL-6—interleukin-6; TF—tissue factor; VII—factor VII of blood coagulation; TM—thrombomodulin; EPCR—endothelial protein C receptor; PAI-1—plasminogen activator inhibitor-1; t-PA—tissue-type plasminogen activator.

1. When Kupffer cells recognize LPS, they trigger the following processes:
    - Release of inflammatory cytokines: TNF-α, IL-1β, and IL-6 are secreted and promote recruitment of other immune cells (e.g., neutrophils and lymphocytes) [19,47].
    - Production of chemokines: chemokines such as CCL2 and CXCL1 mobilize immune cells from the surrounding blood vessels to the liver [48].
    - Reactive oxygen species and nitrogen oxide production: Reactive oxygen species (ROS) and nitrogen oxides damage hepatocytes and endothelial cells via oxidative stress [49,50]. Oxidative stress induces hepatocyte death through peroxidation of cell membrane lipids and DNA damage, thereby exacerbating the progression of liver injury [51].
2. Hepatic stellate cells normally function as vitamin A storage cells but are activated by LPS stimulation [52]. This activation promotes the overproduction of collagen and other extracellular matrix components that cause fibrosis. Chronic LPS stimulation may accelerate liver fibrosis that eventually progresses to cirrhosis.
3. Vascular endothelial cells that form sinusoids are directly damaged by LPS, causing a decrease in thrombomodulin and an increase in tissue factor [53,54]. Furthermore, it inhibits tissue-type plasminogen activator by inducing the expression of plasminogen activator inhibitor-1 [55]. This results in the formation of a thrombus. In addition, by inducing changes in vascular permeability, local oxygen supply is restricted and hepatocytes are further damaged [56].

Furthermore, the activation of Kupffer cells produces ROS and nitric oxide, which cause oxidative stress [49]. Oxidative stress induces hepatocyte death through the peroxidation of cell membrane lipids and DNA damage, thereby exacerbating the progression of liver injury [51]. Moreover, LPS activates hepatic stellate cells, which may increase collagen production and promote fibrosis. This is an important pathological change in the progression of chronic liver disease [52].

3.3. LPS-Induced Hepatocyte and Sinusoidal Endothelial Cell Changes

LPS-induced inflammation has a wide variety of effects on hepatocytes and endothelial cells, which are the major components of the liver, and leads to various changes in the hepatic microenvironment.

1. Induction of hepatocyte death (apoptosis and necrosis): The release of inflammatory cytokines (especially TNF-α and Fas ligands) by LPS induces apoptosis in hepatocytes [57,58]. Apoptosis is energy-dependent and is characterized by cell shrinkage and nuclear fragmentation; however, at high concentrations of inflammatory cytokines, hepatocytes may enter uncontrolled necrosis [59]. Necrosis leads to further damage to surrounding tissues, forming a vicious cycle that leads to chronic inflammation.
2. Impairment of sinusoidal endothelial cells and blood flow: Sinusoids function as sites of blood circulation and metabolite exchange within the liver, but LPS directly injures sinusoidal endothelial cells [60]. This results in altered vascular permeability and accumulation of inflammatory cells and platelets in the sinusoids. This leads to microthrombus formation and localized blood flow obstruction, resulting in deficient oxygen supply to the liver. Such an impaired blood flow may induce further tissue necrosis and fibrosis.
3. Abnormal platelet aggregation and thrombus formation: LPS activates vascular endothelial cells and platelets and induces platelet aggregation [61,62]. This leads to the formation of microthrombi in portal veins and sinusoids. These thrombi not

only interfere with normal blood flow to the liver but also cause additional oxygen deprivation to the surrounding hepatocytes, contributing to chronic damage.

4. Diagnosis of Hepatitis

In LPS-induced hepatitis, various blood biochemistry tests are clinically performed. The presence or absence of inflammation and the degree of inflammation can be inferred from the results of these tests, which are useful diagnostic tools. LPS plays an important role in the progression of hepatitis by activating the immune system and stimulating the production of inflammatory mediators. Inflammatory markers, oxidative stress, and chemokine dynamics are indicators of the severity and chronicity of hepatitis. The roles of inflammatory markers and their relevance in LPS-induced hepatitis, including their clinical significance, are discussed below.

4.1. Blood Biochemistry Test

Blood biochemistry tests performed in hospitals to diagnose and manage LPS-induced hepatitis are primarily general blood tests that evaluate liver function, inflammation, and coagulation status.

1. Liver function tests:
    - Alanine aminotransferase: specific marker of hepatocyte damage.
    - Aspartate aminotransferase: it reflects damage to the liver and other organs (the heart and muscle).
    - Bilirubin (total bilirubin, direct bilirubin): used to assess bile stasis and liver dysfunction.
    - Albumin: Used to assess hepatic synthetic function. Its levels decrease during chronic inflammation and severe liver injury.
    - Alkaline phosphatase: used to evaluate bile stasis and abnormal bone metabolism.
    - γ-Glutamyltransferase: used to assess liver damage and bile duct disease.
2. Inflammatory markers:
    - CRP level reflects the degree of inflammation and spikes in LPS-induced inflammation.
    - White blood cell count is an indicator of infection and inflammation. Usually, it increases with LPS stimulation.
    - Erythrocyte sedimentation rate is useful to assess chronic inflammation and elevation of LPS-induced hepatitis.
3. Coagulation test:
    - Prothrombin time and international normal ratio: used to assess coagulation capacity and the risk of impaired liver synthesis and disseminated intravascular coagulation.
    - D-dimer: Index of thrombus formation and lysis. It increases during LPS-induced thrombus formation.
    - Platelet count: it is related to thrombus formation and may fluctuate after LPS stimulation.
4. Additional inspection:
    - Abdominal ultrasonography: Used to evaluate liver size, blood flow, and lesions. It is useful for detecting clots and biliary stasis.
    - Blood culture: confirmation of infection by LPS-producing bacteria (e.g., Gram-negative bacteria).
    - Endotoxin measurement: it assesses the amount of LPS (often used in the research phase).

These tests help diagnose and assess the progression of LPS-induced hepatitis and monitor the treatment response. Additional tests may be performed as needed depending on the patient's condition.

4.2. Role and Clinical Significance of Inflammatory Markers

1. C-reactive protein: CRP is widely used as an early indicator of the acute inflammatory response; upon LPS stimulation, CRP is rapidly produced, primarily through IL-6 induction in the liver [63,64]. Elevated blood CRP levels reflect systemic inflammation and indicate the progression of local inflammation in the liver. CRP is an important biomarker for assessing the severity of hepatitis and serves as an early warning for the severity of acute hepatitis and the possibility of sepsis [64].
2. Tumor necrosis factor-α: TNF-α is one of the primary proinflammatory cytokines secreted after LPS stimulation [38,65]. Kupffer cells and macrophages produce TNF-α, which acts on surrounding hepatocytes and endothelial cells to initiate the inflammatory cascade and is involved in increasing vascular permeability, leukocyte infiltration, induction of cell death, and amplification of liver damage. Chronic increases in TNF-α also contribute to the development of liver fibrosis [66].
3. Interleukin-1β: IL-1β is an important proinflammatory cytokine secreted by inflammasomes activated by LPS stimulation [67]. This molecule is a major contributor to the inflammatory environment in the liver and causes damage to the vascular endothelial cells and hepatocytes. In addition, IL-1β induces the production of other cytokines (e.g., IL-6 and IL-8), which contribute to the spread of inflammation [68].

4.3. Association Between Increased CRP Levels and Hepatitis Severity After LPS Stimulation

Once LPS enters the bloodstream, an acute phase response is rapidly initiated, and CRP concentrations increase markedly [63,64]. This increase coincides with the peak production of inflammatory mediators and reflects the severity of local inflammation in the liver.

1. CRP as an indicator of hepatitis progression: A rapid increase in CRP levels can be used as a real-time indicator of inflammatory activity. In animal models, CRP concentrations have been observed to increase rapidly within hours of LPS administration and subsequently correlate with inflammatory changes in liver tissue [33]. Therefore, CRP measurement plays an important role in the early diagnosis of LPS-induced hepatitis and monitoring therapeutic efficacy.
2. High CRP levels and risk of chronicity: Persistently high CRP levels suggest uncontrolled inflammation, which increases the risk of hepatitis becoming chronic and transitioning into liver fibrosis. In such cases, the introduction of anti-inflammatory therapy and additional diagnostic procedures is warranted.

4.4. Role of Oxidative Stress (Reactive Oxygen Species) and Its Progression to Liver Fibrosis

Oxidative stress is a core pathophysiological process in LPS-induced hepatitis. After LPS stimulation, reactive oxygen species (ROS) production markedly increases, thereby damaging cell membranes, proteins, and DNA. The detailed role of oxidative stress is as follows [69]:

1. Production of ROS and induction of hepatocyte death: Kupffer cells and hepatocytes activated by LPS overproduce ROS, causing hepatocytes to undergo apoptosis or necrosis [16]. This process is closely associated with damage to the mitochondrial membrane and decreased ATP production. Oxidative stress disrupts intercellular signaling pathways and contributes to a vicious cycle of inflammation.
2. Endothelial cell damage and thrombus formation: ROS damage sinusoidal endothelial cells, increase vascular permeability, and induce thrombus formation [70]. Consequently, impaired blood flow and local oxygen deprivation are exacerbated, leading to progressive liver dysfunction.
3. Transition to hepatic fibrosis: Chronic oxidative stress promotes the activation of astrocytes (the major effector cells of intrahepatic fibrosis) [71,72]. Activated astrocytes

produce excess extracellular matrix and destroy the normal structure of the liver. This contributes to the development of fibrosis and, eventually, cirrhosis.

4.5. Role of Chemokines

Chemokine production after LPS stimulation is important for the mobilization of inflammatory cells and development of liver fibrosis.

1. Types of chemokines and their functions: Chemokines (e.g., MCP-1, CXCL8) are responsible for inducing immune cells such as neutrophils, monocytes, and macrophages to the liver [73,74]. These cells further exacerbate inflammation and damage liver tissue through the production of inflammatory mediators and ROS.
2. Contribution to fibrosis: Chemokines are also involved in astrocyte activation and abnormal signaling between hepatocytes. MCP-1 overexpression promotes the maintenance of chronic inflammation and fibrotic processes [73].

4.6. Histopathological Features of LPS-Induced Hepatitis

LPS-induced hepatitis causes significant changes in liver structure and function. These changes can be clearly observed via histological and ultrastructural analyses, allowing a detailed understanding of pathogenesis. Therefore, histopathological features of LPS-induced hepatitis may be described in terms of histological changes and ultrastructural analysis using electron microscopy [33].

4.6.1. Details of Histological Changes

As shown in Figure 3, LPS administration results in multiple pathological abnormalities in the liver tissue. These abnormalities are observed in hepatocytes, vascular structures, and sinusoidal endothelial cells and reflect a progressive inflammatory process.

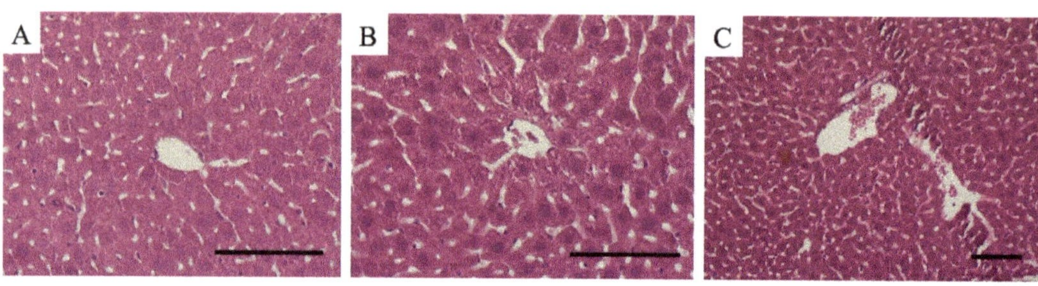

Figure 3. (**A**,**B**) Hematoxylin and eosin-stained images of the liver lobules; (**C**) hematoxylin and eosin-stained image of the intrahepatic vein. (**A**) Normal mouse liver; (**B**,**C**) mouse liver 24 h after LPS administration. As shown in (**A**), the hepatocytes of normal mice are healthy, and staining is even. The sinusoids are normal and few hemocytes are observed in the sinusoids. As shown in (**B**), the hepatocytes in the LPS-treated group are not uniformly stained and are damaged. The sinusoids are swollen. There is an increase in the number of erythrocytes and leukocytes in sinusoids. In (**C**), a large number of blood cells are stored in the vein. All scale bars are 100 μm.

1. Hepatocyte atrophy and decreased eosin staining: Hepatocyte atrophy is caused by a decrease in cellular metabolic activity and disruption of the cytoskeleton. Trophied cells are characterized by decreased eosin staining. Decreased eosin staining indicates degeneration or degradation of cytoplasmic proteins and reflects the progression of apoptosis (programmed cell death) or necrosis. This phenomenon is closely related to cell death mechanisms associated with the release of inflammatory cytokines (particularly TNF-α and IL-1β).

2. Dilation of sinusoids and increased vascular permeability: Dilatation of the sinusoids indicates a change in hemodynamics. LPS stimulation increases the internal pressure of the sinusoids, resulting in the blockage of blood flow, which is observed in sinusoids. In addition, inflammatory cytokines and ROS damage sinusoidal endothelial cells and increase vascular permeability. This increased permeability allows plasma components and immune cells to leak into surrounding tissues, further expanding the inflammatory response within the liver.
3. Blood cell accumulation and thrombus formation: The accumulation of blood cells in the blood vessels and sinusoids is an important indicator of inflammation progression. This accumulation primarily consists of leukocytes (especially neutrophils) and erythrocytes. These cells form microthrombi by interacting with the endothelial cells and activated platelets. The thrombus causes vascular occlusion, resulting in localized ischemia and an inadequate oxygen supply. This results in accelerated necrosis of hepatocytes and progressive dysfunction of the entire liver.
4. Immune cell infiltration: LPS stimulation causes neutrophils and macrophages to infiltrate liver tissue. These cells release inflammatory mediators and ROS that amplify inflammation and cause tissue damage. In particular, neutrophil overactivation causes direct damage to the surrounding cells and tissues through degranulation.

4.6.2. Details of Scanning Electron Microscope Analysis

In addition to optical microscopy, electron microscopy is an extremely important tool for analyzing the ultrastructure of lesions and elucidating the molecular mechanisms of LPS-induced hepatitis. Transmission electron microscopy is useful for detailed intracellular examinations; however, scanning electron microscopy is useful for observing blood cells and thrombi because the sections are extremely thin (Figure 4).

Figure 4. Scanning electron microscopy images of intrahepatic veins 24 h after LPS administration. (**A**,**B**) show different vascular images; (**B′**) magnified image of the red frame in (**B**). A thrombus is observed in the veins. All scale bars are 100 μm.

1. Detailed structural observation of blood cell accumulation: Scanning electron microscopy provides a detailed view of leukocyte and red blood cell aggregation in blood vessels and sinusoids. Red blood cell aggregation not only increases the viscosity of the blood stream but also decreases its oxygen-carrying capacity, contributing to systemic hypoxia. Platelets are also observed, suggesting that fibrin formation is an important component of the thrombi.
2. Endothelial cell damage and changes in surface structure: Endothelial cells undergo coarsening of the cell surface under the direct influence of LPS and the action of inflammatory mediators. This coarsening implies the loss of microvilli on the endothelial cell surface and the degradation of intercellular junctions. This results in abnormally increased vascular permeability and accelerated inflammatory progres-

sion. In addition, endothelial cells often show signs of apoptosis or necrosis, resulting in the loss of vascular structural integrity.

4.6.3. Details of Transmission Electron Microscope Analysis

Intracellular organelles and cell membrane changes and immunoelectron microscopy can be used to reveal detailed intracellular LPS-induced pathological changes.

1. Changes in intracellular organelles: LPS stimulation of hepatocytes reveals morphological and functional abnormalities in mitochondria. Specifically, mitochondrial swelling, disruption of the cristae structure, and matrix rarefaction are observed. These changes reflect increased oxidative stress and a decreased ability to produce ATP, which promotes cell death. Structural disruption of the Golgi apparatus and rough endoplasmic reticulum also indicates impaired protein synthesis and abnormal stress responses.
2. Localization of inflammatory mediators: Electron microscopy combined with immunogold labeling techniques reveal the localization of inflammatory cytokines (e.g., TNF-α and IL-1β). These cytokines are detected at high concentrations, primarily in Kupffer cells and activated endothelial cells, suggesting that they play a central role in inflammation.

4.6.4. Usefulness of Correlative Light and Electron Microscopy Methods for Pathological Analysis

Histopathological analysis using optical microscopy is frequently performed in clinical practice. However, analysis using electron microscopy is limited to a small number of cases because of the difficulty and time required to prepare samples. Nevertheless, electron microscopy can be used to prove findings that are difficult to confirm using optical microscopy. Moreover, the correlative light and electron microscopy (CLEM) method, in which the same specimen is analyzed using electron microscopy, may be useful (Figure 5). The CLEM method combines the advantages of optical and electron microscopy to analyze cellular and tissue structures in detail. It is thought to be much less expensive and more effective than conventional electron microscopic analysis and can be used for diagnosis and research. The histopathological analysis of LPS-induced hepatitis has several advantages [33].

1. Structural analysis at high resolution
    - The pattern of inflammation and damage throughout the liver tissue in LPS-induced hepatitis is observed using an optical microscope. Specifically, histological changes such as dilation of sinusoids, accumulation of blood cells, and atrophy of hepatocytes can be observed. Using electron microscopy on the same liver tissue specimen, ultrafine structures (e.g., coarse endothelial cells and morphological changes of platelets and leukocytes) that cannot be observed with optical microscopy can be analyzed in detail.
2. Specific localization analysis at the cellular level
    - The site of inflammation can be identified using optical microscopy, and the morphology and interactions of the cells (Kupffer cells, leukocytes, platelets, etc.) present at the site can be confirmed using scanning electron microscopy.
    - The CLEM method using transmission electron microscopy allows a detailed analysis of LPS-stimulated intracellular structural changes (e.g., lysosomal hypertrophy, nuclear fragmentation, and endothelial cell damage).
3. Detailed analysis of thrombus formation
    - Accumulation of platelets, leukocytes, and red blood cells in the sinusoids can be observed and analyzed to determine their involvement in thrombus formation.

- Electron microscopy can be used to observe the morphological characteristics of the fibrin network and platelet aggregation that underlies the thrombus.

4. Correlation analysis of pathological changes and molecular functions
    - Detailed analysis of abnormal regions observed via optical and electron microscopy can link pathological changes to abnormalities at the cellular level.
    - The distribution of specific proteins (e.g., tissue factors and inflammatory cytokines) can be confirmed using optical techniques such as immunostaining, and the sites can be further analyzed using electron microscopy.
5. Quantitative analysis
    - By quantifying the number of inflammatory cells, degree of sinusoidal dilation, and frequency of thrombus formation, the progression of LPS-induced hepatitis can be objectively evaluated.
6. Elucidation of complex interactions
    - LPS-stimulated changes in cell–cell adhesion and extracellular matrix can be visualized to reveal the details of inflammation and coagulation processes.

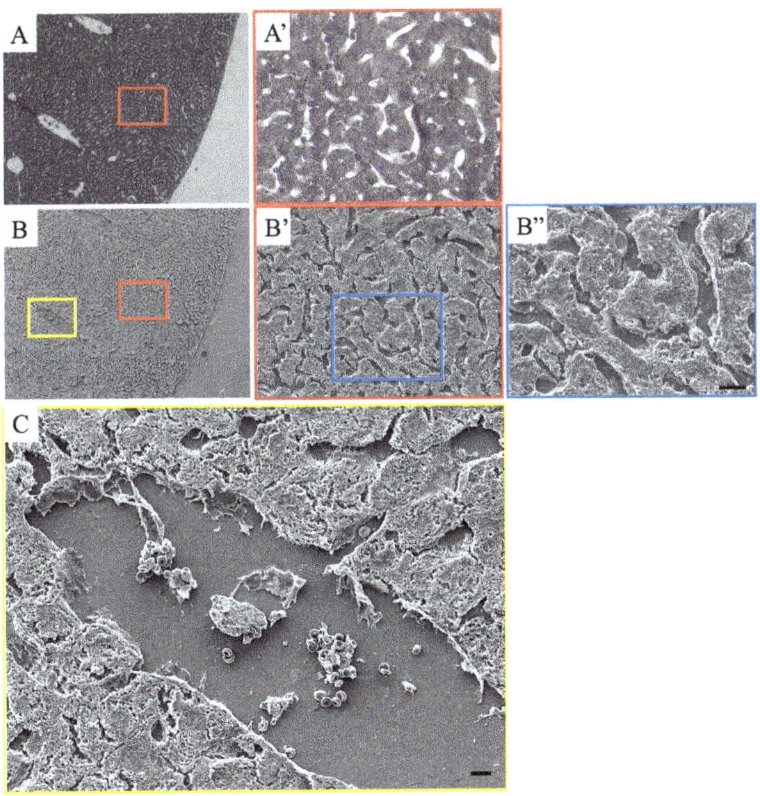

Figure 5. Identical histopathological images obtained using correlative light and electron microscopy. (**A**,**A′**) Hematoxylin-stained images; (**B**–**B″**) scanning electron microscopy images of the same field of view; (**A′**) magnified image of the red frame in (**A**); (**B′**) magnified image of the red frame in (**B**); (**B″**) magnified image of the blue frame in (**B′**); (**C**) magnified image of the yellow frame in (**B**). The morphology of the blood cells is observed using an optical microscope and confirmed using a scanning electron microscope. It is possible to observe blood cells and thrombi in the blood vessels in detail. Scale bar is 10 μm.

The CLEM method enables multifaceted analysis from the tissue level to the ultrastructural level in LPS-induced hepatitis and contributes to the elucidation of the mechanisms of inflammation and thrombus formation. We believe that this method can provide a basis for the detailed analysis of changes in hepatocytes and endothelial cells and for elucidating new therapeutic targets for inflammatory diseases.

5. Therapeutic Approach: Anti-Inflammatory and Antithrombotic

To control the progression of LPS-induced hepatitis, a multifaceted approach that simultaneously suppresses the inflammatory responses and thrombus formation is required. This section focuses on aspirin, a widely used anti-inflammatory and antithrombotic agent, and details its mechanism of action and therapeutic efficacy. In addition, the development status of new treatment methods and the perspective of prevention through lifestyle modification are discussed in depth.

5.1. Effects of Aspirin and Its Mechanism of Action

Aspirin is a leading nonsteroidal anti-inflammatory drug, particularly known for its inhibition of the cyclooxygenase (COX) enzyme, which is responsible for prostaglandin synthesis. Prostaglandins play an important role in amplifying inflammatory responses and platelet aggregation [75]. Aspirin inhibits both COX-1 and COX-2 but irreversibly inhibits COX-1, effectively suppressing platelet aggregation and exerting an antithrombotic effect [76]. In contrast, the inhibition of COX-2 suppresses the production of prostaglandin E2 and prostacyclin, resulting in an anti-inflammatory effect that inhibits the secretion of inflammatory cytokines (e.g., TNF-α and IL-1β) [77,78].

Experiments using an LPS-induced hepatitis model have shown that aspirin treatment has a marked effect. Specifically, aspirin treatment significantly reduces the levels of inflammatory markers such as CRP, IL-1β, and TNF-α [33]. These results suggest that aspirin may not only inhibit the progression of inflammation but also prevent the severity of liver injury caused by LPS. Previous studies have reported a decrease in platelet aggregation in the blood vessels and sinusoids, indicating that aspirin has the potential to reduce hepatocyte necrosis and apoptosis through improved blood flow [79,80]. Furthermore, the effect of aspirin is not limited to suppressing inflammatory responses but also involves the control of LPS-induced oxidative stress [81]. Because oxidative stress contributes to hepatocyte death and fibrosis progression, its control is important in the management of LPS-induced hepatitis. Therefore, aspirin is clinically useful as a therapeutic agent with multifaceted actions.

Quantitative data on aspirin therapy have been widely reported in the areas of cardiovascular disease prevention and inflammation control [82–85]. The data play an important role in the evaluation of therapeutic efficacy and risk of adverse effects. First, many clinical trials have been conducted on the effects of aspirin in cardiovascular disease prevention, with the ASPREE trial being the leading report on primary prevention. This trial examined the effect of aspirin in 19,114 elderly patients and found that the rate of cardiovascular events in the aspirin group was 10.7 events per 10,000 person-years, not significantly different from the placebo group. However, the risk of bleeding was increased in the aspirin group (8.6 events per 10,000 person-years). In secondary prevention, on the other hand, the ISIS-2 trial was conducted in patients with acute myocardial infarction and showed a 23% reduction in vascular death with aspirin (160 mg/day). In the suppression of inflammation, studies have also been conducted on LPS-induced inflammation models. In mouse models, aspirin administration (50–100 mg/kg) has been shown to reduce blood levels of TNF-α and IL-1β by more than 50% and reduce levels of CRP (C-reactive protein) by approximately 30–40% [32,33].

5.2. Potential for New Treatment Methods

In the treatment of LPS-induced hepatitis, new therapies are being developed in addition to conventional drugs such as aspirin. Of particular interest are therapies that directly target LPS and its associated signaling pathways. One approach is to regulate LPS-binding proteins (LBPs), which bind to LPS in the blood and are responsible for transferring LPS to TLR4 receptors. Regulation of this process can inhibit TLR4 signaling and prevent excessive inflammatory responses. Therapeutic approaches that utilize soluble TLR4 have also been investigated [86,87]. Soluble TLR4 competes with cell-surface TLR4 for binding to LPS, thus preventing LPS from activating intracellular signaling. This approach has the potential to fundamentally inhibit LPS-induced inflammation.

Additionally, biologics and molecular-targeted therapies are promising. Specifically, monoclonal antibodies that directly inhibit inflammatory cytokines (e.g., TNF-α and IL-1β) and small-molecule drugs that inhibit chemokine signaling have been investigated. These novel therapies are expected to be effective for reducing chronic inflammation and fibrosis in LPS-induced hepatitis.

5.3. Perspectives on Lifestyle Improvement and Prevention

Improvement of the intestinal environment plays an important role in the prevention of LPS-induced hepatitis because LPS leaks into the bloodstream owing to an imbalance in the intestinal microbiota or disruption of the intestinal barrier. Regulation of intestinal microbiota can reduce the burden of LPS [88]. Probiotics (lactobacilli and bifidobacteria) and prebiotics (dietary fiber) increase the diversity of intestinal bacteria and decrease intestinal permeability [89,90]. In addition, the intake of zinc and vitamin D, which strengthen the intestinal barrier, may reduce the risk of LPS migration into the bloodstream [91,92].

Nutritional approaches have also attracted considerable attention. Foods containing omega-3 fatty acids and polyphenols contribute to liver health maintenance through their anti-inflammatory and antioxidant effects [93,94]. For example, omega-3 fatty acids inhibit the production of TNF-α and IL-1β, thereby reducing the progression of chronic inflammation [95]. Polyphenols found in green tea and fruits may reduce oxidative stress and prevent LPS-induced hepatocellular damage caused by LPS [96].

6. Future Prospects

Although much progress has been made in the study of LPS-induced hepatitis in recent years, several areas remain unexplored. The inflammatory responses induced by LPS and the mechanisms of hepatitis progression caused by these responses are extremely complex, and a deep understanding of these mechanisms is important for future research. In addition, the development of biomarkers, novel therapeutic agents, and public health prevention strategies is essential to establish effective diagnostic and therapeutic strategies. This section details future challenges and prospects for the control of LPS-induced hepatitis.

6.1. Further Elucidation of Signaling Pathways in LPS-Induced Inflammation

LPS activates innate immunity via TLR4 and triggers the secretion of inflammatory cytokines (e.g., TNF-α and IL-1β) through various signaling pathways. However, many aspects of these signaling pathways remain unexplored. For example, the crosstalk between MyD88-dependent and MyD88-independent pathways downstream of TLR4 and their involvement in hepatocyte death and fibrosis are not yet fully understood. The roles of LPS-induced oxidative stress and endoplasmic reticulum stress in hepatitis progression require further investigation.

Recent studies have focused on how LPS modulates inflammatory responses through interactions with the gut microbiota. Elucidating how changes in the intestinal environment affect blood LPS levels and the extent to which they contribute as an inflammatory stimulus to the liver may lead to the development of new therapeutic strategies based on the gut–liver axis.

6.2. Development of a Predictive Biomarker for LPS-Induced Hepatitis Progression

The development of reliable biomarkers is necessary to accurately assess the progression of LPS-induced hepatitis and determine the timing of appropriate therapeutic interventions. Currently, inflammatory markers such as CRP, IL-1β, and TNF-α are widely used to evaluate hepatitis, but these markers may lack specificity. Therefore, new biomarkers that reflect the hepatocellular damage, thrombus formation, and oxidative stress caused by LPS are needed. Variations in metabolites and microRNAs are potential biomarkers for LPS-induced hepatitis. MicroRNAs play important roles in the regulation of inflammatory responses and cell death and their expression profiles are closely related to the pathophysiology of the liver. Comprehensive proteomic and metabolomic analyses are underway to identify molecular patterns specific to LPS-induced hepatitis.

6.3. Evaluation of LPS Inhibitors for Clinical Application

Therapeutic agents that act directly on LPS or target their associated pathways have high potential clinical value. However, cost-effectiveness and safety evaluations are essential for widespread clinical application. Agents targeting soluble TLR4 and LBPs, which are currently under development, are promising new therapeutic tools to reduce LPS-induced inflammation; however, their production costs and side-effect profiles have not yet been fully evaluated.

In particular, the safety aspects require careful consideration of the impact of LPS suppression on the overall immune response. As LPS plays a critical role in the host defense against pathogens, excessive suppression may increase the risk of infection. Therefore, optimizing the dosage and timing of drug administration is a challenge.

6.4. Developing Hepatitis Prevention Strategies at the Public Health Level

Public health efforts beyond individual treatments are important to control LPS-induced hepatitis. In particular, lifestyle modifications to prevent hepatitis are an effective way to reduce the inflammatory burden caused by LPS. For example, maintaining balanced intestinal microflora through dietary modifications can reduce blood LPS levels. Because high-fat diets and the excessive intake of refined sugars may worsen the intestinal environment, an active intake of dietary fiber and fermented foods is recommended. In addition, maintaining an exercise routine has been shown to improve intestinal permeability and prevent LPS leakage.

Moreover, it is important to promote early hepatitis diagnosis and appropriate management through health education and screening programs at the community level. It is hoped that by spreading awareness of LPS-related inflammatory diseases, early intervention and preventive measures can be effectively implemented.

6.5. Potential Side Effects of Aspirin

Aspirin is widely used as an antiplatelet agent, but its use is associated with several potential side effects and challenges in long-term use. It is important to fully understand these issues and manage them appropriately. The most common side effect is gastrointestinal distress [97,98]. Aspirin is a nonselective inhibitor of cyclooxygenase, resulting in the decreased production of prostaglandins, which are necessary for gastric mucosal protection. This results in an increased risk of gastric ulcers and gastric bleeding, which is of particular concern in the elderly and those infected with H. pylori. It can also cause a range of symptoms from mild abdominal pain to severe gastrointestinal perforation.

Next, the risk of bleeding is also a serious issue. Aspirin prolongs hemostasis time by inhibiting thromboxane A2 production and platelet aggregation [99]. This action increases the tendency for systemic bleeding and requires careful management before and after surgery and in trauma patients. Additionally, the risk of cerebral hemorrhage may be increased in hypertensive patients. Other side effects associated with aspirin use include allergic reactions and renal dysfunction [100,101]. In particular, a condition called aspirin asthma can cause a narrowing of the airways and can lead to asthma attacks [102]. Cases have also been reported in which the long-term use of aspirin has decreased renal blood flow and worsened renal dysfunction. Although very rare, hepatotoxicity due to high-dose use is also a concern [103].

Challenges in long-term use include careful identification of indications, thorough management of side effects, monitoring with periodic blood tests, and consideration of alternative therapies if necessary. The use of proton pump inhibitors (PPIs) to prevent gastrointestinal disturbances and at lower doses is recommended. Switching to COX-2 selective inhibitors or other antiplatelet agents (e.g., clopidogrel) is also an option. While aspirin offers significant benefits, the management of side effects is important. It must be used safely and effectively by assessing the risks and benefits for each patient and developing an appropriate treatment plan. In light of these side effects, the effectiveness of thromboprophylaxis in the body's organs, including the liver, should be monitored.

7. Conclusions

LPS is an important virulence factor derived from Gram-negative bacteria and plays a central role in the development of hepatitis. LPS activates the innate immune system and affects the hepatocytes, Kupffer cells, and sinusoidal endothelial cells. This induces the excessive production of inflammatory cytokines, thereby leading to increased oxidative stress, thrombus formation, and fibrosis. These factors accelerate the progression of hepatitis from an acute to a chronic disease.

Although blood biochemical tests are the foundation for the diagnosis of hepatitis, histopathological examinations play an important role. Electron microscopic analysis is particularly useful for observing the localization of inflammatory cells, endothelial cell damage, and detailed structure of thrombus formation. This allows us to elucidate the pathogenesis of LPS-induced hepatitis at the molecular and cellular levels. Aspirin, a nonsteroidal anti-inflammatory drug, has received attention as a treatment option. Aspirin inhibits platelet aggregation by inhibiting COX and reducing the production of inflammatory cytokines. Aspirin has been shown to reduce hepatitis and thrombus formation in LPS-induced hepatitis models. However, owing to the limitations of existing therapies, the development of novel therapeutic agents targeting soluble TLR4 and LBP is required.

Furthermore, lifestyle modifications, such as the adjustment of intestinal microflora and dietary habits, are effective in preventing LPS-induced hepatitis. The intake of fermented foods and dietary fiber improves intestinal permeability and inhibits LPS leakage into the systemic circulation. Early diagnosis and educational activities at the public health level are also key to preventing hepatitis progression. Research on the mechanism of action of LPS and its inhibition strategies has the potential to revolutionize the treatment and prevention of inflammatory liver diseases. Biomarker development and the integration of new diagnostic technologies for personalized medicine are expected to improve the quality of life of patients and public health.

Author Contributions: Conceptualization, H.S., K.K. and K.N.; methodology, H.S., M.S., F.M., N.M., N.I., K.T., K.K. and K.N.; formal analysis, H.S., M.S., F.M., N.M., N.I. and K.T.; writing—original draft preparation, K.K. and K.N.; writing—review and editing, K.K. and K.N.; funding acquisition, K.N. All authors have read and agreed to the published version of the manuscript.

Funding: This work was supported by JSPS KAKENHI Grant Number 23K08620 (K.N.).

Institutional Review Board Statement: This study was reviewed and approved by the relevant committees of Meiji Pharmaceutical University (No. 2704, 1 April 2023–2024).

Informed Consent Statement: Not applicable.

Data Availability Statement: The datasets used and/or analyzed during this study are available from the corresponding author upon reasonable request.

Conflicts of Interest: The authors declare no conflicts of interest.

Abbreviations

The following abbreviations are used in this manuscript:

LPS	Lipopolysaccharide
TLR4	Toll-like receptor 4
TNF	Tumor necrosis factor
IL	Interleukin
CRP	C-reactive protein
ROS	Reactive oxygen species
CLEM	Correlative light and electron microscopy
COX	Cyclooxygenase
LBP	LPS-binding proteins

References

1. Benavent, S.; Carlos, S.; Reina, G. *Rocahepevirus ratti* as an Emerging Cause of Acute Hepatitis Worldwide. *Microorganisms* **2023**, *11*, 2996. [CrossRef] [PubMed]
2. Riazi, K.; Azhari, H.; Charette, J.H.; Underwood, F.E.; King, J.A.; Afshar, E.E.; Swain, M.G.; Congly, S.E.; Kaplan, G.G.; Shaheen, A.A. The prevalence and incidence of NAFLD worldwide: A systematic review and meta-analysis. *Lancet Gastroenterol. Hepatol.* **2022**, *7*, 851–861. [CrossRef] [PubMed]
3. Tai, C.M.; Yu, M.L. Hepatitis C virus micro-elimination in people who inject drugs: Challenges and chance in Taiwan and worldwide. *Kaohsiung J. Med. Sci.* **2024**, *40*, 112–118. [CrossRef]
4. Huang, M.; Chen, H.; Wang, H.; Wang, X.; Wang, D.; Li, Y.; Zhou, Q.; Zhang, D.; Li, M.; Ma, L. Worldwide burden of liver cancer due to metabolic dysfunction-associated steatohepatitis from 1990 to 2019: Insights from the Global Burden of Disease study. *Front. Oncol.* **2024**, *14*, 1424155. [CrossRef] [PubMed]
5. Gavilan, P.; Gavilan, J.C.; Arnedo, R.; Clavijo, E.; Viciana, I.; Gonzalez-Correa, J.A. Prediction model of hepatocellular carcinoma development in chronic hepatitis B virus infection in a Spanish cohort. *Med. Clin.* **2024**, *163*, 609–616. [CrossRef] [PubMed]
6. Su, W.C.; Chen, T.T.; Wu, C.H.; Shih, L.N.; Liu, C.K.; Hsueh, Y.A. The association between comprehensive patient-centered care and health-related quality of life (HRQoL) for patients with chronic viral hepatitis-A pathway analysis. *J. Formos. Med Assoc.* **2024**, *123*, 1057–1063. [CrossRef] [PubMed]
7. Lu, M.; Rupp, L.B.; Melkonian, C.; Trudeau, S.; Daida, Y.G.; Schmidt, M.A.; Gordon, S.C. Persistent pruritus associated with worse quality of life in patients with chronic hepatitis. *Liver Int.* **2024**, *44*, 577–588. [CrossRef]
8. Jimenez-Castillo, R.A.; Gonzalez-Martinez, C.E.; Tovar-Bojorquez, E.M.; Cortez-Hernandez, C.A.; Jaquez-Quintana, J.O.; Gonzalez-Gonzalez, J.A.; Gonzalez-Gonzalez, J.G.; Maldonado-Garza, H.J. Prognostic value of relative adrenal insufficiency in patients with severe alcohol-associated hepatitis-a prospective clinical study. *Gastroenterol. Hepatol.* **2024**, 502322. [CrossRef]
9. Sharma, N.R.; Siriya, P.; Sandhu, S.P.S.; Pokhrel, M.; Ching, D. Complex Diagnostic Dilemma of Viral Hepatitis vs. Drug-Induced Hepatitis: A Case Report. *Cureus* **2023**, *15*, e49191. [CrossRef]
10. Alberti, K.G.; Ilic, V.; Record, C.; Williamson, D.H.; Wright, R. Metabolic abnormalities in patients with active chronic hepatitis. *Clin. Sci.* **1971**, *41*, 4P. [CrossRef] [PubMed]
11. Patil, V.S.; Harish, D.R.; Sampat, G.H.; Roy, S.; Jalalpure, S.S.; Khanal, P.; Gujarathi, S.S.; Hegde, H.V. System Biology Investigation Revealed Lipopolysaccharide and Alcohol-Induced Hepatocellular Carcinoma Resembled Hepatitis B Virus Immunobiology and Pathogenesis. *Int. J. Mol. Sci.* **2023**, *24*, 11146. [CrossRef]
12. Nien, H.C.; Hsu, S.J.; Su, T.H.; Yang, P.J.; Sheu, J.C.; Wang, J.T.; Chow, L.P.; Chen, C.L.; Kao, J.H.; Yang, W.S. High Serum Lipopolysaccharide-Binding Protein Level in Chronic Hepatitis C Viral Infection Is Reduced by Anti-Viral Treatments. *PLoS ONE* **2017**, *12*, e0170028. [CrossRef]

13. Chen, Z.; Liu, H.; Lei, S.; Zhao, B.; Xia, Z. LY294002 prevents lipopolysaccharide-induced hepatitis in a murine model by suppressing IκB phosphorylation. *Mol. Med. Rep.* **2016**, *13*, 811–816. [CrossRef] [PubMed]
14. Sun, X.; Widdison, W.; Mayo, M.; Wilhelm, S.; Leece, B.; Chari, R.; Singh, R.; Erickson, H. Design of antibody-maytansinoid conjugates allows for efficient detoxification via liver metabolism. *Bioconjugate Chem.* **2011**, *22*, 728–735. [CrossRef]
15. Holen, E.; Araujo, P.; Xie, S.; Softeland, L.; Espe, M. Resveratrol inhibited LPS induced transcription of immune genes and secretion of eicosanoids in Atlantic salmon (*Salmo salar*), comparing mono-, co- and a novel triple cell culture model of head kidney leukocytes, liver cells and visceral adipocyte tissue. *Comp. Biochem. Physiol. C Toxicol. Pharmacol.* **2019**, *224*, 108560. [CrossRef]
16. Yuan, S.; Wang, Y.; Yang, J.; Tang, Y.; Wu, W.; Meng, X.; Jian, Y.; Lei, Y.; Liu, Y.; Tang, C.; et al. Treadmill exercise can regulate the redox balance in the livers of APP/PS1 mice and reduce LPS accumulation in their brains through the gut-liver-kupffer cell axis. *Aging* **2024**, *16*, 1374–1389. [CrossRef]
17. Thomas, P.; Lazure, D.A.; Moussa, R.; Bajenova, O.; Burke, P.A.; Ganguly, A.; Forse, R.A. Identification of two novel LPS-binding proteins in Kupffer cells: Implications in TNF-α production. *J. Endotoxin Res.* **2006**, *12*, 352–357. [CrossRef]
18. Kojima, S.; Negishi, Y.; Tsukimoto, M.; Takenouchi, T.; Kitani, H.; Takeda, K. Purinergic signaling via P2 × 7 receptor mediates IL-1β production in Kupffer cells exposed to silica nanoparticle. *Toxicology* **2014**, *321*, 13–20. [CrossRef]
19. Han, R.; Zhang, F.; Wan, C.; Liu, L.; Zhong, Q.; Ding, W. Effect of perfluorooctane sulphonate-induced Kupffer cell activation on hepatocyte proliferation through the NF-κB/TNF-α/IL-6-dependent pathway. *Chemosphere* **2018**, *200*, 283–294. [CrossRef] [PubMed]
20. Gu, C.; Du, W.; Chai, M.; Jin, Z.; Zhou, Y.; Guo, P.; Zhou, Y.; Tan, W.S. Human umbilical cord-derived mesenchymal stem cells affect urea synthesis and the cell apoptosis of human induced hepatocytes by secreting IL-6 in a serum-free co-culture system. *Biotechnol. J.* **2022**, *17*, e2100096. [CrossRef] [PubMed]
21. Dai, J.; Sun, C.; Yao, Z.; Chen, W.; Yu, L.; Long, M. Exposure to concentrated ambient fine particulate matter disrupts vascular endothelial cell barrier function via the IL-6/HIF-1α signaling pathway. *FEBS Open Bio* **2016**, *6*, 720–728. [CrossRef] [PubMed]
22. Smith, S.G.; Kleinnijenhuis, J.; Netea, M.G.; Dockrell, H.M. Whole Blood Profiling of Bacillus Calmette-Guerin-Induced Trained Innate Immunity in Infants Identifies Epidermal Growth Factor, IL-6, Platelet-Derived Growth Factor-AB/BB, and Natural Killer Cell Activation. *Front. Immunol.* **2017**, *8*, 644. [CrossRef]
23. Edamitsu, S.; Matsukawa, A.; Ohkawara, S.; Takagi, K.; Nariuchi, H.; Yoshinaga, M. Role of TNF α, IL-1, and IL-1ra in the mediation of leukocyte infiltration and increased vascular permeability in rabbits with LPS-induced pleurisy. *Clin. Immunol. Immunopathol.* **1995**, *75*, 68–74. [CrossRef]
24. Nerurkar, S.S.; McDevitt, P.J.; Scott, G.F.; Johanson, K.O.; Willette, R.N.; Yue, T.L. Lipopolysaccharide (LPS) contamination plays the real role in C-reactive protein-induced IL-6 secretion from human endothelial cells in vitro. *Arterioscler. Thromb. Vasc. Biol.* **2005**, *25*, e136. [CrossRef] [PubMed]
25. Ng, P.M.; Jin, Z.; Tan, S.S.; Ho, B.; Ding, J.L. C-reactive protein: A predominant LPS-binding acute phase protein responsive to Pseudomonas infection. *J. Endotoxin Res.* **2004**, *10*, 163–174. [CrossRef] [PubMed]
26. Wu, H.; Wang, Y.; Zhang, Y.; Xu, F.; Chen, J.; Duan, L.; Zhang, T.; Wang, J.; Zhang, F. Breaking the vicious loop between inflammation, oxidative stress and coagulation, a novel anti-thrombus insight of nattokinase by inhibiting LPS-induced inflammation and oxidative stress. *Redox Biol.* **2020**, *32*, 101500. [CrossRef]
27. Zhang, Y.Y.; Yao, Y.D.; Cheng, Q.Q.; Huang, Y.F.; Zhou, H. Establishment of a High Content Image Platform to Measure NF-κB Nuclear Translocation in LPS-induced RAW264.7 Macrophages for Screening Anti-inflammatory Drug Candidates. *Curr. Drug Metab.* **2022**, *23*, 394–414. [CrossRef]
28. Moon, K.M.; Lee, B.; Jeong, J.W.; Kim, D.H.; Park, Y.J.; Kim, H.R.; Park, J.Y.; Kim, M.J.; An, H.J.; Lee, E.K.; et al. Thio-barbiturate-derived compounds are novel antioxidants to prevent LPS-induced inflammation in the liver. *Oncotarget* **2017**, *8*, 91662–91673. [CrossRef]
29. Chorawala, M.R.; Chauhan, S.; Patel, R.; Shah, G. Cell Wall Contents of Probiotics (*Lactobacillus species*) Protect Against Lipopolysaccharide (LPS)-Induced Murine Colitis by Limiting Immuno-inflammation and Oxidative Stress. *Probiotics Antimicrob. Proteins* **2021**, *13*, 1005–1017. [CrossRef] [PubMed]
30. Li, C.; Chen, P.; Vadasz, B.; Ma, L.; Zhou, H.; Lang, S.; Freedman, J.; Ni, H. Co-stimulation with LPS or Poly I:C markedly enhances the anti-platelet immune response and severity of fetal and neonatal alloimmune thrombocytopenia. *Thromb. Haemost.* **2013**, *110*, 1250–1258. [CrossRef] [PubMed]
31. He, C.L.; Yi, P.F.; Fan, Q.J.; Shen, H.Q.; Jiang, X.L.; Qin, Q.Q.; Song, Z.; Zhang, C.; Wu, S.C.; Wei, X.B.; et al. Xiang-Qi-Tang and its active components exhibit anti-inflammatory and anticoagulant properties by inhibiting MAPK and NF-κB signaling pathways in LPS-treated rat cardiac microvascular endothelial cells. *Immunopharmacol. Immunotoxicol.* **2013**, *35*, 215–224. [CrossRef]
32. Liu, X.; Tao, Q.; Shen, Y.; Liu, X.; Yang, Y.; Ma, N.; Li, J. Aspirin eugenol ester ameliorates LPS-induced inflammatory responses in RAW264.7 cells and mice. *Front. Pharmacol.* **2023**, *14*, 1220780. [CrossRef] [PubMed]

33. Saitoh, H.; Sakaguchi, M.; Miruno, F.; Muramatsu, N.; Ito, N.; Tadokoro, K.; Kawakami, K.; Nakadate, K. Histopathological Analysis of Lipopolysaccharide-Induced Liver Inflammation and Thrombus Formation in Mice: The Protective Effects of Aspirin. *Curr. Issues Mol. Biol.* **2024**, *46*, 14291–14303. [CrossRef]
34. Vell, M.S.; Krishnan, A.; Wangensteen, K.; Serper, M.; Seeling, K.S.; Hehl, L.; Rendel, M.D.; Zandvakili, I.; Vujkovic, M.; Scorletti, E.; et al. Aspirin is associated with a reduced incidence of liver disease in men. *Hepatol. Commun.* **2023**, *7*, e0268. [CrossRef]
35. Christiansen, D.; Brekke, O.L.; Stenvik, J.; Lambris, J.D.; Espevik, T.; Mollnes, T.E. Differential effect of inhibiting MD-2 and CD14 on LPS- versus whole *E. coli* bacteria-induced cytokine responses in human blood. *Adv. Exp. Med. Biol.* **2012**, *946*, 237–251. [CrossRef] [PubMed]
36. Vives-Pi, M.; Somoza, N.; Fernandez-Alvarez, J.; Vargas, F.; Caro, P.; Alba, A.; Gomis, R.; Labeta, M.O.; Pujol-Borrell, R. Evidence of expression of endotoxin receptors CD14, toll-like receptors TLR4 and TLR2 and associated molecule MD-2 and of sensitivity to endotoxin (LPS) in islet β cells. *Clin. Exp. Immunol.* **2003**, *133*, 208–218. [CrossRef] [PubMed]
37. Hsieh, W.-T.; Hsu, M.-H.; Lin, W.-J.; Xiao, Y.-C.; Lyu, P.-C.; Liu, Y.-C.; Lin, W.-Y.; Kuo, Y.-H.; Chung, J.-G. Ergosta-7, 9 (11), 22-trien-3β-ol Interferes with LPS Docking to LBP, CD14, and TLR4/MD-2 Co-Receptors to Attenuate the NF-κB Inflammatory Pathway In Vitro and *Drosophila*. *Int. J. Mol. Sci.* **2021**, *22*, 6511. [CrossRef] [PubMed]
38. Kim, M.W.; Kang, J.H.; Jung, H.J.; Park, S.Y.; Hwang, J.I.; Seong, J.K.; Yoon, Y.S.; Oh, S.H. Deficiency of Ninjurin1 attenuates LPS/D-galactosamine-induced acute liver failure by reducing TNF-α-induced apoptosis in hepatocytes. *J. Cell Mol. Med.* **2022**, *26*, 5122–5134. [CrossRef] [PubMed]
39. Song, Y.; Shi, Y.; Ao, L.H.; Harken, A.H.; Meng, X.Z. TLR4 mediates LPS-induced HO-1 expression in mouse liver: Role of TNF-α and IL-1β. *World J. Gastroenterol.* **2003**, *9*, 1799–1803. [CrossRef]
40. Treffkorn, L.; Scheibe, R.; Maruyama, T.; Dieter, P. PGE2 exerts its effect on the LPS-induced release of TNF-α, ET-1, IL-1α, IL-6 and IL-10 via the EP2 and EP4 receptor in rat liver macrophages. *Prostaglandins Other Lipid Mediat.* **2004**, *74*, 113–123. [CrossRef] [PubMed]
41. Khan, H.U.; Aamir, K.; Jusuf, P.R.; Sethi, G.; Sisinthy, S.P.; Ghildyal, R.; Arya, A. Lauric acid ameliorates lipopolysaccharide (LPS)-induced liver inflammation by mediating TLR4/MyD88 pathway in Sprague Dawley (SD) rats. *Life Sci.* **2021**, *265*, 118750. [CrossRef]
42. Wang, Y.; Chen, H.; Chen, Q.; Jiao, F.Z.; Zhang, W.B.; Gong, Z.J. The Protective Mechanism of CAY10683 on Intestinal Mucosal Barrier in Acute Liver Failure through LPS/TLR4/MyD88 Pathway. *Mediators Inflamm.* **2018**, *2018*, 7859601. [CrossRef] [PubMed]
43. Zhang, W.; Liu, K.; Ren, G.M.; Wang, Y.; Wang, T.; Liu, X.; Li, D.X.; Xiao, Y.; Chen, X.; Li, Y.T.; et al. BRISC is required for optimal activation of NF-κB in Kupffer cells induced by LPS and contributes to acute liver injury. *Cell Death Dis.* **2023**, *14*, 743. [CrossRef] [PubMed]
44. Momen-Heravi, F.; Bala, S.; Kodys, K.; Szabo, G. Exosomes derived from alcohol-treated hepatocytes horizontally transfer liver specific miRNA-122 and sensitize monocytes to LPS. *Sci. Rep.* **2015**, *5*, 9991. [CrossRef] [PubMed]
45. Frink, M.; Hsieh, Y.C.; Thobe, B.M.; Choudhry, M.A.; Schwacha, M.G.; Bland, K.I.; Chaudry, I.H. TLR4 regulates Kupffer cell chemokine production, systemic inflammation and lung neutrophil infiltration following trauma-hemorrhage. *Mol. Immunol.* **2007**, *44*, 2625–2630. [CrossRef]
46. Lukkari, T.A.; Jarvelainen, H.A.; Oinonen, T.; Kettunen, E.; Lindros, K.O. Short-term ethanol exposure increases the expression of Kupffer cell CD14 receptor and lipopolysaccharide binding protein in rat liver. *Alcohol. Alcohol.* **1999**, *34*, 311–319. [CrossRef] [PubMed]
47. Toki, Y.; Takenouchi, T.; Harada, H.; Tanuma, S.; Kitani, H.; Kojima, S.; Tsukimoto, M. Extracellular ATP induces P2 × 7 receptor activation in mouse Kupffer cells, leading to release of IL-1β, HMGB1, and PGE2, decreased MHC class I expression and necrotic cell death. *Biochem. Biophys. Res. Commun.* **2015**, *458*, 771–776. [CrossRef]
48. Su, L.; Li, N.; Tang, H.; Lou, Z.; Chong, X.; Zhang, C.; Su, J.; Dong, X. Kupffer cell-derived TNF-α promotes hepatocytes to produce CXCL1 and mobilize neutrophils in response to necrotic cells. *Cell Death Dis.* **2018**, *9*, 323. [CrossRef]
49. Marangoni, A.; Accardo, S.; Aldini, R.; Guardigli, M.; Cavrini, F.; Sambri, V.; Montagnani, M.; Roda, A.; Cevenini, R. Production of reactive oxygen species and expression of inducible nitric oxide synthase in rat isolated Kupffer cells stimulated by Leptospira interrogans and Borrelia burgdorferi. *World J. Gastroenterol.* **2006**, *12*, 3077–3081. [CrossRef] [PubMed]
50. Curran, R.D.; Billiar, T.R.; Stuehr, D.J.; Hofmann, K.; Simmons, R.L. Hepatocytes produce nitrogen oxides from L-arginine in response to inflammatory products of Kupffer cells. *J. Exp. Med.* **1989**, *170*, 1769–1774. [CrossRef]
51. Li, B.; Li, W.; Tian, Y.; Guo, S.; Qian, L.; Xu, D.; Cao, N. Selenium-Alleviated Hepatocyte Necrosis and DNA Damage in Cyclophosphamide-Treated Geese by Mitigating Oxidative Stress. *Biol. Trace Elem. Res.* **2020**, *193*, 508–516. [CrossRef] [PubMed]
52. Ho, C.H.; Huang, J.H.; Sun, M.S.; Tzeng, I.S.; Hsu, Y.C.; Kuo, C.Y. Wild Bitter Melon Extract Regulates LPS-Induced Hepatic Stellate Cell Activation, Inflammation, Endoplasmic Reticulum Stress, and Ferroptosis. *Evid. Based Complement. Altern. Med.* **2021**, *2021*, 6671129. [CrossRef] [PubMed]

53. Arai, M.; Mochida, S.; Ohno, A.; Ogata, I.; Obama, H.; Maruyama, I.; Fujiwara, K. Blood coagulation equilibrium in rat liver microcirculation as evaluated by endothelial cell thrombomodulin and macrophage tissue factor. *Thromb. Res.* **1995**, *80*, 113–123. [CrossRef]
54. Usui, M.; Kuriyama, N.; Kisawada, M.; Hamada, T.; Mizuno, S.; Sakurai, H.; Tabata, M.; Imai, H.; Okamoto, K.; Uemoto, S.; et al. Tissue factor expression demonstrates severe sinusoidal endothelial cell damage during rejection after living-donor liver transplantation. *J. Hepatobiliary Pancreat. Surg.* **2009**, *16*, 513–520. [CrossRef]
55. Fang, W.F.; Chen, Y.M.; Lin, C.Y.; Huang, H.L.; Yeh, H.; Chang, Y.T.; Huang, K.T.; Lin, M.C. Histone deacetylase 2 (HDAC2) attenuates lipopolysaccharide (LPS)-induced inflammation by regulating PAI-1 expression. *J. Inflamm.* **2018**, *15*, 3. [CrossRef]
56. Dai, S.; Liu, F.; Qin, Z.; Zhang, J.; Chen, J.; Ding, W.X.; Feng, D.; Ji, Y.; Qin, X. Kupffer cells promote T-cell hepatitis by producing CXCL10 and limiting liver sinusoidal endothelial cell permeability. *Theranostics* **2020**, *10*, 7163–7177. [CrossRef]
57. Zhou, R.; Li, S.; Wang, Q.; Bi, Y.; Li, X.; Wang, Q. Silencing of GDF11 suppresses hepatocyte apoptosis to relieve LPS/D-GalN acute liver failure. *J. Biochem. Mol. Toxicol.* **2024**, *38*, e23577. [CrossRef]
58. Zhao, S.; Liu, Y.; Pu, Z. Bone marrow mesenchymal stem cell-derived exosomes attenuate D-GalN/LPS-induced hepatocyte apoptosis by activating autophagy in vitro. *Drug Des. Dev. Ther.* **2019**, *13*, 2887–2897. [CrossRef]
59. Treon, S.P.; Thomas, P.; Broitman, S.A. Lipopolysaccharide (LPS) processing by Kupffer cells releases a modified LPS with increased hepatocyte binding and decreased tumor necrosis factor-α stimulatory capacity. *Proc. Soc. Exp. Biol. Med.* **1993**, *202*, 153–158. [CrossRef]
60. Merkel, S.M.; Kamoun, W.; Karaa, A.; Korneszczuk, K.; Schrum, L.W.; Clemens, M.G. LPS inhibits endothelin-1-mediated eNOS translocation to the cell membrane in sinusoidal endothelial cells. *Microcirculation* **2005**, *12*, 433–442. [CrossRef] [PubMed]
61. Brusilovskaya, K.; Hofer, B.S.; Simbrunner, B.; Eichelberger, B.; Lee, S.; Bauer, D.J.M.; Mandorfer, M.; Schwabl, P.; Panzer, S.; Reiberger, T.; et al. Platelet Function Decreases with Increasing Severity of Liver Cirrhosis and Portal Hypertension-A Prospective Study. *Thromb. Haemost.* **2023**, *123*, 1140–1150. [CrossRef]
62. Kusakabe, J.; Hata, K.; Miyauchi, H.; Tajima, T.; Wang, Y.; Tamaki, I.; Kawasoe, J.; Okamura, Y.; Zhao, X.; Okamoto, T.; et al. Complement-5 Inhibition Deters Progression of Fulminant Hepatitis to Acute Liver Failure in Murine Models. *Cell Mol. Gastroenterol. Hepatol.* **2021**, *11*, 1351–1367. [CrossRef] [PubMed]
63. Kar, E.; Kar, F.; Can, B.; Cakir Gundogdu, A.; Ozbayer, C.; Kocak, F.E.; Senturk, H. Prophylactic and Therapeutic Efficacy of Boric Acid on Lipopolysaccharide-Induced Liver and Kidney Inflammation in Rats. *Biol. Trace Elem. Res.* **2024**, *202*, 3701–3713. [CrossRef] [PubMed]
64. Berkoz, M.; Unal, S.; Karayakar, F.; Yunusoglu, O.; Ozkan-Yilmaz, F.; Ozluer-Hunt, A.; Aslan, A. Prophylactic effect of myricetin and apigenin against lipopolysaccharide-induced acute liver injury. *Mol. Biol. Rep.* **2021**, *48*, 6363–6373. [CrossRef] [PubMed]
65. Ajmal, M.R.; Yaccha, M.; Malik, M.A.; Rabbani, M.U.; Ahmad, I.; Isalm, N.; Abdali, N. Prevalence of nonalcoholic fatty liver disease (NAFLD) in patients of cardiovascular diseases and its association with hs-CRP and TNF-α. *Indian. Heart J.* **2014**, *66*, 574–579. [CrossRef] [PubMed]
66. Mou, W.L.; Chen, S.R.; Wu, Z.T.; Hu, L.H.; Zhang, J.Y.; Chang, H.J.; Zhou, H.; Liu, Y. LPS-TLR4/MD-2-TNF-α signaling mediates alcohol-induced liver fibrosis in rats. *J. Toxicol. Pathol.* **2022**, *35*, 193–203. [CrossRef] [PubMed]
67. Liang, W.; Lindeman, J.H.; Menke, A.L.; Koonen, D.P.; Morrison, M.; Havekes, L.M.; van den Hoek, A.M.; Kleemann, R. Metabolically induced liver inflammation leads to NASH and differs from LPS- or IL-1β-induced chronic inflammation. *Lab. Investig.* **2014**, *94*, 491–502. [CrossRef] [PubMed]
68. Zhao, S.; Yang, X.; He, Y.; Yu, Q.; Liu, L.M. Kupffer-Cell-Targeted Carboxylesterase 1f Knockdown Deteriorates Lipopolysaccharide/D-Galactosamine-Induced Acute Liver Failure Through Regulating Cellular Polarization in Mice. *Can. J. Gastroenterol. Hepatol.* **2024**, *2024*, 6410484. [CrossRef]
69. Li, J.; Jiang, L.; Zhao, K.; Tang, Y.; Yuan, X.; Xu, Y. Myeloid-Derived Tlr4-Trif Signaling Pathway Mediates Oxidative Stress in Lps/D-Galn-Induced Acute Liver Failure. *Shock* **2024**, *62*, 582–587. [CrossRef] [PubMed]
70. Chen, H.J.; Liang, T.M.; Lee, I.J.; Huang, Y.T.; Lin, Y.L. Scutellariae radix suppresses LPS-induced liver endothelial cell activation and inhibits hepatic stellate cell migration. *J. Ethnopharmacol.* **2013**, *150*, 835–842. [CrossRef] [PubMed]
71. Han, C.; Wu, X.; Zou, N.; Zhang, Y.; Yuan, J.; Gao, Y.; Chen, W.; Yao, J.; Li, C.; Hou, J.; et al. *Cichorium pumilum* Jacq Extract Inhibits LPS-Induced Inflammation via MAPK Signaling Pathway and Protects Rats From Hepatic Fibrosis Caused by Abnormalities in the Gut-Liver Axis. *Front. Pharmacol.* **2021**, *12*, 683613. [CrossRef]
72. Schippers, M.; Post, E.; Eichhorn, I.; Langeland, J.; Beljaars, L.; Malo, M.S.; Hodin, R.A.; Millán, J.L.; Popov, Y.; Schuppan, D.; et al. Phosphate Groups in the Lipid A Moiety Determine the Effects of LPS on Hepatic Stellate Cells: A Role for LPS-Dephosphorylating Activity in Liver Fibrosis. *Cells* **2020**, *9*, 2708. [CrossRef] [PubMed]
73. Zahner, G.; Schaper, M.; Panzer, U.; Kluger, M.; Stahl, R.A.; Thaiss, F.; Schneider, A. Prostaglandin EP2 and EP4 receptors modulate expression of the chemokine CCL2 (MCP-1) in response to LPS-induced renal glomerular inflammation. *Biochem. J.* **2009**, *422*, 563–570. [CrossRef] [PubMed]

74. Gouwy, M.; Struyf, S.; Verbeke, H.; Put, W.; Proost, P.; Opdenakker, G.; Van Damme, J. CC chemokine ligand-2 synergizes with the nonchemokine G protein-coupled receptor ligand fMLP in monocyte chemotaxis, and it cooperates with the TLR ligand LPS via induction of CXCL8. *J. Leukoc. Biol.* **2009**, *86*, 671–680. [CrossRef]
75. Roca-Ferrer, J.; Perez-Gonzalez, M.; Garcia-Garcia, F.J.; Pereda, J.; Pujols, L.; Alobid, I.; Mullol, J.; Picado, C. Low prostaglandin E2 and cyclooxygenase expression in nasal mucosa fibroblasts of aspirin-intolerant asthmatics. *Respirology* **2013**, *18*, 711–717. [CrossRef]
76. Mirabito Colafella, K.M.; Neuman, R.I.; Visser, W.; Danser, A.H.J.; Versmissen, J. Aspirin for the prevention and treatment of pre-eclampsia: A matter of COX-1 and/or COX-2 inhibition? *Basic. Clin. Pharmacol. Toxicol.* **2020**, *127*, 132–141. [CrossRef] [PubMed]
77. Brox, R.; Hackstein, H. Physiologically relevant aspirin concentrations trigger immunostimulatory cytokine production by human leukocytes. *PLoS ONE* **2021**, *16*, e0254606. [CrossRef] [PubMed]
78. Majkic, T.; Bekvalac, K.; Beara, I. Plantain (*Plantago* L.) species as modulators of prostaglandin E_2 and thromboxane A_2 production in inflammation. *J. Ethnopharmacol.* **2020**, *262*, 113140. [CrossRef]
79. van Zijverden, L.M.; Schutte, M.H.; Madsen, M.C.; Bonten, T.N.; Smulders, Y.M.; Wiepjes, C.M.; van Diemen, J.J.K.; Thijs, A. The efficacy of aspirin to inhibit platelet aggregation in patients hospitalised with a severe infection: A multicentre, open-label, randomised controlled trial. *Clin. Exp. Med.* **2023**, *23*, 3501–3508. [CrossRef] [PubMed]
80. Byrnes, J.J.; Moake, J.L. Thrombotic thrombocytopenic purpura and the haemolytic-uraemic syndrome: Evolving concepts of pathogenesis and therapy. *Clin. Haematol.* **1986**, *15*, 413–442. [CrossRef] [PubMed]
81. Tjahjono, Y.; Caroline; Foe, K.; Wijaya, H.; Dewi, B.D.N.; Karnati, S.; Esar, S.Y.; Karel, P.; Partana, F.R.; Henrikus, M.A.; et al. 2-(3-(Chloromethyl)benzoyloxy)benzoic Acid reduces prostaglandin E-2 concentration, NOX2 and NFKB expression, ROS production, and COX-2 expression in lipopolysaccharide-induced mice. *Prostaglandins Other Lipid Mediat.* **2024**, *174*, 106866. [CrossRef] [PubMed]
82. Stocks, N.P.; Gonzalez-Chica, D.A.; Woods, R.L.; Lockery, J.E.; Wolfe, R.S.J.; Murray, A.M.; Kirpach, B.; Shah, R.C.; Nelson, M.R.; Reid, C.M.; et al. Quality of Life for 19,114 participants in the ASPREE (ASPirin in Reducing Events in the Elderly) study and their association with sociodemographic and modifiable lifestyle risk factors. *Qual. Life Res.* **2019**, *28*, 935–946. [CrossRef]
83. Lockery, J.E.; Collyer, T.A.; Abhayaratna, W.P.; Fitzgerald, S.M.; McNeil, J.J.; Nelson, M.R.; Orchard, S.G.; Reid, C.; Stocks, N.P.; Trevaks, R.E.; et al. Recruiting general practice patients for large clinical trials: Lessons from the Aspirin in Reducing Events in the Elderly (ASPREE) study. *Med. J. Aust.* **2019**, *210*, 168–173. [CrossRef] [PubMed]
84. Wolfe, R.; Murray, A.M.; Woods, R.L.; Kirpach, B.; Gilbertson, D.; Shah, R.C.; Nelson, M.R.; Reid, C.M.; Ernst, M.E.; Lockery, J.; et al. The aspirin in reducing events in the elderly trial: Statistical analysis plan. *Int. J. Stroke* **2018**, *13*, 335–338. [CrossRef]
85. Margolis, K.L.; Mahady, S.E.; Nelson, M.R.; Ives, D.G.; Satterfield, S.; Britt, C.; Ekram, S.; Lockery, J.; Schwartz, E.C.; Woods, R.L.; et al. Development of a standardized definition for clinically significant bleeding in the ASPirin in Reducing Events in the Elderly (ASPREE) trial. *Contemp. Clin. Trials Commun.* **2018**, *11*, 30–36. [CrossRef] [PubMed]
86. Meng, J.; Zou, Y.; Chen, J.; Qin, F.; Chen, X.; Chen, X.; Dai, S. sTLR4/sMD-2 complex alleviates LPS-induced acute lung injury by inhibiting pro-inflammatory cytokines and chemokine CXCL1 expression. *Exp. Ther. Med.* **2018**, *16*, 4632–4638. [CrossRef] [PubMed]
87. Zou, Y.; Qin, F.; Chen, J.; Meng, J.; Wei, L.; Wu, C.; Zhang, Q.; Wei, D.; Chen, X.; Wu, H.; et al. sTLR4/MD-2 complex inhibits colorectal cancer in vitro and in vivo by targeting LPS. *Oncotarget* **2016**, *7*, 52032–52044. [CrossRef] [PubMed]
88. Yan, A.; Ding, H.; Liu, J.; Bi, C.; Han, Z.; Wang, Z.; Nawaz, S.; Shen, Y.; Liu, S. Black *Lycium barbarum* polysaccharide attenuates LPS-induced intestine damage via regulation gut microbiota. *Front. Microbiol.* **2022**, *13*, 1080922. [CrossRef] [PubMed]
89. Ayob, N.; Muhammad Nawawi, K.N.; Mohamad Nor, M.H.; Raja Ali, R.A.; Ahmad, H.F.; Oon, S.F.; Mohd Mokhtar, N. The Effects of Probiotics on Small Intestinal Microbiota Composition, Inflammatory Cytokines and Intestinal Permeability in Patients with Non-Alcoholic Fatty Liver Disease. *Biomedicines* **2023**, *11*, 640. [CrossRef]
90. Liu, Q.; Tian, H.; Kang, Y.; Tian, Y.; Li, L.; Kang, X.; Yang, H.; Wang, Y.; Tian, J.; Zhang, F.; et al. Probiotics alleviate autoimmune hepatitis in mice through modulation of gut microbiota and intestinal permeability. *J. Nutr. Biochem.* **2021**, *98*, 108863. [CrossRef]
91. Tran, C.D.; Hawkes, J.; Graham, R.D.; Kitchen, J.L.; Symonds, E.L.; Davidson, G.P.; Butler, R.N. Zinc-fortified oral rehydration solution improved intestinal permeability and small intestinal mucosal recovery. *Clin. Pediatr.* **2015**, *54*, 676–682. [CrossRef] [PubMed]
92. Yang, K.; Zhu, J.; Wu, J.; Zhong, Y.; Shen, X.; Petrov, B.; Cai, W. Maternal Vitamin D Deficiency Increases Intestinal Permeability and Programs Wnt/β-Catenin Pathway in BALB/C Mice. *JPEN J. Parenter. Enteral Nutr.* **2021**, *45*, 102–114. [CrossRef]
93. Pewan, S.B.; Otto, J.R.; Kinobe, R.T.; Adegboye, O.A.; Malau-Aduli, A.E.O. Nutritional Enhancement of Health Beneficial Omega-3 Long-Chain Polyunsaturated Fatty Acids in the Muscle, Liver, Kidney, and Heart of Tattykeel Australian White MARGRA Lambs Fed Pellets Fortified with Omega-3 Oil in a Feedlot System. *Biology* **2021**, *10*, 912. [CrossRef]

94. Deng, W.; Yang, Q.-N.; Wu, D.-T.; Li, J.; Liu, H.-Y.; Hu, Y.-C.; Zou, L.; Gan, R.-Y.; Yan, H.-L.; Huang, J.-W. Comparison of Protective Effects of Polyphenol-Enriched Extracts from Thinned Immature Kiwifruits and Mature Kiwifruits against Alcoholic Liver Disease in Mice. *Foods* **2024**, *13*, 3072. [CrossRef] [PubMed]
95. Hu, L.; Zeng, X.; Yang, K.; Peng, H.; Chen, J. n-3 polyunsaturated fatty acids improve depression-like behavior by inhibiting hippocampal neuroinflammation in mice via reducing TLR4 expression. *Immun. Inflamm. Dis.* **2022**, *10*, e707. [CrossRef]
96. Lebda, M.A.; Elmassry, I.H.; Taha, N.M.; Elfeky, M.S. Nanocurcumin alleviates inflammation and oxidative stress in LPS-induced mastitis via activation of Nrf2 and suppressing TLR4-mediated NF-κB and HMGB1 signaling pathways in rats. *Environ. Sci. Pollut. Res. Int.* **2022**, *29*, 8294–8305. [CrossRef]
97. Masutani, N.; Ogawa, H.; Soejima, H.; Okada, S.; Masuda, I.; Waki, M.; Jinnouchi, H.; Saito, Y.; Morimoto, T. Long-Term Effects of Low-Dose Aspirin on Gastrointestinal Symptoms and Bleeding Complications in Patients with Type 2 Diabetes. *Am. J. Cardiovasc. Drugs* **2024**. [CrossRef]
98. Ide, Y.; Morikawa, G.; Yoshida, K.; Takano, Y.; Kubota, K.; Okazawa, K.; Yasu, T. The risk of upper gastrointestinal bleeding associated with concomitant proton pump inhibitor administration during dual antiplatelet therapy with aspirin and prasugrel: A retrospective single-center study. *J. Pharm. Health Care Sci.* **2024**, *10*, 76. [CrossRef] [PubMed]
99. Suzuki, R.; Kozuma, Y.; Inoue, C.; Tanabe, K.; Noboruo, I.; Arao, H.; Kawaguchi, T.; Shimizu, N.; Yamamoto, T. Artificial cerebrospinal fluid restores aspirin-inhibited physiological hemostasis through recovery of platelet aggregation function. *Acta Neurochir.* **2023**, *165*, 1269–1276. [CrossRef]
100. Huang, J.; White, A.A. Biologics in Aspirin-Exacerbated Respiratory Disease and Allergic Bronchopulmonary Aspergillosis. *Immunol. Allergy Clin. N. Am.* **2024**, *44*, 673–692. [CrossRef] [PubMed]
101. Kim, C.H.; Oh, H.J.; Kim, Y.S.; Kim, Y.-L.; Chang, J.H.; Ryu, D.-R. The Effect of Aspirin on Preventing Vascular Access Dysfunction in Incident Hemodialysis Patients: A Prospective Cohort Study in Korean Clinical Research Centers for End-Stage Renal Disease (CRC for ESRD). *J. Clin. Med.* **2019**, *8*, 677. [CrossRef] [PubMed]
102. Lundberg, M.; Voutilainen, H.; Lyly, A.; Karjalainen, J.; Huhtala, H.; Laidlaw, T.M.; Lee, S.E.; Nuutinen, M.; Toppila-Salmi, S. Adult-onset asthma, allergy, and aspirin hypersensitivity associate with self-reported food avoidance. *Eur. Clin. Respir. J.* **2024**, *11*, 2347073. [CrossRef]
103. Altay, D.; Pamukcu, O.; Baykan, A.; Uzum, K.; Arslan, D. Aspirin-induced hepatotoxicity and anemia in children with acute rheumatic fever. *Turk. J. Pediatr.* **2021**, *63*, 193–199. [CrossRef]

current issues in molecular biology

MDPI

Review

Advanced In Vitro Models for Preclinical Drug Safety: Recent Progress and Prospects

Dileep G. Nair [1,2,*] and Ralf Weiskirchen [1]

1 Institute of Molecular Pathobiochemistry, Experimental Gene Therapy and Clinical Chemistry (IFMPEGKC), University Hospital Aachen, D-52074 Aachen, Germany; rweiskirchen@ukaachen.de
2 Keliomics Inc., 4640 S Macadam Ave #270, Portland, OR 97239, USA
* Correspondence: grdileep@gmail.com

Abstract: The majority of drugs are typically orally administered. The journey from drug discovery to approval is often long and expensive, involving multiple stages. A major challenge in the drug development process is drug-induced liver injury (DILI), a condition that affects the liver, the organ responsible for metabolizing most drugs. Traditionally, identifying DILI risk has been difficult due to the poor correlation between preclinical animal models and in vitro systems. Differences in physiology between humans and animals or cell lines contribute to the failure of many drug programs during clinical trials. The use of advanced in vitro systems that closely mimic human physiology, such as organ-on-a-chip models like gut–liver-on-a-chip, can be crucial in improving drug efficacy while minimizing toxicity. Additionally, the adaptation of these technologies has the potential to significantly reduce both the time and cost associated with obtaining safe drug approvals, all while adhering to the 3Rs principle (replacement, reduction, refinement). In this review, we discuss the significance, current status, and future prospects of advanced platforms, specifically organ-on-a-chip models, in supporting preclinical drug discovery.

Keywords: OOC; organ-on-a-chip; DILI; drug-induced liver injury; AI/ML; artificial intelligence and machine learning; HTS; high-throughput screening; MSFLD; metabolic dysfunction-associated fatty liver disease (MAFLD); iPSC; induced pluripotent stem cells; MPS; microphysiological systems

1. Introduction

The current drug discovery process is costly and plagued by several bottlenecks, including a prolonged timeline and a high failure rate of approximately 90% from Phase 1 trials to market. The primary reasons for this elevated failure rate are the lack of efficacy and toxicity issues that often emerge at later stages of drug development. Drug-induced liver injury (DILI) significantly contributes to these high attrition rates, sometimes leading to the withdrawal of approved drugs [1]. A major challenge in addressing these adverse effects is the inability to identify them using traditional animal models.

There is an urgent demand for efficient, human-relevant in vitro models that can screen drug candidates utilizing human-based liver systems to detect potential hepatotoxicity. These advanced models offer advantages such as high-throughput capability, cost-effectiveness, and reduced resource requirements compared to conventional in vivo approaches, all while adhering to the 3Rs principle (replacement, reduction, refinement).

Approximately 80% of best-selling drugs are orally administered [2]. Therefore, developing a combined ex vivo model that integrates gut and liver systems would be highly beneficial. By incorporating ex vivo models into a microfluidics platform, characterized by ease of control, high resolution, rapid processing, and low cost, it is possible to create a system that enhances throughput and efficiency. Additionally, individual efficient models (either gut or liver) have significant potential for studying disease progression and serving as advanced disease models for drug discovery.

However, optimizing these ex vivo organ systems involves complex procedures with numerous parameters to consider—including media composition, oxygen and nutrient gradients, growth factors/cytokines essential for proper maintenance, shear stress conditions, extracellular matrix composition, and rigidity. The integration of artificial intelligence (AI) and machine learning (ML) could play a pivotal role in this optimization process. When combined with advanced in vitro models, AI/ML can enhance preclinical studies by addressing factors such as environmental influences, lifestyle variations, genetic diversity, nutrient supply dynamics, and drug metabolism.

2. Animal Models for Predicting Toxicity

Animal models have been used to study human physiology and pathophysiology. The progress in drug discovery and development requires advancements in animal models associated with therapeutic areas. The complexity of human physiology can be mimicked to an extent using specialized animal models [3]. Additionally, toxicology studies and studies requiring multiple organ-level manipulations are typically conducted in animal models before advancing to human clinical trials. Some examples of the significance of animal models are described below.

Rat and mice models are the most commonly used animal models to investigate the overall effects of therapeutics or treatments on the body. The availability of extensive databases containing research data from animal studies is also an advantage of animal models. The development of 'humanized mice' has emerged as a major tool for accurately modeling human immune responses [4–6]. Other commonly used murine model types include genetically engineered mouse models and xenograft models [7]. These models have been successfully utilized to model human immunity, discover and develop therapeutic agents, and conduct toxicology studies. However, these models may have several interspecies incompatibilities of cytokines and cytokine receptors.

Non-human primates, due to their similarity in biochemical and phylogenetic aspects, have been extensively used for vaccine development, orthopedic surgical techniques or devices, and Parkinson's disease [8]. Studies involving primates can be long-term and have benefitted diseases like Parkinson's disease. Disease conditions such as human immunodeficiency virus (HIV), Zika virus, and human tuberculosis (TB) are modeled in primates as murine models lack the ability to become infected or mimic human pathology. Most recently, SARS-CoV-2 (COVID-19) research has extensively utilized non-human primates, including rhesus macaques, to understand the disease and develop vaccines and treatment options [9]. The availability of previous research data and the similarity of immune responses to humans make rhesus macaques a reliable model.

Over the past two decades, more than USD 28 billion per year has been spent on irreproducible preclinical research in the United States [10]. Animal testing has traditionally been employed to predict drug safety in humans during preclinical studies. However, this process is often time-consuming and costly, contributing to the high failure rates currently observed in drug development. Additionally, pharmacogenomic differences between humans and animals can lead to discrepancies in drug efficacy and toxicity.

A comprehensive review by Gail Van Norman highlights two critical decisions regarding animal testing of drug candidates: the safe tagging of a toxic drug and the toxic tagging of a beneficial drug [11]. A notable example of a misclassified drug is Vioxx (rofecoxib) from Merck, which was linked to numerous cases of myocardial infarction and stroke, resulting in over USD 8.5 billion in legal settlements [12]. Siramshetty et al. reported that nearly half of the 578 drugs withdrawn or discontinued post-approval in the United States and Europe were due to toxicity issues [13]. Accessing human-relevant data before advancing to clinical trials may provide greater confidence than relying solely on animal testing data. Positive results from preliminary screenings using humanized models can facilitate project progression that might otherwise be terminated based solely on animal data.

In light of these inherent challenges associated with animal-based data, the FDA Modernization Act 2.0 has recently been approved, allowing for alternatives to animal testing for drug and biological product applications [14]. These alternatives include advanced in vitro models such as organoids, OOC systems, and human-induced pluripotent stem cells (iPSCs), as well as AI and ML methods for assessing drug metabolism and toxicity [14]. Similarly, the European Union (EU) has implemented a complete ban on cosmetic products developed using animal models, followed by countries such as Canada, Brazil, Colombia, and Mexico [15]. The growing acceptance of alternative preclinical models has spurred efforts to develop efficient and humanized systems.

Recent advances in in vitro platforms are discussed below with an emphasis on toxicity evaluation—a significant bottleneck in drug discovery.

3. In Vitro Cellular Models for Drug Discovery

The process of drug discovery and development is complex, costly, and time-consuming, typically taking between 10 and 15 years and requiring billions of dollars [16] (Figure 1). In vitro cellular models play a crucial role in high-throughput screening (HTS), serving as disease models for efficacy assessments and safety evaluations but requiring further advancements [17]. Over the years, these models have evolved into rapid and reproducible tools for studying both efficacy and toxicity. Below are some of the major applications and advancements in this field. Elfawal et al. recently reported a high-throughput screening platform where over 30,000 compounds were screened to identify 55 compounds with broad anthelmintic activity against gastrointestinal nematodes [18].

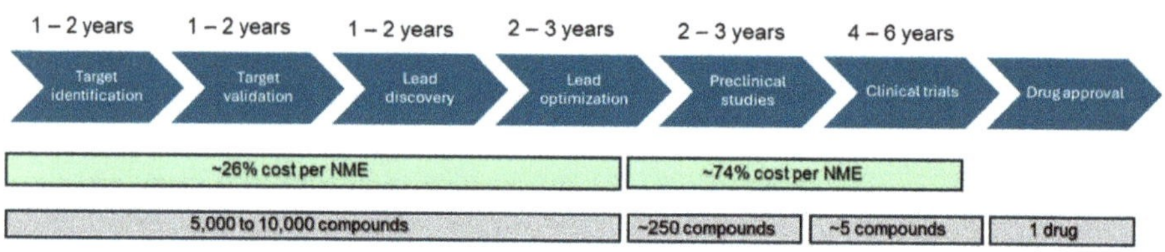

Figure 1. Schematic representation of the drug discovery and development pipeline. The first step is target identification, followed by target validation. Lead discovery is the next step and is generally pursued using high-throughput screening. Lead optimization is carried out before preclinical studies. The drug candidates selected from preclinical studies are taken to clinical trials, and a successful drug is approved for clinical usage.

3.1. *In Vitro Cellular Models for High-Throughput Screening*

Cell-based in vitro models are widely used for screening and account for nearly half of HTS efforts, especially for orally available drugs. Unlike biochemical or acellular assays, cellular assays reflect molecular interactions within a cellular environment, involving multiple biochemical pathways and their interactions and effects. Furthermore, cellular assays provide deeper insights into toxicity profiles, impacts on various signaling pathways or receptors, and overall cellular effects.

Compared to in vivo counterparts, cell-based in vitro assays offer advantages in scalability, cost-effectiveness, and reproducibility. Establishing effective in vitro assays can also reduce the number of animals used and sacrificed to obtain data related to drug efficacy. By using optimized in vitro screening methods, thousands of compounds can be evaluated using fluorometric, luminescent, and colorimetric platforms across 384-well to 1036-well formats. HTS methodologies typically involve miniaturized assays with minimal reagent requirements and have significant potential for multiparameter measurements from individual wells [19].

However, major challenges include optimizing assay conditions to develop and validate effective protocols that meet statistical acceptance criteria. With advancements in AI, liquid handling technologies, as well as enhanced readouts and data processing capabilities, the development of assays targeting novel pathways or approaches has become increasingly common. Using cell-based HTS platforms allows researchers to screen large numbers of compounds to identify the most active molecules, conduct hit-to-lead studies, and assess potential stability and bioavailability using advanced in vitro models as described in the sections below.

3.2. Drug Toxicity Testing Using In Vitro Models

Precision in determining the toxicity of drug candidates is crucial for their development, approval, and ultimate use. Incorporating toxicity assessments during key stages of drug discovery, such as hit-to-lead optimization, structure-activity relationship (SAR) studies, and ADME (absorption, distribution, metabolism, and excretion) processes, can help minimize costs and reduce potential failure risks [20]. Traditional drug toxicology approaches include computational models, biochemical assays, cellular in vitro assays, and in vivo studies. While computational screening is commonly used for initial HTS projects during the drug discovery phase, it may lack comprehensive outcomes on a global physiological scale compared to advanced cellular assays. Specialized cell-based assays are used to identify organ-specific toxicities.

Routine 2D cell cultures can offer basic insights into cytotoxicity. However, they often do not replicate the tissue-specific mechanical and biochemical characteristics of target organs. Current in vitro models for assessing liver hepatotoxicity utilize various cell sources—including genetically engineered cells, hepatoma cell lines, stem cell-derived hepatocyte-like cells, primary hepatocytes, and co-cultures of hepatocytes with non-parenchymal cells (NPCs). While co-cultures of hepatocytes and NPCs provide the most physiologically relevant model, maintaining consistency can be challenging [21]. Integrating a co-culture model into a 3D culture system could enhance predictions of in vivo hepatotoxicity and hepatic clearance of orally available drugs in a high-throughput manner.

To address these challenges, organ-specific cell-based assays, including co-cultures, are increasingly used for HTS assays. However, incorporating more physiological parameters such as extracellular matrix components and organ-specific cell-to-cell interactions can significantly improve the identification of potential toxicity. In both preclinical and clinical phases, toxicology studies focus on risk identification, management, and mitigation. The findings from these studies inform proper clinical trial design and execution.

Identification of an active and selective compound is not enough to consider it a potential candidate for drug development. Poor pharmacokinetics and toxicological profiles are responsible for the majority of drug program failures (63% of all preclinical candidates) [22,23]. Absorption, distribution, metabolism, and excretion (ADME) studies are crucial in drug development, with 40% of drug candidates failing solely due to poor human pharmacokinetics before formal implementation in the drug discovery process [24]. Several ADME assays are available in high-throughput formats and are typically initiated in the early stages of discovery. Some examples include cytochrome inhibition assays using non-fluorescent probe substrates [25] and Caco-2 cell permeability assays using LC-MS (liquid chromatography/mass spectrometry) [26]. Recent approaches to expedite ADME data collection include automated sample preparation, fast chromatography, automated data processing, and sample pooling [23].

Perception of structure–activity relationships (SARs) data is crucial for drug development. Small molecule profiling typically involves characterizing compounds through multiple biological measurements [27]. SAR studies aid in understanding the mechanism of action and provide detailed information on biological effects [27]. Data can be obtained by incorporating information from various assays such as binding studies, cell-line studies for drug sensitivity, and other high-throughput assays [28–30]. The availability of public databases like Binding DB, ChEMBL, and ChemBank helps to expedite and streamline the process [31,32]. Furthermore, advancements in high-throughput assay platforms, like high-throughput gene expression signature analysis and multiplex cytological profiling assays for measuring diverse cellular states, have enabled the characterization of SAR analysis in a high-throughput manner [33]. Progress in computational methods also plays a significant role in deducing SAR data from high-dimensional profiling data [27].

Recently, there has been a strong push to establish human-relevant ex vivo models due to the limitations associated with traditional animal models. Some notable advanced in vitro systems include 3D cell culture models such as organoids and OOC or microphysiological systems (MPSs).

3.3. *Advantages of Complex Cell Culture Models for Drug Discovery and Development*

Advanced 3D cell culture models can effectively maintain cell differentiation, as well as RNA and protein expression levels that closely resemble those found in actual organs [34]. These models are widely utilized in drug discovery projects and encompass various formats, including spheroids, organoids, and individual cells suspended within an extracellular matrix [35].

Spheroids are multicellular cultures compatible with a range of cell types, including stem cells, neuronal cells, epithelial cells, and hepatocytes. The advantages of spheroid cultures include their capacity to develop co-cultures and their ability to replicate gradients of nutrients, oxygen, and metabolites. Additionally, they facilitate interactions between cells as well as between cells and the extracellular matrix, mimicking physiological or pathophysiological conditions [36].

Organoids consist of organ-specific cell types and can be classified as tissue-derived or stem cell-derived organoids based on their origin. Stem cells used in organoid formation may be either organ-restricted adult stem cells or pluripotent induced/embryonic stem cells. Generally, organoids contain both stem cells and differentiated cells, allowing them to generate multicellular structures that exhibit tissue-level architecture and function.

Intestinal organoids have significant applications, particularly in oral drug screening and delivery, studying gut microbiome interactions, disease modeling, and developmental research (Figure 2).

One of the earliest studies conducted by Sato et al. in 2009 demonstrated that the leucine-rich repeat-containing G protein-coupled receptor 5 (Lgr5) stem cells possess the ability to develop crypt-villus structures with native tissue-like functions [37]. Qu and his team developed an efficient intestinal organoid model for injury-induced intestinal regeneration, which has the potential to identify candidates that could promote this regeneration [37]. Mature epithelial cell differentiation into enterocytes, goblet cells, Paneth cells, and enteroendocrine cells is associated with the expression of specific markers such as villin, Mucin 2, lysozyme, and chromogranin A, respectively [38].

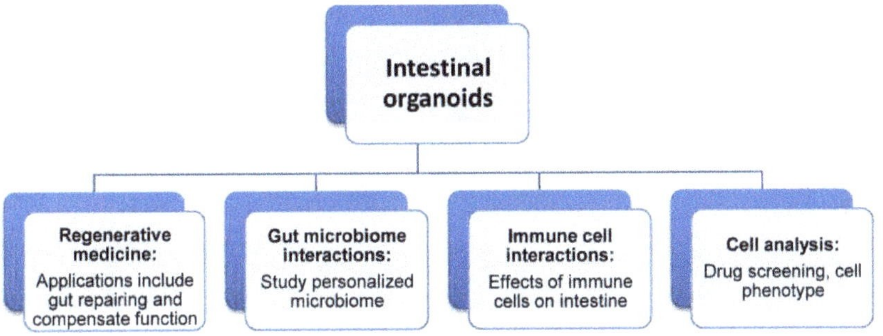

Figure 2. Uses of intestinal organoids. Intestinal organoids have a variety of common applications, including replacing or compensating for functional structures, mimicking host–microbe interactions, studying interactions with immune cell types, and investigating drug toxicity or developmental biology.

Liver organoids can be derived from both embryonic or pluripotent stem cells and adult stem cells. Cholangiocyte organoids, which are based on cholangiocytes that line the biliary tree, play a role in parenchymal fibrosis. Huch and his team have successfully established cholangiocyte organoids from both human and mouse sources [39–41]. These organoids are valuable for studying the molecular aspects of cholangiopathies and for developing personalized therapeutic strategies.

Hepatocytes constitute the majority of liver cells, and hepatic organoids have been utilized to model various liver diseases [42]. Regenerative hepatocytes resulting from liver injury or other pathophysiological conditions can repopulate to compensate for cell loss. The advantages of liver organoids include their potential for gene editing, biobanking, and HTS. Personalized or patient-centric organoids can be created to recreate specific pathophysiological conditions. Both pluripotent stem cell (PSC)-derived and adult stem cell (ASC)-derived organoids are widely used in various studies [43].

Marsee and her team reported a systematic nomenclature for organoids, including hepatic organoids categorized by their source cells [44]. The secretion of albumin and the activity of cytochromes are commonly used criteria to assess the mature hepatic phenotype. However, one major challenge in establishing these organoids is the lengthy cell differentiation process, along with complex media requirements that can affect the functional lifespan of the models and lead to batch-to-batch variability [45]. Recently, Kamishibahara et al. reported using Laminin-511 (LN) direct coating technology to successfully differentiate human induced pluripotent stem cells (hiPSCs) into hepatocytes, yielding functional hepatic organoids [46]. Nevertheless, more complex systems with more human relevance are needed to detect efficacy and potential toxicity in drug candidates (Figure 3).

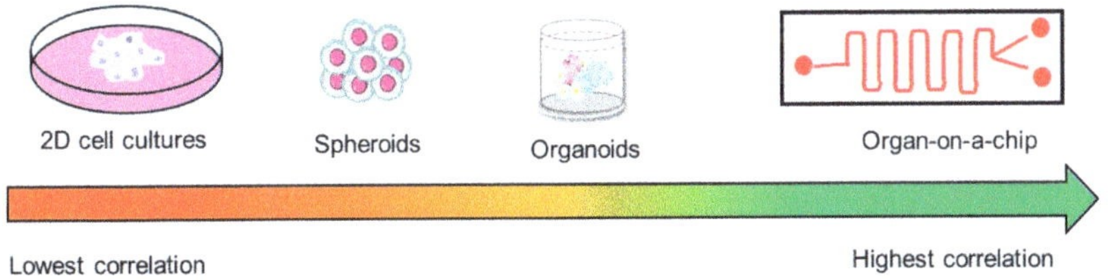

Figure 3. Clinically relevant in vitro models that demonstrate increased correlation to human biology. This progression is evident through the advancement from 2D cell culture models to 3D models like spheroids or organoids and ultimately to organ-on-a-chip models.

3.4. Organ-on-a-Chip Platform for Drug Development

Organ-on-a-chip (OOC) technology enables the modeling of complex organs using a microfluidic design that incorporates cellular microenvironments, simulating the physiological, mechanical, and chemical properties of the tissue of origin. This innovative technology has the potential to address many challenges associated with traditional cell culturing and in vivo models while providing greater insights into human physiology at both single-organ and multi-organ levels. Key components involved in OOC systems include microfluidics, cellular microenvironments, cell models, and sensing elements. The successful integration of these components is crucial for the efficiency and robustness of OOC systems.

Since the pioneering work by Huh et al. in 2010 on the 'breathing lung' model, diverse OOC models for various organs—including lung, liver, brain, gut, and heart—have been reported [47,48]. Multi-organ OOCs that simulate physiological interactions between two or more organs are also being explored. Major applications of OOC technology include drug safety evaluation, pharmacokinetic/pharmacodynamic (PK/PD) modeling, disease modeling, and personalized medicine.

Approximately 80% of drugs are administered orally, with the liver playing a critical role in metabolizing most of these compounds. Traditional 2D cell cultures have been extensively utilized to study drug availability; Caco-2 monolayers are particularly popular due to their simplicity and efficiency in analyzing transepithelial drug transport [49]. The inclusion of a mucus layer is often achieved by co-culturing with goblet-cell-like HT29-MTX cells [49]. However, the absence of other intestinal epithelial cells—such as Paneth cells and endocrine cells—along with differences in transporter expression and enzyme distribution can lead to inconsistent data compared to in vivo results [49].

Intestinal organoids have emerged as a promising next step in drug development; however, many models lack fluid flow, oxygen gradients, and mechanical forces necessary for physiological relevance. Gut-on-a-chip models can overcome these limitations by utilizing microfluidic platforms that enable gut-like functions. Cremer et al. reported an advanced in vitro model that mimics gut microbiota under physiological flow conditions [50]. By incorporating these additional layers of functionality, it may be possible to develop models that yield more reliable predictions regarding drug absorption and metabolism as well as disease pathology and developmental studies.

Liver-on-a-chip systems can replicate complex properties of the liver more effectively than hepatocyte organoids by addressing issues such as polarization deficiency while enhancing oxygen/nutrient supply and metabolic activity through shear force simulation. Using pumps and microfluidic channels allows for the recreation and maintenance of oxygen gradients and glucose concentrations while preserving appropriate cell-to-cell ratios and metabolite concentrations. Stern et al. recently reviewed the significance of OOC liver models for addressing idiosyncratic drug-induced liver injury (iDILI), which involves complex disease development processes that are generally challenging to replicate [51]. By incorporating non-parenchymal cells such as Kupffer cells, stellate cells, and dendritic cells into these model systems, hepatic inflammation—a key feature of iDILI—can be effectively modeled [51].

Haque et al. demonstrated that hepatocytes cultured within low-volume microfluidic channels maintained their differentiated state for up to 21 days compared to just 7 days in conventional cultures. Additionally, cells within these microfluidic environments produced hepato-inductive signals alongside pro-fibrogenic factors that sustained their phenotype and function [52].

A combined gut and liver-on-a-chip model can effectively mimic in vivo conditions, where there is a direct link between the absorption of drugs in the gut and their metabolism in the liver. Personalized models incorporating both gut and liver OOC systems can enhance toxicity studies, paralleling clinical trial conditions. Additionally, these models allow for the collection of cells or co-cultures for detailed genomic or proteomic analyses, providing further insights into toxicity profiles. The significance of growth factors and cytokines is often underestimated in many ex vivo disease or organ models. For example, growth factors such as platelet-derived growth factor (PDGF) and various cytokines have been shown to influence gastrointestinal inflammatory diseases and hepatic fibrosis. Therefore, including similar factors can help replicate physiological or pathophysiological conditions [53]. Regulating the supply of these factors will contribute to developing a sustainable OOC platform.

In addition to cell source, cell type, and circulating media, the extracellular matrix (ECM) used for cell support and the materials employed for the chip are critical determinants of successful OOC models. The priority lies in using a biocompatible ECM that sustains cells while maintaining their native physiological characteristics across multiple organs. Pairing this with a biocompatible chip interface, such as chitin or alginate, that possesses antibacterial and anti-inflammatory properties while facilitating the delivery of growth factors and cytokines can be highly beneficial [54].

Yang et al. reported on an integrated gut–liver-on-a-chip platform designed to model metabolic dysfunction-associated fatty liver disease (MAFLD) [55]. In this study, Caco-2 cells representing gut epithelial cells were connected via microfluidic channels to HepG2 cells mimicking liver cells. The gut–liver axis was shown to have a profound effect on MAFLD development. Free fatty acids (FFAs) were administered to the system over 1 and 7 days to initiate and progress MAFLD [55]. Furthermore, this model provided insights into MAFLD-like gene expression patterns, suggesting its potential use as an alternative to animal models in drug discovery (Figure 4).

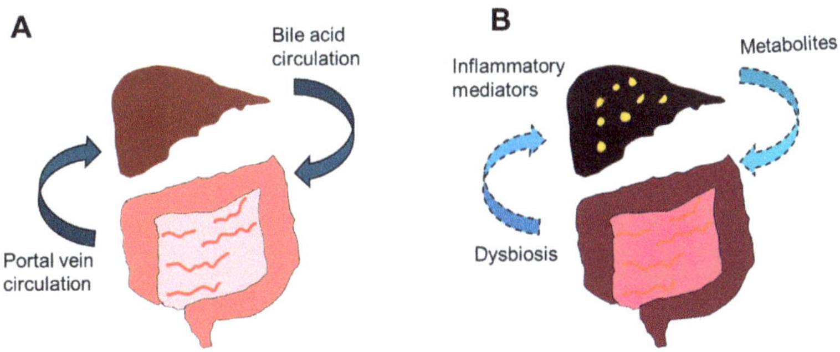

Figure 4. Gut–liver axis during normal conditions compared to conditions with metabolic dysfunction-associated fatty liver disease (MAFLD). (**A**) In normal conditions, there is bidirectional communication between the intestines and liver, known as the gut–liver axis, which helps maintain homeostasis. (**B**) However, under MAFLD conditions, factors originating in the gut, such as dysbiosis, leaky gut, and inflammatory cytokines, can impact the liver. Additionally, altered bile acid metabolites can affect gut function. An efficient organ-on-a-chip model can be used to represent this scenario [50].

3.5. Significance of Extracellular Matrix (ECM) on OOC and Similar Platforms

The ECM plays a major role in regulating cellular processes through structural support, biochemical signaling, and cellular signal regulation. It promotes healing and preserves tissue balance, which is important for the proper functioning of cells, tissue engineering, and developing ex vivo models [56]. Polysaccharide hydrogels are widely used to mimic the ECM due to their biocompatibility and similarity to the natural ECM. Tissue constructs can now be fabricated using controlled biomaterial composition, shape, and cellular components through 3D bioprinting. The hydrogels are sourced from diverse sources, including marine, plant/seed, starch, pectin, bacteria, and animal/human origins. They have differences in polysaccharide composition based on their types and origins and are being extensively characterized to improve tissue-like properties [56].

Similar to hydrogel-based ECM, a collagen-based scaffold has been used to model difficult-to-establish tumors for studying disease pathology and drug discovery. Miserocchi et al. reported a biomimetic collagen-based scaffold for oropharyngeal squamous cell carcinomas (OSCCs) and studied the impact of the ECM on the phenotypic and genotypic properties of the cells [57]. They also studied the impact of the scaffold on the cell line and primary cell responses to standard of care drugs [57]. De Vita et al. used collagen-based 3D models to characterize soft tissue sarcomas (STSs) and the activity of trabectedin on them [58]. They used primary cells derived from 10 patients and established a 3D culture system using collagen to analyze the activity of trabectedin on the cells. The models showed high morphological and genomic similarity to in vivo samples [58]. Broen et al. used primary human white adipose tissue sandwiched between adipose-derived stem cells to create a microphysiological system (MPS) to establish breast cancer models [59]. The sandwich model contained native ECM, mature adipocytes, fibroblasts, and immune cells and was stable for up to 14 days for studying cancer–ECM interactions and for drug-screening studies [59].

3.6. Recent Progress in Organ-on-a-Chip Platforms

Deng et al. recently published an extensive review on OOC platforms designed to model complex female reproductive pathophysiological conditions [60]. Compared to existing 2D models and in vivo systems, OOC platforms demonstrate the ability to replicate various physiological and pathophysiological states. Some notable models include placenta-on-a-chip, which represents placental barrier function; OOCs that study the invasive capabilities of placental trophoblast cells; and OOCs designed for investigating endometriosis and cancer [60].

Recent advances in 3D printing technology have significantly impacted the fabrication of microfluidic chips and tissue construction using bioinks. For instance, 3D printing technology has been explored for personalized treatments of Crohn's disease [61]. Common materials used for 3D printing microfluidic chips include polydimethylsiloxane (PDMS), hydrogels, and polyethylene terephthalate (PET) [62]. The selection of materials is critical for establishing efficient OOC systems. Modified resins with no residual cytotoxicity, ideal transparency, and biocompatibility are employed in constructing OOC systems [63].

Many OOC projects utilize PDMS due to its cellular compatibility, high transparency, and low cost; however, it also presents drawbacks such as poor printability along with hydrophobic and porous characteristics. Villegas et al. demonstrated the application of a lubricant-infused coating on PDMS molds using fluorinated silane to reduce surface roughness and enhance optical properties [64]. Considerable progress has also been made with alternative materials like PET and thermoplastic polyurethane (TPU) for creating microfluidic chips through direct or indirect fabrication methods tailored to specific system needs.

Bioinks—comprising biomaterials, cells, and bioactive compounds—are essential for bioprinting tissues onto microfluidic chips in a process known as 3D bioprinting. Natural hydrogels commonly used as bioinks include gelatin, chitosan, and decellularized extracellular matrix-derived materials (dECMs) [62]. Widely adopted 3D bioprinting techniques encompass inkjet-based, extrusion-based, or digital light/laser-polymerization methods [65]. Cell sources can be derived from stem cells, such as induced pluripotent stem cells (iPSCs), cell lines, or cancer cells; diverse organ-specific OOC platforms have been established.

Kang et al. reported successful 3D bioprinting of liver OOC using extrusion bioprinting technology to construct hepatic lobular arrays that incorporate vascularization alongside various hepatic cell types [66]. The extruded hepatic lobules were functional while maintaining structural integrity, demonstrating potential applications in liver tissue engineering. Ma et al. showcased rapid 3D bioprinting of liver tissue, utilizing a hydrogel-based triculture that included hiPSC-derived hepatic cells along with vascular endothelial cells and adipose-derived stem cells [67]. This generated tissue exhibited high liver-specific gene expression levels alongside increased cytochrome P450 production and metabolite generation, underscoring its significance for drug screening and disease modeling experiments.

In addition to incorporating a vascular system into their models, Lee et al. integrated a biliary system into their bioprinted liver OOC platform. A decellularized ECM bioink was utilized to establish a 3D microenvironment featuring both vascular and biliary fluidic channels; this approach resulted in a model displaying liver-specific gene expression profiles and functional capabilities [68]. Furthermore, this model demonstrated effective drug responses, suggesting its potential utility for drug screening assays.

There has been enormous progress in the OOC platforms with several new organotypic models emerging [69]. Some of the major OOC models established for organ-level research include a second-generation lung-on-a-chip model [70]. The model reconstitutes the air–blood barrier using a biological, biodegradable, and stretchable membrane made of elastin and collagen, along with primary lung alveolar epithelial cells and primary lung alveolar endothelial cells. The generated biomembrane reproduced the composition and in vivo functionality of the lung alveolar barrier, and these properties were preserved for up to 3 weeks. Ross et al. described a novel microfluidic platform called the lymph node slice-on-a-chip to analyze local events in the lymph node, a structurally complex organ with high significance for human health. The lymph node slices maintained cellular organization, were accessible for local stimulation, and have the potential for the development of novel immunotherapies [71].

One of the premier studies in developing intestinal OOCs was conducted by Sung et al., who created the first 3D hydrogel-based platform to simulate human intestinal villi [72]. They utilized a combination of laser ablation and sacrificial molding techniques with calcium alginate to fabricate complex structures. Through this technology, they successfully produced microscale collagen structures that mimic human intestinal villi. In another significant study, Jalili-Firoozinezhad et al. established a complex human gut microbiome in an anaerobic intestine-on-a-chip [73]. The composition and function of the human microbiome, particularly the gut microbiome, have a profound impact on human health [74]. The microfluidic intestine OOC developed by Jalili-Firoozinezhad et al. allowed for the prolonged co-culture of human intestinal epithelium with stable communities of aerobic and anaerobic human gut microbiota. By creating a hypoxia gradient in the OOC, they were able to enhance barrier function and maintain a diverse microbial population, including over 200 unique taxonomic units from 11 different genera [73]. This OOC, which closely resembles the human gut, has the potential to aid in predicting drug absorption, developing probiotics, nutraceuticals, and microbiome-related therapeutics.

The filtration and reabsorption functions of kidney cells, or nephrons, consisting of the glomerulus, renal capsule, and renal tubule, were successfully modeled by several groups. Jang et al. established a kidney OOC lined by living human epithelial cells exposed to fluidic flow [75]. The microfluidic chamber mimics an in vivo system with a luminal channel and interstitial space, showing enhanced epithelial cell polarizations, primary cilia formation, increased albumin transport, and glucose reabsorption. Similarly, Sakolish et al. reported a kidney-based reusable microfluidic device [76]. The usage of human-induced pluripotent stem cells to successfully generate human islet organoids was reported by Tao et al. [77].

Zhang et al. reported a 3D tissue-engineered cell culture model on a chip for the heart, aimed at determining cardiac drug efficacy. This model utilized fibrin-based media and a high-speed impedance detection technique [78]. Schneider et al. demonstrated the use of heart OOCs with pluripotent stem cells [79]. The system is user-friendly and can generate multiple physiologically relevant tissues, with a simple readout using bright-field video microscopy. Some of the studies establishing OOC models are summarized below in Table 1.

Table 1. Some of the organ-on-a-chip studies summarized below with the ECM/platforms used.

Sl No.	Organ Focused	ECM/Platform Used	Reference
1	Lung	Elastin and collagen membrane	[70]
2	Lymph node	Tissue slice-on-a-chip	[71]
3	Intestine	Hydrogel-based	[72]
4	Gut (anaerobic microbiome)	Matrigel and collagen	[74]
5	Kidney	Microfluidic flow systems	[75–77]
6	Heart	Fibrin/impedance/microscopy	[78,79]

Multi-OOC platforms involve cells from different organs cultured simultaneously and connected by a fluidic network. These platforms are useful for recreating the complexity, functions, and integrity of organ functions [80]. Tsamandouras reported a two-organ gut–liver OOC model using a novel mesofluidic multi-MPS platform coupled with computational modeling to connect a gut MPS with a liver MPS. A three-tissue OOC was reported by Skardal et al., where a liver, heart, and lung OOC was established with relative proportions [81]. The multi-organ OOC was used for drug and toxicology screening. Furthermore, multi-organ OOC models involving four to ten organs were reported [82,83].

4. Future Prospects

One of the major bottlenecks in the profitable commercialization of advanced in vitro models, such as spheroids or OOC platforms, is the lack of automated systems to support these processes. Recently, Lindner et al. demonstrated the development of an automated system utilizing affordable components, including a 3D bioprinter, to produce perfusion-ready OOC devices [84]. In comparison to existing alternatives like the BioAssembly Platform by Advanced Solutions Life Sciences and HUMIMIC AutoLab by TissUse GmbH, Berlin, Germany the system presented by Lindner et al. is notably adaptable, scalable, transferable, and customizable [85].

To establish robust hepatic screening platforms for assessing aflatoxin B1 toxicity, Schmidt et al. employed a mixture of gelatin, alginate, and Matrigel to maximize cell viability, printability, and stability over a duration of 21 days [86]. They utilized 3D-printed liver spheroids with in situ image-based quantification for screening drug-induced toxicity [81]. HepG2 liver spheroids were cultivated in mini-fabricated hydrogels using a 3D printer and analyzed for drug-induced mitochondrial permeability transition (MPT), apoptosis, and cytosolic calcium levels. Advances in imaging technologies combined with embedded sensors enhanced by AI/ML capabilities may facilitate the acquisition of multi-parametric data at higher resolutions for improved toxicity prediction.

Incorporating vascular networks into OOC platforms is critical for ensuring an adequate supply of nutrients and oxygen; however, it presents significant challenges. In liver OOCs specifically, adding bile-transporting microtubes or ducts has been shown to enhance physiological functions. Tronolone et al. employed deep learning techniques to analyze chemical and mechanical factors influencing MPS architecture in previously established models [87]. A chained neural network was trained using common morphological metrics to generate a vascular network quality index (VNQI), which can be applied across various MPS platforms. Furthermore, the VNQI algorithm was validated using vascularized islet chips.

Most OOC models rely on fluorescence image-based data outputs; however, data processing can be tedious, time-consuming, and prone to errors. Recent advances in AI technology now make efficient high-throughput data analysis feasible. A summary of recent reports on OOC or MPS platforms integrated with AI/ML technologies is provided below in Table 2.

Table 2. Examples of organ-on-a-chip integrated with artificial intelligence for data analysis in various applications.

Sl No.	Platform Tested	Target/Therapeutic Area	Reference
1	Microscopic images	Red blood cell deformability	[88]
2	2D time-lapse imaging	cancer cell cycle dynamics, motility, etc.	[89]
3	Lens-free imaging and quantitative agglutination	SARS-CoV-2 sensing	[90]
4	Label-free efficacy testing using tumor spheroids	Tumor spheroids for cancer	[91]
5	Live cell imaging	Brain metastasis	[92]
6	Smartphone-based imaging	Fentanyl detection	[93]

Molecular mechanisms of drug toxicity include the generation of reactive oxygen species (ROS), apoptosis or necrosis pathway activation, alanine aminotransferase (ALT) secretion, and cytochrome P450 enzymatic activity, and metabolite profiling can be explored more for establishing predictive toxicity modeling. Iachetta et al. recently demonstrated a non-invasive optoporation-based chronic cardiotoxicity assessment using iPSC-derived cardiomyocytes [94]. Compared to traditional models, the optoporation-based method was able to read for up to 35 days in vitro [94]. Real-time monitoring of liver fibrosis development was carried out using embedded electrochemical sensors [95]. The microfluidic system was established using a 3D printer with TEER (trans-epithelial electrical resistance) and ROS sensors incorporated using chemical vapor deposition (CVD) and 3D printing technology. Liver fibrosis was induced by adding TGF-β1, resulting in the decrease in the TEER value and increase in ROS production. The liver MPS also helped to optimize ECM for ideal functioning. Similarly, Wang et al. reviewed kidney-on-a-chip models incorporating mechanical stimulation and integrated sensors, especially for measuring TEER [96]. Several molecular pathways contribute to the complexity of cancer, and tracking these complex pathways is cumbersome for establishing tumor-on-a-chip platforms. However, with the recent progress in the field of fluorescent nano- and microparticles capable of sensing diverse microenvironments, we can support developing efficient platforms. Fluorescent ratiometric nanoparticles are capable of sensing pH, oxygen content, ROS, and inorganic cations and anions like Ca^{2+}, Na^{+}, K^{+}, and Cl^{-}, and can serve as probes for biomarkers like matrix metalloprotease-2 (MMP-2) [97]. This fluorescent dye-based tracking can benefit not only the cancer field but also other therapeutic areas, including diagnostics platforms, fibrotic diseases, and inflammatory diseases. The merging of advanced molecular biology and biochemistry technologies, like multi-omics, including single-cell analysis, advanced imaging platforms, and mass spectrometry, is needed to track parameters like metabolites, cellular phenotype, and cell health in real time to develop efficient and predictable in vitro models.

5. Conclusions

Drug discovery is a lengthy and costly process, traditionally relying on animal models for toxicity and efficacy studies, which significantly contribute to drug failures due to their inherent limitations. However, recent advancements in in vitro models, particularly organoids and OOC systems, coupled with 3D bioprinting and AI/ML technologies, represent a significant leap forward in this field. These innovative platforms offer the potential to closely mimic human physiological and pathophysiological conditions, thereby enhancing the relevance of toxicity and efficacy studies while minimizing reliance on animal testing. Given that the majority of drugs are administered orally, developing robust gut–liver OOC models that accurately reflect physiological aspects, mechanistic inputs, and human-relevant cellular phenotypes can have a substantial impact on drug discovery processes by providing insights that align more closely with human biology. Despite challenges associated with incorporating complex features such as vascular networks, extracellular matrix components, and standardization of OOC systems, ongoing research continues to address these issues [98]. The successful application of deep learning techniques to analyze microphysiological system architecture, along with advances in sensing molecular pathways involved, exemplifies the progress being made in this area. As we move toward a future where personalized medicine becomes increasingly attainable, the development of advanced in vitro models will play a crucial role in transforming drug discovery into a more efficient, ethical, and effective process. Regulatory measures, patient consent, data security, and safety measures should be incorporated before implementing AI technologies on patient samples to create personalized OOC models [99]. Strict and far-reaching policies with proper training datasets to avoid bias and discrimination will minimize the issues associated with AI and can increase patient confidence to pursue similar research in an efficient way [100].

Author Contributions: Conceptualization, D.G.N.; resources, R.W.; writing—original draft preparation, D.G.N.; writing—review and editing, D.G.N. and R.W.; visualization, D.G.N.; supervision, R.W.; project administration, R.W.; funding acquisition, R.W. All authors have read and agreed to the published version of the manuscript.

Funding: R.W. is supported by grants from the German Research Foundation (projects WE2554/13-1, WE2554/15-1, and WE2554/17-1), the Deutsche Krebshilfe (grant 70115581), and the Interdisciplinary Centre for Clinical Research within the Faculty of Medicine at the RWTH Aachen University (grant PTD 1-5). The APC was funded by the RWTH Aachen University.

Institutional Review Board Statement: Not applicable.

Informed Consent Statement: Not applicable.

Data Availability Statement: No new data were created or analyzed in this study. Data sharing is not applicable to this article.

Conflicts of Interest: Author Dileep G. Nair was employed by the company Keliomics, Inc. The remaining authors declare that the research was conducted in the absence of any commercial or financial relationships that could be construed as a potential conflict of interest.

References

1. Serras, A.S.; Rodrigues, J.S.; Cipriano, M.; Rodrigues, A.V.; Oliveira, N.G.; Miranda, J.P. A Critical Perspective on 3D Liver Models for Drug Metabolism and Toxicology Studies. *Front. Cell Dev. Biol.* **2021**, *9*, 626805. [CrossRef] [PubMed]
2. Alqahtani, M.S.; Kazi, M.; Alsenaidy, M.A.; Ahmad, M.Z. Advances in Oral Drug Delivery. *Front. Pharmacol.* **2021**, *12*, 618411. [CrossRef] [PubMed]
3. Mukherjee, P.; Roy, S.; Ghosh, D.; Nandi, S.K. Role of Animal Models in Biomedical Research: A Review. *Lab. Anim. Res.* **2022**, *38*, 18. [CrossRef] [PubMed]
4. Yu, H.; Borsotti, C.; Schickel, J.-N.; Zhu, S.; Strowig, T.; Eynon, E.E.; Frleta, D.; Gurer, C.; Murphy, A.J.; Yancopoulos, G.D.; et al. A Novel Humanized Mouse Model with Significant Improvement of Class-Switched, Antigen-Specific Antibody Production. *Blood* **2017**, *129*, 959–969. [CrossRef]
5. Herndler-Brandstetter, D.; Shan, L.; Yao, Y.; Stecher, C.; Plajer, V.; Lietzenmayer, M.; Strowig, T.; de Zoete, M.R.; Palm, N.W.; Chen, J.; et al. Humanized mouse model supports development, function, and tissue residency of human natural killer cells. *Proc. Natl. Acad. Sci. USA* **2017**, *114*, E9626–E9634. [CrossRef]
6. Rongvaux, A.; Willinger, T.; Martinek, J.; Strowig, T.; Gearty, S.V.; Teichmann, L.L.; Saito, Y.; Marches, F.; Halene, S.; Palucka, A.K.; et al. Development and function of human innate immune cells in a humanized mouse model. *Nat. Biotechnol.* **2014**, *32*, 364–372. [CrossRef]
7. Wagar, L.E.; DiFazio, R.M.; Davis, M.M. Advanced model systems and tools for basic and translational human immunology. *Genome Med.* **2018**, *10*, 73. [CrossRef]
8. Andersen, M.L.; Winter, L.M.F. Animal Models in Biological and Biomedical Research—Experimental and Ethical Concerns. *An. Acad. Bras. Cienc.* **2019**, *91*, e20170238. [CrossRef]
9. Chang, M.C.; Hild, S.; Grieder, F. Nonhuman Primate Models for SARS-CoV-2 Research: Consider Alternatives to Macaques. *Lab Anim.* **2021**, *50*, 113–114. [CrossRef]
10. Freedman, L.P.; Cockburn, I.M.; Simcoe, T.S. The Economics of Reproducibility in Preclinical Research. *PLoS Biol.* **2015**, *13*, e1002165. [CrossRef]
11. Van Norman, G.A. Limitations of Animal Studies for Predicting Toxicity in Clinical Trials. *JACC Basic Transl. Sci.* **2019**, *4*, 845–854. [CrossRef] [PubMed]
12. Jüni, P.; Nartey, L.; Reichenbach, S.; Sterchi, R.; Dieppe, P.A.; Egger, M. Risk of Cardiovascular Events and Rofecoxib: Cumulative Meta-Analysis. *Lancet* **2004**, *364*, 2021–2029. [CrossRef] [PubMed]
13. Siramshetty, V.B.; Nickel, J.; Omieczynski, C.; Gohlke, B.-O.; Drwal, M.N.; Preissner, R. WITHDRAWN—A Resource for Withdrawn and Discontinued Drugs. *Nucleic Acids Res.* **2016**, *44*, D1080–D1086. [CrossRef] [PubMed]
14. Zushin, P.-J.H.; Mukherjee, S.; Wu, J.C. FDA Modernization Act 2.0: Transitioning beyond Animal Models with Human Cells, Organoids, and AI/ML-Based Approaches. *J. Clin. Investig.* **2023**, *133*, e175824. [CrossRef]
15. Morel, S.; Sapino, S.; Peira, E.; Chirio, D.; Gallarate, M. Regulatory Requirements for Exporting Cosmetic Products to Extra-EU Countries. *Cosmetics* **2023**, *10*, 62. [CrossRef]
16. Wang, L.; Hu, D.; Xu, J.; Hu, J.; Wang, Y. Complex in Vitro Model: A Transformative Model in Drug Development and Precision Medicine. *Clin. Transl. Sci.* **2024**, *17*, e13695. [CrossRef]
17. Workman, P. How Much Gets There and What Does It Do?: The Need for Better Pharmacokinetic and Pharmacodynamic Endpoints in Contemporary Drug Discovery and Development. *Curr. Pharm. Des.* **2003**, *9*, 891–902. [CrossRef]
18. Elfawal, M.A.; Goetz, E.; Kim, Y.; Chen, P.; Savinov, S.N.; Barasa, L.; Thompson, P.R.; Aroian, R.V. High-Throughput Screening of More Than 30,000 Compounds for Anthelmintics against Gastrointestinal Nematode Parasites. *ACS Infect. Dis.* 2024, *online ahead of print*. [CrossRef]
19. Ryoo, H.; Kimmel, H.; Rondo, E.; Underhill, G.H. Advances in High Throughput Cell Culture Technologies for Therapeutic Screening and Biological Discovery Applications. *Bioeng. Transl. Med.* **2024**, *9*, e10627. [CrossRef]
20. Moreau, J.-F.; Pradeu, T.; Grignolio, A.; Nardini, C.; Castiglione, F.; Tieri, P.; Capri, M.; Salvioli, S.; Taupin, J.-L.; Garagnani, P.; et al. The Emerging Role of ECM Crosslinking in T Cell Mobility as a Hallmark of Immunosenescence in Humans. *Ageing Res. Rev.* **2017**, *35*, 322–335. [CrossRef]
21. Nudischer, R.; Renggli, K.; Hierlemann, A.; Roth, A.B.; Bertinetti-Lapatki, C. Characterization of a Long-Term Mouse Primary Liver 3D Tissue Model Recapitulating Innate-Immune Responses and Drug-Induced Liver Toxicity. *PLoS ONE* **2020**, *15*, e0235745. [CrossRef]
22. White, R.E. High-Throughput Screening in Drug Metabolism and Pharmacokinetic Support of Drug Discovery. Annu. *Rev. Pharmacol. Toxicol.* **2000**, *40*, 133–157. [CrossRef] [PubMed]
23. Kassel, D.B. Applications of High-Throughput ADME in Drug Discovery. *Curr. Opin. Chem. Biol.* **2004**, *8*, 339–345. [CrossRef] [PubMed]
24. Siramshetty, V.; Williams, J.; Nguyễn, Đ.-T.; Neyra, J.; Southall, N.; Mathé, E.; Xu, X.; Shah, P. Validating ADME QSAR Models Using Marketed Drugs. *SLAS Discov.* **2021**, *26*, 1326–1336. [CrossRef] [PubMed]

25. Rodrigues, A.D.; Lin, J.H. Screening of Drug Candidates for Their Drug—Drug Interaction Potential. *Curr. Opin. Chem. Biol.* **2001**, *5*, 396–401. [CrossRef]
26. Li, Y.; Shin, Y.G.; Yu, C.; Kosmeder, J.W.; Hirschelman, W.H.; Pezzuto, J.M.; van Breemen, R.B. Increasing the Throughput and Productivity of Caco-2 Cell Permeability Assays Using Liquid Chromatography-Mass Spectrometry: Application to Resveratrol Absorption and Metabolism. *Comb. Chem. High. Throughput Screen.* **2003**, *6*, 757–767. [CrossRef]
27. Wawer, M.J.; Jaramillo, D.E.; Dančík, V.; Fass, D.M.; Haggarty, S.J.; Shamji, A.F.; Wagner, B.K.; Schreiber, S.L.; Clemons, P.A. Automated Structure-Activity Relationship Mining: Connecting Chemical Structure to Biological Profiles. *J. Biomol. Screen.* **2014**, *19*, 738–748. [CrossRef]
28. Petrone, P.M.; Simms, B.; Nigsch, F.; Lounkine, E.; Kutchukian, P.; Cornett, A.; Deng, Z.; Davies, J.W.; Jenkins, J.L.; Glick, M. Rethinking Molecular Similarity: Comparing Compounds on the Basis of Biological Activity. *ACS Chem. Biol.* **2012**, *7*, 1399–1409. [CrossRef]
29. Garnett, M.J.; Edelman, E.J.; Heidorn, S.J.; Greenman, C.D.; Dastur, A.; Lau, K.W.; Greninger, P.; Thompson, I.R.; Luo, X.; Soares, J.; et al. Systematic Identification of Genomic Markers of Drug Sensitivity in Cancer Cells. *Nature* **2012**, *483*, 570–575. [CrossRef]
30. Wagner, B.K.; Clemons, P.A. Connecting Synthetic Chemistry Decisions to Cell and Genome Biology Using Small-Molecule Phenotypic Profiling. *Curr. Opin. Chem. Biol.* **2009**, *13*, 539–548. [CrossRef]
31. Seiler, K.P.; George, G.A.; Happ, M.P.; Bodycombe, N.E.; Carrinski, H.A.; Norton, S.; Brudz, S.; Sullivan, J.P.; Muhlich, J.; Serrano, M.; et al. ChemBank: A Small-Molecule Screening and Cheminformatics Resource Database. *Nucleic Acids Res.* **2007**, *36*, D351–D359. [CrossRef]
32. Wassermann, A.M.; Bajorath, J. BindingDB and ChEMBL: Online Compound Databases for Drug Discovery. *Expert Opin. Drug Discov.* **2011**, *6*, 683–687. [CrossRef] [PubMed]
33. Gustafsdottir, S.M.; Ljosa, V.; Sokolnicki, K.L.; Anthony Wilson, J.; Walpita, D.; Kemp, M.M.; Petri Seiler, K.; Carrel, H.A.; Golub, T.R.; Schreiber, S.L.; et al. Multiplex Cytological Profiling Assay to Measure Diverse Cellular States. *PLoS ONE* **2013**, *8*, e80999. [CrossRef] [PubMed]
34. Lin, C.Q.; Bissell, M.J. Multi-faceted Regulation of Cell Differentiation by Extracellular Matrix. *FASEB J.* **1993**, *7*, 737–743. [CrossRef] [PubMed]
35. Fang, Y.; Eglen, R.M. Three-Dimensional Cell Cultures in Drug Discovery and Development. *SLAS Discov.* **2017**, *22*, 456–472. [CrossRef]
36. Sato, T.; Vries, R.G.; Snippert, H.J.; van de Wetering, M.; Barker, N.; Stange, D.E.; van Es, J.H.; Abo, A.; Kujala, P.; Peters, P.J.; et al. Single Lgr5 Stem Cells Build Crypt-Villus Structures in Vitro without a Mesenchymal Niche. *Nature* **2009**, *459*, 262–265. [CrossRef]
37. Qu, M.; Xiong, L.; Lyu, Y.; Zhang, X.; Shen, J.; Guan, J.; Chai, P.; Lin, Z.; Nie, B.; Li, C.; et al. Establishment of Intestinal Organoid Cultures Modeling Injury-Associated Epithelial Regeneration. *Cell Res.* **2021**, *31*, 259–271. [CrossRef]
38. Chen, Y.; Zhou, W.; Roh, T.; Estes, M.K.; Kaplan, D.L. In Vitro Enteroid-Derived Three-Dimensional Tissue Model of Human Small Intestinal Epithelium with Innate Immune Responses. *PLoS ONE* **2017**, *12*, e0187880. [CrossRef]
39. Zhao, Z.; Chen, X.; Dowbaj, A.M.; Sljukic, A.; Bratlie, K.; Lin, L.; Fong, E.L.S.; Balachander, G.M.; Chen, Z.; Soragni, A.; et al. Organoids. *Nat. Rev. Methods Prim.* **2022**, *2*, 94. [CrossRef]
40. Huch, M.; Dorrell, C.; Boj, S.F.; van Es, J.H.; Li, V.S.W.; van de Wetering, M.; Sato, T.; Hamer, K.; Sasaki, N.; Finegold, M.J.; et al. In Vitro Expansion of Single Lgr5+ Liver Stem Cells Induced by Wnt-Driven Regeneration. *Nature* **2013**, *494*, 247–250. [CrossRef]
41. Huch, M.; Gehart, H.; van Boxtel, R.; Hamer, K.; Blokzijl, F.; Verstegen, M.M.A.; Ellis, E.; van Wenum, M.; Fuchs, S.A.; de Ligt, J.; et al. Long-Term Culture of Genome-Stable Bipotent Stem Cells from Adult Human Liver. *Cell* **2015**, *160*, 299–312. [CrossRef]
42. Nair, D.G.; Weiskirchen, R. Recent Advances in Liver Tissue Engineering as an Alternative and Complementary Approach for Liver Transplantation. *Curr. Issues Mol. Biol.* **2023**, *46*, 262–278. [CrossRef] [PubMed]
43. Liu, Y.; Sheng, J.-Y.; Yang, C.-F.; Ding, J.; Chan, Y.-S. A Decade of Liver Organoids: Advances in Disease Modeling. *Clin. Mol. Hepatol.* **2023**, *29*, 643–669. [CrossRef] [PubMed]
44. Marsee, A.; Roos, F.J.M.; Verstegen, M.M.A.; Gehart, H.; de Koning, E.; Lemaigre, F.; Forbes, S.J.; Peng, W.C.; Huch, M.; Takebe, T.; et al. Building Consensus on Definition and Nomenclature of Hepatic, Pancreatic, and Biliary Organoids. *Cell Stem Cell* **2021**, *28*, 816–832. [CrossRef] [PubMed]
45. Shrestha, S.; Acharya, P.; Kang, S.-Y.; Vanga, M.G.; Lekkala, V.K.R.; Liu, J.; Yang, Y.; Joshi, P.; Lee, M.-Y. Regenerative Human Liver Organoids (HLOs) in a Pillar/Perfusion Plate for Hepatotoxicity Assays. *bioRxiv* **2024**. [CrossRef]
46. Kamishibahara, Y.; Okamoto, S.; Ohkuma, T.; Taniguchi, H. Stabilized Generation of Human IPSC-Derived Liver Organoids Using a Modified Coating Approach. *Biol. Methods Protoc.* **2023**, *8*, bpac034. [CrossRef]
47. Huh, D.; Matthews, B.D.; Mammoto, A.; Montoya-Zavala, M.; Hsin, H.Y.; Ingber, D.E. Reconstituting Organ-Level Lung Functions on a Chip. *Science* **2010**, *328*, 1662–1668. [CrossRef]
48. Valiei, A.; Aminian-Dehkordi, J.; Mofrad, M.R.K. Gut-on-a-Chip Models for Dissecting the Gut Microbiology and Physiology. *APL Bioeng.* **2023**, *7*, 011502. [CrossRef]

49. Cheng, Y.; Watanabe, C.; Ando, Y.; Kitaoka, S.; Egawa, Y.; Takashima, T.; Matsumoto, A.; Murakami, M. Caco-2 Cell Sheet Partially Laminated with HT29-MTX Cells as a Novel In Vitro Model of Gut Epithelium Drug Permeability. *Pharmaceutics* **2023**, *15*, 2338. [CrossRef]
50. Cremer, J.; Segota, I.; Yang, C.; Arnoldini, M.; Sauls, J.T.; Zhang, Z.; Gutierrez, E.; Groisman, A.; Hwa, T. Effect of Flow and Peristaltic Mixing on Bacterial Growth in a Gut-like Channel. *Proc. Natl. Acad. Sci. USA* **2016**, *113*, 11414–11419. [CrossRef]
51. Stern, S.; Wang, H.; Sadrieh, N. Microphysiological Models for Mechanistic-Based Prediction of Idiosyncratic DILI. *Cells* **2023**, *12*, 1476. [CrossRef]
52. Haque, A.; Gheibi, P.; Gao, Y.; Foster, E.; Son, K.J.; You, J.; Stybayeva, G.; Patel, D.; Revzin, A. Cell Biology Is Different in Small Volumes: Endogenous Signals Shape Phenotype of Primary Hepatocytes Cultured in Microfluidic Channels. *Sci. Rep.* **2016**, *6*, 33980. [CrossRef] [PubMed]
53. Nair, D.G.; Miller, K.G.; Lourenssen, S.R.; Blennerhassett, M.G. Inflammatory Cytokines Promote Growth of Intestinal Smooth Muscle Cells by Induced Expression of PDGF-Rβ. *J. Cell Mol. Med.* **2014**, *18*, 444–454. [CrossRef] [PubMed]
54. Nair, D.G.; Weiskirchen, R.; Al-Musharafi, S.K. The Use of Marine-Derived Bioactive Compounds as Potential Hepatoprotective Agents. *Acta Pharmacol. Sin.* **2015**, *36*, 158–170. [CrossRef] [PubMed]
55. Yang, J.; Hirai, Y.; Iida, K.; Ito, S.; Trumm, M.; Terada, S.; Sakai, R.; Tsuchiya, T.; Tabata, O.; Kamei, K. Integrated-Gut-Liver-on-a-Chip Platform as an in Vitro Human Model of Non-Alcoholic Fatty Liver Disease. *Commun. Biol.* **2023**, *6*, 310. [CrossRef]
56. Damiri, F.; Fatimi, A.; Liu, Y.; Musuc, A.M.; Fajardo, A.R.; Gowda, B.H.J.; Vora, L.K.; Shavandi, A.; Okoro, O.V. Recent Advances in 3D Bioprinted Polysaccharide Hydrogels for Biomedical Applications: A Comprehensive Review. *Carbohydr. Polym.* **2025**, *348*, 122845. [CrossRef]
57. Miserocchi, G.; Cocchi, C.; De Vita, A.; Liverani, C.; Spadazzi, C.; Calpona, S.; Di Menna, G.; Bassi, M.; Meccariello, G.; De Luca, G.; et al. Three-Dimensional Collagen-Based Scaffold Model to Study the Microenvironment and Drug-Resistance Mechanisms of Oropharyngeal Squamous Cell Carcinomas. *Cancer Biol. Med.* **2021**, *18*, 502–516. [CrossRef]
58. De Vita, A.; Recine, F.; Miserocchi, G.; Pieri, F.; Spadazzi, C.; Cocchi, C.; Vanni, S.; Liverani, C.; Farnedi, A.; Fabbri, F.; et al. The Potential Role of the Extracellular Matrix in the Activity of Trabectedin in UPS and L-Sarcoma: Evidences from a Patient-derived Primary Culture Case Series in Tridimensional and Zebrafish Models. *J. Exp. Clin. Cancer Res.* **2021**, *40*, 165. [CrossRef]
59. Brown, L.M.; Hebert, K.L.; Gurrala, R.R.; Byrne, C.E.; Burow, M.; Martin, E.C.; Lau, F.H. Modeling Breast Cancer in Human Breast Tissue Using a Microphysiological System. *J. Vis. Exp.* **2021**, *170*, e62009. [CrossRef]
60. Deng, Z.-M.; Dai, F.-F.; Wang, R.-Q.; Deng, H.-B.; Yin, T.-L.; Cheng, Y.-X.; Chen, G.-T. Organ-on-a-Chip: Future of Female Reproductive Pathophysiological Models. *J. Nanobiotechnol.* **2024**, *22*, 455. [CrossRef]
61. Mishra, Y.; Mishra, V.; Aljabali, A.A.A.; El-Tanani, M.; Naikoo, G.A.; Charbe, N.; Chava, S.R.; Tambuwala, M.M. 3D Printed Personalized Colon-Targeted Tablets: A Novel Approach in Ulcerative Colitis Management. *Curr. Drug Deliv.* **2024**, *21*, 1211–1225. [CrossRef]
62. Wu, X.; Shi, W.; Liu, X.; Gu, Z. Recent Advances in 3D-Printing-Based Organ-on-a-Chip. *EngMedicine* **2024**, *1*, 100003. [CrossRef]
63. Fritschen, A.; Bell, A.K.; Königstein, I.; Stühn, L.; Stark, R.W.; Blaeser, A. Investigation and Comparison of Resin Materials in Transparent DLP-Printing for Application in Cell Culture and Organs-on-a-Chip. *Biomater. Sci.* **2022**, *10*, 1981–1994. [CrossRef] [PubMed]
64. Villegas, M.; Cetinic, Z.; Shakeri, A.; Didar, T.F. Fabricating Smooth PDMS Microfluidic Channels from Low-Resolution 3D Printed Molds Using an Omniphobic Lubricant-Infused Coating. *Anal. Chim. Acta* **2018**, *1000*, 248–255. [CrossRef] [PubMed]
65. Xiang, Y.; Miller, K.; Guan, J.; Kiratitanaporn, W.; Tang, M.; Chen, S. 3D Bioprinting of Complex Tissues in Vitro: State-of-the-Art and Future Perspectives. *Arch. Toxicol.* **2022**, *96*, 691–710. [CrossRef]
66. Kang, D.; Hong, G.; An, S.; Jang, I.; Yun, W.; Shim, J.; Jin, S. Bioprinting of Multiscaled Hepatic Lobules within a Highly Vascularized Construct. *Small* **2020**, *16*, e1905505. [CrossRef]
67. Ma, X.; Qu, X.; Zhu, W.; Li, Y.-S.; Yuan, S.; Zhang, H.; Liu, J.; Wang, P.; Lai, C.S.E.; Zanella, F.; et al. Deterministically Patterned Biomimetic Human IPSC-Derived Hepatic Model via Rapid 3D Bioprinting. *Proc. Natl. Acad. Sci. USA* **2016**, *113*, 2206–2211. [CrossRef]
68. Lee, H.; Chae, S.; Kim, J.Y.; Han, W.; Kim, J.; Choi, Y.; Cho, D.-W. Cell-Printed 3D Liver-on-a-Chip Possessing a Liver Microenvironment and Biliary System. *Biofabrication* **2019**, *11*, 025001. [CrossRef]
69. Wu, Q.; Liu, J.; Wang, X.; Feng, L.; Wu, J.; Zhu, X.; Wen, W.; Gong, X. Organ-on-a-Chip: Recent Breakthroughs and Future Prospects. *Biomed. Eng. Online* **2020**, *19*, 9. [CrossRef]
70. Zamprogno, P.; Wüthrich, S.; Achenbach, S.; Thoma, G.; Stucki, J.D.; Hobi, N.; Schneider-Daum, N.; Lehr, C.-M.; Huwer, H.; Geiser, T.; et al. Second-Generation Lung-on-a-Chip with an Array of Stretchable Alveoli Made with a Biological Membrane. *Commun. Biol.* **2021**, *4*, 168. [CrossRef]
71. Ross, A.E.; Belanger, M.C.; Woodroof, J.F.; Pompano, R.R. Spatially Resolved Microfluidic Stimulation of Lymphoid Tissue Ex Vivo. *Analyst* **2017**, *142*, 649–659. [CrossRef]

72. Sung, J.H.; Yu, J.; Luo, D.; Shuler, M.L.; March, J.C. Microscale 3-D Hydrogel Scaffold for Biomimetic Gastrointestinal (GI) Tract Model. *Lab Chip* **2011**, *11*, 389–392. [CrossRef] [PubMed]
73. Jalili-Firoozinezhad, S.; Gazzaniga, F.S.; Calamari, E.L.; Camacho, D.M.; Fadel, C.W.; Bein, A.; Swenor, B.; Nestor, B.; Cronce, M.J.; Tovaglieri, A.; et al. A Complex Human Gut Microbiome Cultured in an Anaerobic Intestine-on-a-Chip. *Nat. Biomed. Eng.* **2019**, *3*, 520–531. [CrossRef] [PubMed]
74. Cho, I.; Blaser, M.J. The Human Microbiome: At the Interface of Health and Disease. *Nat. Rev. Genet.* **2012**, *13*, 260–270. [CrossRef]
75. Jang, K.-J.; Mehr, A.P.; Hamilton, G.A.; McPartlin, L.A.; Chung, S.; Suh, K.-Y.; Ingber, D.E. Human Kidney Proximal Tubule-on-a-Chip for Drug Transport and Nephrotoxicity Assessment. *Integr. Biol.* **2013**, *5*, 1119–1129. [CrossRef]
76. Sakolish, C.M.; Philip, B.; Mahler, G.J. A Human Proximal Tubule-on-a-Chip to Study Renal Disease and Toxicity. *Biomicrofluidics* **2019**, *13*, 014107. [CrossRef]
77. Nieskens, T.T.G.; Sjögren, A.-K. Emerging In Vitro Systems to Screen and Predict Drug-Induced Kidney Toxicity. *Semin. Nephrol.* **2019**, *39*, 215–226. [CrossRef]
78. Zhang, D.; Shadrin, I.Y.; Lam, J.; Xian, H.-Q.; Snodgrass, H.R.; Bursac, N. Tissue-Engineered Cardiac Patch for Advanced Functional Maturation of Human ESC-Derived Cardiomyocytes. *Biomaterials* **2013**, *34*, 5813–5820. [CrossRef]
79. Schneider, O.; Zeifang, L.; Fuchs, S.; Sailer, C.; Loskill, P. User-Friendly and Parallelized Generation of Human Induced Pluripotent Stem Cell-Derived Microtissues in a Centrifugal Heart-on-a-Chip. *Tissue Eng. Part A* **2019**, *25*, 786–798. [CrossRef]
80. Lee, S.; Sung, J. Microtechnology-Based Multi-Organ Models. *Bioengineering* **2017**, *4*, 46. [CrossRef]
81. Skardal, A.; Murphy, S.V.; Devarasetty, M.; Mead, I.; Kang, H.-W.; Seol, Y.-J.; Shrike Zhang, Y.; Shin, S.-R.; Zhao, L.; Aleman, J.; et al. Multi-Tissue Interactions in an Integrated Three-Tissue Organ-on-a-Chip Platform. *Sci. Rep.* **2017**, *7*, 8837. [CrossRef]
82. Oleaga, C.; Bernabini, C.; Smith, A.S.T.; Srinivasan, B.; Jackson, M.; McLamb, W.; Platt, V.; Bridges, R.; Cai, Y.; Santhanam, N.; et al. Multi-Organ Toxicity Demonstration in a Functional Human in Vitro System Composed of Four Organs. *Sci. Rep.* **2016**, *6*, 20030. [CrossRef] [PubMed]
83. Edington, C.D.; Chen, W.L.K.; Geishecker, E.; Kassis, T.; Soenksen, L.R.; Bhushan, B.M.; Freake, D.; Kirschner, J.; Maass, C.; Tsamandouras, N.; et al. Interconnected Microphysiological Systems for Quantitative Biology and Pharmacology Studies. *Sci. Rep.* **2018**, *8*, 4530. [CrossRef] [PubMed]
84. Lindner, N.; Mejia-Wille, A.; Fritschen, A.; Blaeser, A. Development of a Robotic-Assisted Handling and Manipulation System for the High-Scale Bioproduction of 3D-Bioprinted Organ-on-a-Chip Devices. *HardwareX* **2024**, *19*, e00572. [CrossRef]
85. Schmidt, K.; Berg, J.; Roehrs, V.; Kurreck, J.; Al-Zeer, M.A. 3D-Bioprinted HepaRG Cultures as a Model for Testing Long Term Aflatoxin B1 Toxicity in Vitro. *Toxicol. Rep.* **2020**, *7*, 1578–1587. [CrossRef]
86. Hong, S.; Song, J.M. A 3D Cell Printing-Fabricated HepG2 Liver Spheroid Model for High-Content in Situ Quantification of Drug-Induced Liver Toxicity. *Biomater. Sci.* **2021**, *9*, 5939–5950. [CrossRef]
87. Tronolone, J.J.; Mathur, T.; Chaftari, C.P.; Sun, Y.; Jain, A. Machine Learning Chained Neural Network Analysis of Oxygen Transport Amplifies the Physiological Relevance of Vascularized Microphysiological Systems. *Bioeng. Transl. Med.* **2023**, *8*, e10582. [CrossRef]
88. Lamoureux, E.S.; Islamzada, E.; Wiens, M.V.J.; Matthews, K.; Duffy, S.P.; Ma, H. Assessing red blood cell deformability from microscopy images using deep learning. *Lab Chip* **2021**, *22*, 26–39. [CrossRef]
89. Taïeb, H.M.; Herment, G.; Robinson, T.; Cipitria, A. Effect of capillary fluid flow on single cancer cell cycle dynamics, motility, volume and morphology. *Lab Chip* **2022**, *23*, 92–105. [CrossRef]
90. Potter, C.J.; Hu, Y.; Xiong, Z.; Wang, J.; McLeod, E. Point-of-care SARS-CoV-2 sensing using lens-free imaging and a deep learning-assisted quantitative agglutination assay. *Lab Chip* **2022**, *22*, 3744–3754. [CrossRef]
91. Zhang, Z.; Chen, L.; Wang, Y.; Zhang, T.; Chen, Y.C.; Yoon, E. Label-Free Estimation of Therapeutic Efficacy on 3D Cancer Spheres Using Convolutional Neural Network Image Analysis. *Anal. Chem.* **2019**, *91*, 14093–14100. [CrossRef]
92. Oliver, C.R.; Altemus, M.A.; Westerhof, T.M.; Cheriyan, H.; Cheng, X.; Dziubinski, M.; Wu, Z.; Yates, J.; Morikawa, A.; Heth, J.; et al. A platform for artificial intelligence based identification of the extravasation potential of cancer cells into the brain metastatic niche. *Lab Chip* **2019**, *19*, 1162–1173. [CrossRef] [PubMed]
93. Chen, H.; Kim, S.; Hardie, J.M.; Thirumalaraju, P.; Gharpure, S.; Rostamian, S.; Udayakumar, S.; Lei, Q.; Cho, G.; Kanakasabapathy, M.K.; et al. Deep learning-assisted sensitive detection of fentanyl using a bubbling-microchip. *Lab Chip* **2022**, *22*, 4531–4540. [CrossRef] [PubMed]
94. Iachetta, G.; Melle, G.; Colistra, N.; Tantussi, F.; De Angelis, F.; Dipalo, M. Long-Term in Vitro Recording of Cardiac Action Potentials on Microelectrode Arrays for Chronic Cardiotoxicity Assessment. *Arch. Toxicol.* **2023**, *97*, 509–522. [CrossRef] [PubMed]
95. Farooqi, H.M.U.; Kang, B.; Khalid, M.A.U.; Salih, A.R.C.; Hyun, K.; Park, S.H.; Huh, D.; Choi, K.H. Real-Time Monitoring of Liver Fibrosis through Embedded Sensors in a Microphysiological System. *Nano Converg.* **2021**, *8*, 3. [CrossRef] [PubMed]
96. Wang, D.; Gust, M.; Ferrell, N. Kidney-on-a-Chip: Mechanical Stimulation and Sensor Integration. *Sensors* **2022**, *22*, 6889. [CrossRef]

97. Grasso, G.; Colella, F.; Forciniti, S.; Onesto, V.; Iuele, H.; Siciliano, A.C.; Carnevali, F.; Chandra, A.; Gigli, G.; del Mercato, L.L. Fluorescent Nano- and Microparticles for Sensing Cellular Microenvironment: Past, Present and Future Applications. *Nanoscale Adv.* **2023**, *5*, 4311–4336. [CrossRef]
98. Alver, C.; Drabbe, E.; Ishahak, M.; Agarwal, A. Roadblocks Confronting Widespread Dissemination and Deployment of Organs on Chips. *Nat. Commun.* **2024**, *15*, 5118. [CrossRef]
99. Murdoch, B. Privacy and Artificial Intelligence: Challenges for Protecting Health Information in a New Era. *BMC Med. Ethics* **2021**, *22*, 122. [CrossRef]
100. Li, Y.; Li, Y.; Wei, M.; Li, G. Innovation and Challenges of Artificial Intelligence Technology in Personalized Healthcare. *Sci. Rep.* **2024**, *14*, 18994. [CrossRef]

current issues in molecular biology

MDPI

Review

Treatment Options for Patients with Non-Small Cell Lung Cancer and Liver Metastases

Vesna Ćeriman Krstić [1,2,*], Natalija Samardžić [2], Milija Gajić [2], Milan Savić [1,3], Biljana Šeha [4], Marina Roksandić Milenković [5] and Dragana Jovanović [6]

[1] Faculty of Medicine, University of Belgrade, 11000 Belgrade, Serbia; drmilansavic@gmail.com
[2] Clinic for Pulmonology, University Clinical Center of Serbia, 11000 Belgrade, Serbia; natalis.dm@gmail.com (N.S.); milijagajic@gmail.com (M.G.)
[3] Clinic for Thoracic Surgery, University Clinical Center of Serbia, 11000 Belgrade, Serbia
[4] Clinic for Neurosurgery, University Clinical Center of Serbia, 11000 Belgrade, Serbia; biljana.seha@yahoo.com
[5] Municipal Institute for Lung Diseases and TB, 11000 Belgrade, Serbia; dr.marr@gmail.com
[6] Internal Medicine Clinic „Akta Medica", 11000 Belgrade, Serbia; draganajv@yahoo.com
* Correspondence: ceriman.vesna@gmail.com

Abstract: Lung cancer represents the most common cause of cancer-related death. Patients with non-small cell lung cancer (NSCLC) and liver metastases have worse prognosis, with an overall survival (OS) from three to six months. The majority of them have a poor response to chemotherapy, and the data are controversial regarding the response to immunotherapy. This could be because the liver is considered to be an immune-tolerant organ, which is characterized by T-cell anergy and immunosuppressive signals. This review evaluates current treatment options for patients with NSCLC and liver metastases. Combination therapies might be a better treatment option for this subgroup of patients. The addition of radiotherapy to immunotherapy could also be an option in selected patients. The resection of single liver metastasis should also be considered.

Keywords: NSCLC; liver metastases; immunotherapy; combination therapies; radiotherapy; surgery

1. Introduction

Lung cancer is the most common cause of death among all cancers [1]. Non-small cell lung cancer (NSCLC) represents around 85% of all lung cancers [1]. In more than half of patients, the disease is diagnosed in the locally advanced or metastatic stage [1]. In the past, the standard of care for those patients was platinum doublet, with a 5-year survival rate of less than 10% [1].

Patients with NSCLC and liver metastases (LMs) have worse prognosis, and they make up about 20% of patients [2]. The studies have shown that these patients live approximately three to six months [2]. The results of one large study which involved more than 20,000 lung cancer patients showed that patients with liver metastases had a 53% higher risk of death compared to patients with brain metastases [3,4]. Similar results were found in the other studies, which also found that LM was associated with worse outcomes [5,6].

A large cohort study investigated which metastatic site was the most prevalent in patients with NSCLC, median time to develop metastasis and prognosis by metastatic site [7]. It was found that the most prevalent metastatic site was lungs. Regarding LM, it was found that those patients had a lower 6-month survival rate compared to patients with brain metastases [7].

Further, Akamine et al. [8] conducted a study to estimate post-recurrence rate, time to recurrence, and the impact of the site of recurrence on post-recurrence survival in patients who had radical surgery due to NSCLC. They found that liver was a significantly unfavorable prognostic site and that recurrence occurred usually within 6 months after surgery [8]. Also, it was found that LM was associated with poor post-recurrence survival; HR 2.20 [8].

In other study conducted by Takeuchi et al. [9], LM, ECOG PS ≥ 2, and non-smoking history were identified as early mortality factors in patients with advanced or metastatic NSCLC treated with mono-immunotherapy.

Deng et al. [10] found that patients with NSCLC and liver metastases had worse organ-specific response to immunotherapy compared to patients with NSCLC and pulmonary metastases. Interestingly, one third of patients had mixed responses to immunotherapy, and among them, the majority had a response in the lungs, but progression in the liver, and others had the opposite response [10]. Further, 27 paired tumor samples from pulmonary metastases and liver metastases were collected to evaluate organ specific differences. There were no significant difference in the consistency of genetic variation, tumor mutation burden, and different gene mutations [10], but lower levels of immune activation and infiltration were found in the tumor microenvironment of the liver metastases [10].

The majority of patients with liver metastases have a poor response to chemotherapy [2], but the response to immunotherapy is controversial. It has also been shown that the groups of patients with driver mutations have a worse response to targeted therapies compared to patients without liver metastases [4,11,12]. Also, the results of some studies suggested that patients with NSCLC and liver metastases respond to the treatment more similarly to patients with primary liver cancer [10,13].

Since it has been shown that patients with NSCLC and LM have worse prognosis compared to patients with other metastatic sites, even compared to patients with brain metastases, we are going to summarize the possible treatment options for this group of patients. However, the underlying mechanisms as to why these patients have a worse treatment response are sill unclear.

2. The Liver as an Immune-Tolerant Organ

The liver is considered as an immune-tolerant organ, which is characterized by T-cell anergy and immunosuppressive signals [10]. This could be the reason why patients with NSCLC and liver metastases have a worse response to PD-1/PD-L1 inhibitors compared to patients without [2]. Studies also have shown that liver metastases could actively induce the apoptosis of CD8+ T-cells after their interaction with macrophages, which means that liver metastases could take over the mechanisms of peripheral tolerance, leading to acquired resistance to immunotherapy through the removal of CD8+ T-cells [10].

PD-L1 is an immune checkpoint that limits T-cell activity by binding to its receptor PD-1 on activated T-cells [10]. PD-L1 can also be found on tumor cells and, as a result, blocks the ability of the immune system to fight against various cancers [10]. Furthermore, blocking PD-1/PD-L1 leads to the reactivation of exhausted T-cells and to the reactivation of the immune system, enabling it to fight against tumors [10]. In this specific process, tumors and the organ-specific microenvironment play important roles [10]. As was already mentioned, the liver is considered to be an immune-tolerant organ, and it is likely that liver metastases will be resistant to immunotherapy [10].

Vascular endothelial growth factor (VEGF) and its receptors have a crucial role in normal and pathologic angiogenesis. By activating the VEGF/VEGFR axis, other signaling pathways are triggered, which further results in endothelial cell survival, mitogenesis, migration, and differentiation, as well as vascular permeability and the mobilization of endothelial progenitor cells from the bone marrow into the peripheral circulation [14]. Further VEGF leads to increased vascular permeability and thus to the development of malignant effusions. Increased permeability can also lead to the deposition of proteins in the interstitium and, in that way, facilitate angiogenesis [14]. Finally, the overexpression of VEGF has been associated with disease progression and worse prognosis in several tumors, including lung cancer [14]. Thus, VEGF could block antitumor activity by promoting vascularization [14], which is often used by tumors for growth and metastasis and interrupting T-cell infiltration into the tumor [15,16].

It has been suggested that a VEGF inhibitor in combination with chemotherapy could enhance PD-L1 inhibitor-induced T-cell-mediated cancer cell death, by reversing VEGF-induced immunosuppression by promoting the infiltration of T-cells into the tumor microenvironment and enabling T-cell activity against tumor antigens, alongside chemotherapy-induced cancer cell death [13].

3. The Impact of Co-Mutations

Recent studies had shown that the presence of co-mutations also had an impact on immunotherapy efficacy. Sclera et al. [17] investigated whether KEAP1 and TP53 (a tumor suppressor gene) have an impact on immunotherapy efficacy. The KEAP1/NFE2L2 pathway has a role in the protection of eukaryotic cells from oxidative stress and xenobiotics [17]. When cells are exposed to stress, KEAP1 enables NFE2L2 to translocate to the nucleus. Further, NFE2L2 initiates a transcriptional program that leads to a cytoprotective response through the modulation of genes encoding antioxidant proteins, detoxifying enzymes, and multidrug resistance proteins [17]. Cancer cells attack this process, in order to prohibit anticancer treatment [17]. Thus, it is speculated that inadequate KEAP1 function leads to uncontrolled NFE2L2 transcription and enhanced cytoprotection [17]. Accordingly, lung adenocarcinoma-carrying KEAP1 mutations are resistant to chemotherapy and radiotherapy [17]. However, regarding immunotherapy, the data are inconsistent. TP53, along with another tumor suppressor gene, CHEK2, have a role in testing whether DNA mutations in damaged cells can be repaired or whether the cell has to be eliminated [18]. Thus, it was concluded that the presence of other co-mutations may contribute to the immunotherapy response [17,18]. KRAS mutations (among the most common mutations in lung cancer), for example, induce the hyper-activation of downstream signaling, controlling cell proliferation [18]. STK11 has a major role in lung cancer differentiation and metastasis [18].

Scalera et al. [17] found that the presence of KEAP1 and TP53 had different impacts on immunotherapy efficacy. Their results suggested that tumors with KEAP1 mutations had the shortest survival, while tumors with a KEAP1/TP53 or only a TP53 mutation had intermediate survival, and the best prognosis was achieved by tumors without any of these two mutations [17]. Interestingly, it was observed that tumors with KEAP1 and tumors with KEAP1/TP53 were entirely different in terms of mutational repertoire, the degree of intra-tumor heterogeneity, evolutionary trajectories, pathway-level signatures, and immune microenvironment [17].

Further, Wang et al. [18] found that the presence of co-mutations might have an impact on immunotherapy efficacy in patients with NSCLC. The majority of patients were treated with mono-immunotherapy [18]. Thus, they found that patients with a TP53 mutation had significantly longer PFS compared to the TP53 wild-type, and a similar observation was found in a combination group, but the difference was not significant [18]. Also, a longer PFS was found if other co-mutations were found with TP53 [18]. Thus, it was suggested that TP53 could be the dominant mutation, which correlated with longer PFS in patients with NSCLC treated with mono-immunotherapy [18]. It was also suggested that different genes could have different effects when co-mutated with TP53 [18]. Thus, it was found that patients with a co-mutation of TP53 with STK11 had longer PFS compared to the wild-type group, but without statistical significance [18].

FAT3 has a role in encoding large proteins with extracellular Cadherin repeats, EGF-like domains, and Laminin G-like domains [19]. It is involved in tumor suppression and planar cell polarity. Further FAT3 mutations have been linked to prognosis and an elevated TMB level in multiple cancers, including NSCLC [19]. LRP1B, the tumor suppressor gene, encodes the low-density lipoprotein family receptor [19]. LRP1B's correlation with TMB and immunotherapy efficacy has been confirmed in multiple cancers, including lung cancer [19]. One study suggested that the co-mutation of FAT3 and LRP1B led to significantly prolonged PFS in patients with lung adenocarcinoma treated with immunotherapy [19], and Ying et al. [20] found that patients with squamous cell lung carcinoma and TP53 and TTN (one of the most commonly mutated genes; its role in cancer development is unclear) co-mutation had higher tumor-mutation burden levels and a better response to immunotherapy. The data regarding CDKN2A/B (tumor suppressor gene) loss have been controversial. Some results suggested that CDKN2A/B loss was associated with worse outcomes with immunotherapy [21,22], while in the study of Adib et al. [23], an association was not found.

It might also be that the presence of some other co-mutations led to the generation of LM, and thus contributed to resistance to treatment.

Thus, it was found that USP22, ID1, NT5DC2, and AK4, all together, could modulate the expression of proteins which were responsible for proliferation, angiogenesis, and the epithelial-to-mesenchymal transition in cancer cells or the hepatic microenvironment [24–29]. Thus, this modulation led to emergence of LMs and their growth [24–29].

4. Mono-Immunotherapy

In a pooled analysis of CheckMate 017 and 057 [30], approximately 23% of the patients had liver metastases. After a 3-year follow up, it was shown that the patients treated with nivolumab had improved OS compared to patients treated with docetaxel [30]. In the updated five-year analysis, the benefit was still observed in this group of patients; HR 0.67 [31].

Qiao et al. [32] investigated whether the site of metastases and their number were associated with the efficacy of immunotherapy in patients with NSCLC. They showed that patients with liver metastases had the shortest PFS and the shortest OS, numerically, compared to patients with other sites of metastases [32]. Also, they tested tumor specimens from 46 patients for PD-L1 expression, with the CD8+ tumor infiltrating lymphocyte density. They found that patients with liver metastases had a lower proportion of PD-L1+CD8+TIL+ tumors compared to patients without LM [32]. Patients who were treated with immunotherapy-based combination therapy had better PFS and OS (numerically) compared to patients treated with mono-immunotherapy [32].

Tumeh et al. [33] conducted research to evaluate the efficacy of pembrolizumab in patients with melanoma and NSCLC who had liver metastases. The results showed significantly reduced PFS and OS in a group of patients with liver metastases compared to those without liver metastases, in both groups of patients [33]. In the melanoma group, the optional biopsy was performed in 62 patients, and the results showed that patients with liver metastases had reduced CD8+ T-cell density at the invasive tumor margin [33]. Also, fewer CD8+T-cells were found in a group of patients who did not have a response to immunotherapy compared to patients who did have a response to the treatment [33]. Further, patients with LM had significantly lower densities of CD8+ T-cells in the other metastatic sites [33]. However, in the same samples, PD-1 and PD-L1 expression did not significantly differ between patients with and without LM [33].

Xie at al. [34] analyzed the efficacy of PD-1/PD-L1 inhibitors in patients with NSCLC and liver metastases. They found that immunotherapy was effective in patients with liver metastases, but the efficacy was inferior compared to patients without liver metastases [34]. Also, they found that patients with liver metastases had considerably lower CD8+ T-cell infiltration compared to those without LM [34].

Tian et al. [35] conducted a meta-analysis to elucidate the relationship between liver metastases and survival outcomes in patients with different tumor types who were treated with immunotherapy. The results showed worse outcomes in patients with liver metastases [35], and patients with urinary tumors had the worse outcomes, followed by patients with melanoma and NSCLC [35].

Sridhar et al. [36] presented similar findings. They analyzed the results from two trials to investigate the prognostic role of liver metastases in patients with advanced and metastatic NSCLC treated with durvalumab [36]. Liver metastases were found in 31.6% and 17.9% of patients, respectively [36]. They found, in both studies, that liver metastases were a negative independent prognostic factor, regardless of PD-L1 status [36].

Another large meta-analysis, which included 66 studies with previously treated patients with NSCLC, estimated the OS of immunotherapy in the second and further treatment lines [37]. Four studies reported OS in patients with LM, and these studies included patients treated with nivolumab [37]. OS ranged from 3.6 months to 7.5 months, and it was worse compared to patients without LM [37]. Further, a significant association was reported between the presence of LM and worse OS in three out of four studies [37]. Regarding OS rate, a 1-year OS rate was reported in three studies, and there were no data about 2- and 3-year OS rates [37]. The pooled 1-year OS rate was 23.8% [37].

Huang et al. [38] conducted a meta-analysis to evaluate the outcomes in patients with NSCLC who were treated with PD-1 inhibitors, and also to determine whether different metastatic sites had an impact on the outcomes [38]. They found that 17% of patients had liver metastases, and they had worse OS compared to those without LM [38].

The results of another study, which involved 215 patients with NSCLC who were treated with immunotherapy, showed that patients with liver metastases (19.1%) had a lower response to the therapy and also shorter PFS and OS [39].

In a retrospective analysis, Komiya et al. [40] showed that patients with liver metastases do benefit from immunotherapy, but they had inferior outcomes compared to patients without LM.

A large retrospective study included 1470 patients with NSCLC who were treated with targeted therapy, chemotherapy, and immunotherapy [41]. LMs were found in 15.9% of patients [41]. There was no difference in OS between patients with and without liver metastases who were treated with immunotherapy [41]. However, patients with liver metastases treated with immunotherapy had longer OS compared to patients who were treated with chemotherapy [41]. Patients treated with chemotherapy and targeted therapy with liver metastases had worse outcomes compared to those without LM [41].

A retrospective study investigated the efficacy of nivolumab in 201 patients with NSCLC, and among them, 29 had liver metastases [42]. The PFS and response rate were lower in patients with liver metastases [42], and it was also shown that liver metastases were associated with younger age, poorer ECOG PS, and a greater number of metastatic sites [42].

The results of the other study showed that PFS was worse in patients with liver metastases compared to those without LMs who were treated with nivolumab [43]. Similar results were presented in two other studies [44,45], and in a study by Topalian et al. [46], the five-year OS rate for NSCLC patients treated with nivolumab was lower compared to patients with other carcinomas. All patients with liver metastases had decreased survival [46].

Maugeais et al. [47] investigated whether LMs and the number of metastatic sites had an impact on immunotherapy efficacy in patients with different solid tumors. A total of 759 patients were included; among them, 71% had NSCLC, 22% had baseline LMs, and 49% had more than one metastatic site [47]. Patients with LMs had worse PFS and OS compared to patients without LMs, but the outcomes in this group of patients were not worse compared to the group of patients with other metastatic sites such as brain, bone, or lung [47]. The results were consistent in a group of patients with NSCLC [47].

A large multicenter retrospective study enrolled 291 patients with different tumors who were treated with immunotherapy [48]. More than half of patients had NSCLC [48]. It was shown that the presence of LMs and brain metastases was associated with lower PFS and OS, regardless of the histology [48].

Another retrospective study investigated the association between sites of metastatic disease and clinical outcomes in patients treated with immunotherapy [49]. A total of 90 patients were included, and only 20% of patients had lung, head, and neck cancer (there was no information about the percentage of lung cancer patients only) [49]. However, the results were similar to the previous study. Patients with LMs had worse PFS and OS compared to the patients without LMs [49].

Table 1 shows the outcomes for patients with LM treated with mono- immunotherapy in some studies.

Table 1. The overview of the outcomes for patients with liver metastases treated with mono-immunotherapy in some studies.

Study	Type of Study	Applied Treatment	No of Patients; % LM	Findings
CheckMate 017 and 057-pooled analysis [31]	Clinical trial	Nivolumab vs. Docetaxel	427 vs. 427; 23.2% vs. 22%	OS HR: 0.67
Tumeh et al. [33]	Clinical trial	Pembrolizumab	165; 27.9%	LM/no LM: PFS-1.8/4.0 months
Xie et al. [34]	A real-world study	PD-1/PD-L1 inhibitor	648; 9.4%	LM/no LM: PFS-6.4/7.9 months; OS 15.2/20.6 months
Tian et al. [35]	Meta-analysis	Immunotherapy	/	OS HR: 1.81
Huang et al. [38]	Meta-analysis	PD-1 inhibitor	8067; 17%	OS HR: 1.73; PFS HR: 1.4

5. Immunotherapy Combinations

5.1. Immunotherapy + VEGF Inhibitor + Chemotherapy

An ImPower150 study [13] investigated the efficacy of different therapeutic options in patients with metastatic NSCLC in the first-line setting. It was shown that patients with liver metastases had better outcomes if they were treated with bevacizumab (a VEGF inhibitor), atezolizumab (a PD-L1 inhibitor) and chemotherapy compared to those who were treated with bevacizumab and chemotherapy alone (progression-free survival of 7.4 vs. 4.9 months; overall survival of 13.3 vs. 9.4 months) [13]. There was no difference in outcomes for patients who were treated with atezolizumab and chemotherapy compared to those treated with bevacizumab and chemotherapy [13].

It could be assumed that susceptibility to the VEGF inhibitor in patients with NSCLC and liver metastases arises from the reversal of immune suppression by the VEGF inhibitor, which basically has reduced-infiltration CD8+ T-cells, which may further enhance the T-cell-mediated killing of cancer cells by the PD-L1 inhibitor [13]. This efficacy was shown regardless of PD-L1 expression [13]. This could be explained by the activation of both mechanisms—T-cell activation by atezolizumab and interferon, and the γ-mediated induction of PD-L1 expression through the reprograming of the tumor microenvironment into the active immune system by the VEGFR inhibitor [13]. The efficacy of bevacizumab and atezolizumab in combination with chemotherapy in patients with NSCLC and liver metastases is supported by an investigation in patients with unresectable hepatocellular carcinoma who were treated with bevacizumab and atezolizumab in combination with chemotherapy [13]. It seems that patients with NSCLC and liver metastases respond to the treatment more similarly to patients with primary liver cancer [13].

5.2. Immunotherapy + Chemotherapy

A KEYNOTE 189 [50] study was investigating the efficacy of pembrolizumab versus a placebo in combination with pemetrexed–platinum in patients with metastatic NSCLC. The results showed that patients who were treated with pembrolizumab and pemetrexed–platinum had a significantly improved overall survival (OS), progression-free survival (PFS), and overall response rate (ORR) compared to the control group of patients, regardless of PD-L1 expression [50]. This study also evaluated the efficacy of pembrolizumab and pemetrexed–platinum in patients with liver metastases. The results showed improved OS and PFS in the subset of patients with liver metastases who were treated with the combination therapy, and the results were similar to those without liver metastases [50].

Matsumoto et al. [51] evaluated the efficacy of pembrolizumab mono-therapy compared to the combination of pembrolizumab with chemotherapy in patients with unresectable NSCLC [51]. The results of their research suggested that patients with liver metastases may benefit more from pembrolizumab monotherapy [51].

Another retrospective study evaluated the efficacy of mono-immunotherapy compared to the combination therapy—immunotherapy and chemotherapy—in 1133 patients with NSCLC [52]. Six hundred and seventy-five patients were treated in the first-line setting, 13.3% of them had liver metastases, and there were 14.2% patients with LM in the overall study population. There were no differences in PFS and OS between patients treated with mono-immunotherapy compared to patients treated with the combination therapy [52]. In contrast to the previous study, it was found that liver metastases could be a predictor for early disease progression, among others, in a group of patients treated with mono-immunotherapy, and this was not shown in a group of patients treated with the combination therapy [52]. Further, it was found that STK11 alterations were associated with lower PD-L1 expression, and TP53, TERT, and NF1 alterations were associated with high PD-L1 expression [52]. Also, it was found that STK11 loss and JAK2 alterations were potential predictors of early progression on mono-immunotherapy, and CDKN2A alterations were associated with worse long-term outcomes with the combination treatment [52].

Chen et al. [53] evaluated whether liver metastases could be a predictive indicator in patients with various advanced cancers treated with immunotherapy-based therapies. They found that 8.4% of all patients had liver metastases [53]. Patients with liver metastases had significantly shorter OS than those without LMs [53]. In a subgroup analysis, similar results were found in patients treated with mono-immunotherapy, but in patients treated with the combination therapy, OS was numerically lower, but without statistical significance [53]. In a group of patients with NSCLC, there were 18% of patients with liver metastases, and they had lower PFS and OS compared to patients without liver metastases, but the disease control rate was similar in both groups [53]. Finally, they also performed a meta-analysis to compare treatment outcomes in NSCLC patients with and without liver metastases who were treated with immunotherapy. The results showed that immunotherapy-based treatment correlated with better OS and PFS compared to the treatment with chemotherapy in both groups [53]. A subgroup analysis showed that patients with liver metastases treated with mono-immunotherapy did not have prolonged PFS compared to the patients treated with chemotherapy [53]. In contrast to those patients, patients with liver metastases treated with the combination of immunotherapy and chemotherapy did have prolonged PFS compared to patients treated with chemotherapy [53].

Another meta-analysis showed the PFS and OS benefit for patients with and without liver metastases treated with immunotherapy compared to standard therapies [54]. Also, it was shown that patients with liver metastases who were treated with immunotherapy had worse PFS compared to those without LMs, but the significance was not significant regarding OS [54].

Yin et al. [55] found in their meta-analysis that, in patients with NSCLC and liver metastases, treatment with pembrolizumab+ chemotherapy, atezolizumab+ bevacizumab+ chemotherapy, and nivolumab was superior to the other treatments regarding OS. However, regarding PFS, only atezolizumab+ bevacizumab+ chemotherapy and pembrolizumab+ chemotherapy were superior [55].

A meta-analysis conducted by Shi et al. [56] evaluated the efficacy and safety of first-line immunotherapy and chemotherapy in patients with non-squamous NSCLC. Five studies were included in the final analysis. A pooled analysis showed a significantly lower risk of disease progression and death if patients were treated with the combination of immunotherapy and chemotherapy compared to the chemotherapy alone [56]. This benefit was shown in patients with and without LM [56].

Another meta-analysis evaluated the efficacy of immunotherapy as a mono-therapy and in combination in NSCLC patients with liver, brain, or bone metastases [57]. It was found that patients with liver metastases could benefit from mono-immunotherapy as well as immunotherapy in the combination of chemotherapy plus anti-VEGF therapy, while patients with brain metastases could benefit only from immunotherapy+ chemotherapy [57].

Table 2 shows the outcomes for patients with LM treated with immunotherapy combinations in some studies.

Table 2. The overview of the outcomes for patients with liver metastases treated with immunotherapy combinations in some studies.

Study	Type of Study	Applied Treatment	No of Patients; % LM	Findings
IMpower150 [13]	Clinical trial	Atezolizumab+ Bevacizumab+ Paclitaxel/Platinum vs. Atezolizumab+ Paclitaxel/Platinum vs. Bevacizumab + Paclitaxel/Platinum	1202; 10.2%	OS HR: 0.68
KN189 [50]	Clinical trial	Pembrolizumab+ Pemetrexed–Platinum vs. Pemetrexed–Platinum	616; 16.1% vs. 23.8%	OS HR (LM/no LM): 0.62/0.58
Shi et al. [56]	Meta-analysis	Immunotherapy+ chemotherapy vs. chemotherapy	/	PFS (LM/no LM): HR 0.63/HR 0.56
Yang et al. [57]	Meta-analysis	Immunotherapy combinations or mono-immunotherapy vs. chemotherapy	988; 729 pts with LM	OS HR: 0.72; PFS HR: 0.72

5.3. Immunotherapy + Radiotherapy

It was speculated that radiotherapy (RT) may enhance the response to immunotherapy by improving the antigen release and thus further activating CD8+ T-cells [58–60]. Also, it was shown that stereotactic body radiation therapy (SBRT) could activate the innate and the adaptive immune response within an irradiated tumor field and also in distant metastases [61]. This phenomenon is known as "abscopal effect". Also, RT could increase the expression of MHC I on tumor cells and, thus, it could lead to the better recognition of antigens by cytotoxic T-cells [62].

Further, it was suggested that SBRT delivered to LM may cause a greater immune response than SBRT administrated to lung metastases [24]. In combination with CTLA-4 immune-check point inhibitors, SBRT showed clinical benefits in patients with lung or liver metastases [24].

A study evaluated the safety and efficacy of pembrolizumab with and without concurrent radiotherapy for lung and liver metastases in patients with metastatic NSCLC [63]. It was shown that patients treated with immunotherapy and RT had a better PFS, numerically, compared to patients treated with immunotherapy alone, but without statistical significance [63].

In a study which enrolled patients with oligometastatic NSCLC (including patients with liver metastases) who were treated with the locally ablative therapy of all known metastatic sites, patients were treated with pembrolizumab after 4–12 weeks [64]. The locally ablative therapy included surgery, SBRT and chemo-radiotherapy [64]. The results showed improved PFS without new adverse events [64].

In a large study which included 25 centers, 427 patients with different primary tumors and liver metastases were identified, who were then treated with SBRT [65]. Primary lung cancer was found in 12% of patients [65]. The median OS depended on the primary site of the tumor, which was 10 months for patients with lung cancer, and the one-year survival rate was 50% [65]. The median OS was also associated with the volume of metastases and the biologically effective dose [65]. Patients with liver metastases with a volume < 40 m^3 had better OS, which was also linked with the biologically effective dose ≥ 100 Gy [65].

In a Dutch–Belgian registry, 515 patients with liver metastases were identified, who were treated with SBRT [66]. There were 8.9% patients with primary lung cancer [66]. The local control and OS rate was 87% and 84% at 1 year, 75% and 63% at 2 years, and 68% and 44% at 3 years, respectively [66]. However, the results for the local control and OS according to the primary site of the tumor were not shown [66].

6. Surgery

Al Farai et al. [67] detailed very interesting results. They investigated whether the resection of isolated liver metastasis had an impact on OS. The resection was performed in 23 patients, and 4 of them had NSCLC [67]. They found that the OS after the resection was 24 months, without a significant difference regarding the primary tumor site [68]. Also, they reported 1-, 3-, and 5-year survival rates of 70%, 23%, and 23%, respectively [67].

Ishige et al. [68] retrospectively evaluated the outcomes of seven patients with oligo-recurrent NSCLC in the liver, who underwent hepatic resection. The median OS after liver resection was 24 months [68].

Further, Hakoda et al. [69] reported a case of a patient who was alive 41 months after liver resection due to lung cancer metastasis, and Nagashima et al. [70] reported a case of a patient who was alive more than 5 years after liver resection due to NSCLC metastasis.

A large study was conducted in order to evaluate the outcomes in patients with metastatic NSCLC [71]. Some of the patients underwent surgery of a primary tumor or metastasis [71]. Patients with liver metastases had the worst prognosis, with a median OS of 4 months. Further, patients with only bone, brain, liver, or lung metastasis, as well as those with multiple metastases, were divided into groups depending on whether the primary tumor or metastasis was operated on [71]. So, patients with bone, brain, liver, lung metastasis, as well as those with multiple metastases, had a better outcome if they had undergone an operation of the primary tumor compared to patients who did not [71]. Similarly, this group of patients, except for patients with liver metastasis, also had a better outcome if they underwent an operation of the metastatic lesion, compared to patients who did not [71]. However, the difference was not significant in patients with liver metastasis [71].

7. Conclusions

As was shown, patients with liver metastases had worse prognosis compared to other groups of patients, even if they were treated with immunotherapy. As these patients represent a significant number of patients, we need different therapeutic options. It might be that combination therapies (atezolizumab+ bevacizumab+ chemotherapy; immunotherapy+ chemotherapy) represent the best therapeutic option for this subset of patients at this particular moment. Also, the addition of radiotherapy or resection could be considered for selected patients with a limited number of liver metastases.

Author Contributions: Conceptualization, V.Ć.K.; investigation, V.Ć.K., N.S., M.G., M.S., B.Š., M.R.M. and D.J.; writing—original draft preparation, V.Ć.K.; writing—review and editing, V.Ć.K., N.S., M.G., M.S., B.Š., M.R.M. and D.J. All authors have read and agreed to the published version of the manuscript.

Funding: This research received no external funding.

Conflicts of Interest: The authors declare no conflicts of interest.

References

1. Global Data. Non-Small Cell Lung Cancer (NSCLC)—Epidemiology Forecast to 2025. Available online: https://store.globaldata.com/report/gdhcer132-16--epicast-report-non-small-cell-lung-cancer-nsclc-epidemiology-forecast-to-2025/ (accessed on 15 May 2021).
2. Qin, B.D.; Jiao, X.D.; Liu, J.; Liu, K.; He, X.; Wu, Y.; Ling, Y.; Duan, X.P.; Qin, W.X.; Wang, Z.; et al. The effect of liver metastasis on efficacy of immunotherapy plus chemotherapy in advanced lung cancer. *Crit. Rev. Oncol. Hematol.* **2020**, *147*, 102893. [CrossRef] [PubMed]
3. Riihimäki, M.; Hemminki, A.; Fallah, M.; Thomsen, H.; Sundquist, K.; Sundquist, J.; Hemminki, K. Metastatic sites and survival in lung cancer. *Lung Cancer* **2014**, *86*, 78–84. [CrossRef] [PubMed]
4. Qu, F.J.; Zhou, Y.; Wu, S. Progress of immune checkpoint inhibitors therapy for non-small cell lung cancer with liver metastases. *Br. J. Cancer* **2024**, *130*, 165–175. [CrossRef]
5. Tamura, T.; Kurishima, K.; Nakazawa, K.; Kagohashi, K.; Ishikawa, H.; Satoh, H.; Hizawa, N. Specific organ metastases and survival in metastatic non-small-cell lung cancer. *Mol. Clin. Oncol.* **2015**, *3*, 217–221. [CrossRef] [PubMed]

6. Ashour Badawy, A.; Khedr, G.; Omar, A.; Bae, S.; Arafat, W.; Grant, S. Site of Metastases as Prognostic Factors in Unselected Population of Stage IV Non-Small Cell Lung Cancer. *Asian Pac. J. Cancer Prev.* **2018**, *19*, 1907–1910.
7. da Silva, L.M.; da Silva, G.T.; Bergmann, A.; Costa, G.J.; Zamboni, M.M.; Santos Thuler, L.C. Impact of different patterns of metastasis in non-small-cell lung cancer patients. *Future Oncol.* **2021**, *17*, 775–782. [CrossRef] [PubMed]
8. Akamine, T.; Takenaka, T.; Yano, T.; Okamoto, T.; Yamazaki, K.; Hamatake, M.; Kinoshita, F.; Kohno, M.; Shimokawa, M.; Yoshizumi, T.; et al. Impact of timing and initial recurrence site on post-recurrence survival in resected non-small cell lung cancer. *Eur. J. Surg. Oncol.* **2024**, *50*, 108374. [CrossRef]
9. Takeuchi, E.; Kondo, K.; Okano, Y.; Kunishige, M.; Kondo, Y.; Kadota, N.; Machida, H.; Hatakeyama, N.; Naruse, K.; Ogino, H.; et al. Early mortality factors in immune checkpoint inhibitor monotherapy for advanced or metastatic non-small cell lung cancer. *J. Cancer Res. Clin. Oncol.* **2023**, *149*, 3139–3147. [CrossRef] [PubMed]
10. Deng, J.-Y.; Gou, Q.; Yang, L.; Chen, Z.H.; Yang, M.Y.; Yang, X.R.; Yan, H.H.; Wei, X.W.; Liu, J.Q.; Su, J.; et al. Immune suppressive microenvironment in liver metastases contributes to organ-specific response of immunotherapy in advanced non-small cell lung cancer. *J. Immuno Ther. Cancer* **2023**, *11*, e007218. [CrossRef]
11. Jiang, T.; Cheng, R.; Zhang, G.; Su, C.X.; Zhao, C.; Li, X.; Zhang, J.; Wu, F.; Chen, X.; Gao, G.; et al. Characterization of liver metastasis and its effect on targeted therapy in EGFR-mutant NSCLC: A Multicenter Study. *Clin. Lung Cancer* **2017**, *18*, 631–639.e2. [CrossRef]
12. Wu, K.L.; Tsai, M.J.; Yang, C.J.; Chang, W.A.; Hung, J.Y.; Yen, C.J.; Shen, C.H.; Kuo, T.Y.; Lee, J.Y.; Chou, S.H.; et al. Liver metastasis predicts poorer prognosis in stage IV lung adenocarcinoma patients receiving first-line gefitinib. *Lung Cancer* **2015**, *88*, 187–194. [CrossRef] [PubMed]
13. Nogami, N.; Barlesi, F.; Socinski, M.A.; Reck, M.; Thomas, C.A.; Cappuzzo, F.; Mok, T.S.K.; Finley, G.; Aerts, J.G.; Orlandi, F.; et al. IMpower150 Final Exploratory Analyses for Atezolizumab Plus Bevacizumab and Chemotherapy in Key NSCLC Patient Subgroups with EGFR Mutations or Metastases in the Liver or Brain. *J. Thorac. Oncol.* **2022**, *17*, 309–323. [CrossRef]
14. Hicklin, D.J.; Ellis, L.M. Role of the vascular endothelial growth factor pathway in tumor growth and angiogenesis. *J. Clin. Oncol.* **2005**, *23*, 1011–1027. [CrossRef]
15. Kim, J.M.; Chen, D.S. Immune escape to PD-L1/PD-1 blockade: Seven steps to success (or failure). *Ann. Oncol.* **2016**, *27*, 1492–1504. [CrossRef] [PubMed]
16. Terme, M.; Colussi, O.; Marcheteau, E.; Tanchot, C.; Tartour, E.; Taieb, J. Modulation of immunity by antiangiogenic molecules in cancer. *Clin. Dev. Immunol.* **2012**, *2012*, 492920. [CrossRef]
17. Scalera, S.; Mazzotta, M.; Corleone, G.; Sperati, F.; Terrenato, I.; Krasniqi, E.; Pizzuti, L.; Barba, M.; Vici, P.; Gallo, E.; et al. KEAP1 and TP53 Frame Genomic, Evolutionary, and Immunologic Subtypes of Lung Adenocarcinoma with Different Sensitivity to Immunotherapy. *J. Thorac. Oncol.* **2021**, *16*, 2065–2077. [CrossRef] [PubMed]
18. Wang, S.; Jiang, M.; Yang, Z.; Huang, X.; Li, N. The role of distinct co-mutation patterns with TP53 mutation in immunotherapy for NSCLC. *Genes Dis.* **2020**, *9*, 245–251. [CrossRef]
19. Zhu, M.; Zhang, L.; Cui, H.; Zhao, Q.; Wang, H.; Zhai, B.; Jiang, R.; Jiang, Z. Co-Mutation of FAT3 and LRP1B in Lung Adenocarcinoma Defines a Unique Subset Correlated with the Efficacy of Immunotherapy. *Front. Immunol.* **2022**, *12*, 800951. [CrossRef]
20. Ying, K.; Zou, L.; Wang, D.; Wang, R.; Qian, J. Co-mutation of TP53 and TTN is Correlated with the Efficacy of Immunotherapy in Lung Squamous Cell Carcinoma. *Comb. Chem. High Throughput Screen.* **2024**, *27*, 2699–2711. [CrossRef]
21. Han, G.; Yang, G.; Hao, D.; Lu, Y.; Thein, K.; Simpson, B.S.; Chen, J.; Sun, R.; Alhalabi, O.; Wang, R.; et al. 9p21 loss confers a cold tumor immune microenvironment and primary resistance to immune checkpoint therapy. *Nat. Commun.* **2021**, *12*, 5606. [CrossRef]
22. Gutiontov, S.I.; Turchan, W.T.; Spurr, L.F.; Rouhani, S.J.; Chervin, C.S.; Steinhardt, G.; Lager, A.M.; Wanjari, P.; Malik, R.; Connell, P.P.; et al. CDKN2A loss-of-function predicts immunotherapy resistance in non-small cell lung cancer. *Sci. Rep.* **2021**, *11*, 20059. [CrossRef] [PubMed]
23. Adib, E.; Nassar, A.H.; Akl, E.W.; Abou Alaiwi, S.; Nuzzo, P.V.; Mouhieddine, T.H.; Sonpavde, G.; Haddad, R.I.; Mouw, K.W.; Giannakis, M.; et al. CDKN2A Alterations and Response to Immunotherapy in Solid Tumors. *Clin. Cancer Res.* **2021**, *27*, 4025–4035. [CrossRef] [PubMed]
24. Zhu, L.; Yu, X.; Tang, X.; Hu, C.; Wu, L.; Liu, Y.; Zhou, Q. Evolving landscape of treatments targeting the microenvironment of liver metastases in non-small cell lung cancer. *Chin. Med. J.* **2024**, *137*, 1019–1032. [CrossRef] [PubMed]
25. Zhang, K.; Yang, L.; Wang, J.; Sun, T.; Guo, Y.; Nelson, R.; Tong, T.R.; Pangeni, R.; Salgia, R.; Raz, D.J. Ubiquitin-specific protease 22 is critical to in vivo angiogenesis, growth and metastasis of non-small cell lung cancer. *Cell Commun. Signal.* **2019**, *17*, 167. [CrossRef] [PubMed]
26. Jin, X.; Liu, X.; Zhang, Z.; Xu, L. NT5DC2 suppression restrains progression towards metastasis of non-small-cell lung cancer through regulation p53 signaling. *Biochem. Biophys. Res. Commun.* **2020**, *533*, 354–361. [CrossRef]
27. Castañón, E.; Soltermann, A.; López, I.; Román, M.; Ecay, M.; Collantes, M.; Redrado, M.; Baraibar, I.; López-Picazo, J.M.; Rolfo, C.; et al. The inhibitor of differentiation-1 (Id1) enables lung cancer liver colonization through activation of an EMT program in tumor cells and establishment of the pre-metastatic niche. *Cancer Lett.* **2017**, *402*, 43–51. [CrossRef]

28. Chen, B.; Song, Y.; Zhan, Y.; Zhou, S.; Ke, J.; Ao, W.; Zhang, Y.; Liang, Q.; He, M.; Li, S.; et al. Fangchinoline inhibits non-small cell lung cancer metastasis by reversing epithelial-mesenchymal transition and suppressing the cytosolic ROS-related Akt-mTOR signaling pathway. *Cancer Lett.* **2022**, *543*, 215783. [CrossRef]
29. Jan, Y.H.; Lai, T.C.; Yang, C.J.; Lin, Y.F.; Huang, M.S.; Hsiao, M. Adenylate kinase 4 modulates oxidative stress and stabilizes HIF-1α to drive lung adenocarcinoma metastasis. *J. Hematol. Oncol.* **2019**, *12*, 12. [CrossRef]
30. Vokes, E.E.; Ready, N.; Felip, E.; Horn, L.; Burgio, M.A.; Antonia, S.L.; Arén Frontera, O.; Gettinger, S.; Holgado, E.; Spigel, D.; et al. Nivolumab versus docetaxel in previously treated advanced non-small-cell lung cancer (CheckMate 017 and CheckMate 057): 3-year update and outcomes in patients with liver metastases. *Ann. Oncol.* **2018**, *29*, 959–965. [CrossRef]
31. Borghaei, H.; Gettinger, S.; Vokes, E.E.; Chow, L.Q.M.; Burgio, M.A.; de Castro Carpeno, J.; Pluzanski, A.; Arrieta, O.; Frontera, O.A.; Chiari, R.; et al. Five-Year Outcomes from the Randomized, Phase III Trials CheckMate 017 and 057: Nivolumab Versus Docetaxel in Previously Treated Non-Small-Cell Lung Cancer. *J. Clin. Oncol.* **2021**, *39*, 723–733. [CrossRef]
32. Qiao, M.; Zhou, F.; Hou, L.; Li, X.F.; Zhao, C.; Jiang, T.; Gao, G.; Su, C.; Wu, C.; Ren, S.; et al. Efficacy of immune-checkpoint inhibitors in advanced non-small cell lung cancer patients with different metastases. *Ann. Transl. Med.* **2021**, *9*, 34. [CrossRef] [PubMed]
33. Tumeh, P.C.; Hellmann, M.D.; Hamid, O.; Tsai, K.K.; Loo, K.L.; Gubens, M.A.; Rosenblum, M.; Harview, C.L.; Taube, J.M.; Handley, N.; et al. Liver Metastases and Treatment Outcome with Anti-PD1 Monoclonal Antibody in Patients with Melanoma and NSCLC. *Cancer Immunol Res.* **2017**, *5*, 417–424. [CrossRef] [PubMed]
34. Xie, M.; Li, N.; Xu, X.; Xu, Y.; Li, H.; Zhu, L.; Sheng, J.; Zhou, Z.; Fan, Y. The Efficacy of PD-1/PD-L1 Inhibitors in Patients with Liver Metastasis of Non-Small Cell Lung Cancer: A Real-World Study. *Cancers* **2022**, *14*, 4333. [CrossRef] [PubMed]
35. Tian, B.W.; Han, C.L.; Wang, H.C.; Yan, L.J.; Ding, Z.N.; Liu, H.; Mao, X.C.; Tian, J.C.; Xue, J.S.; Yang, L.S.; et al. Effect of liver metastasis on the efficacy of immune checkpoint inhibitors in cancer patients: A systemic review and meta-analysis. *Clin. Exp. Metastasis* **2023**, *40*, 255–287. [CrossRef]
36. Sridhar, S.; Paz-Ares, L.; Liu, H.; Shen, K.; Morehouse, C.; Rizvi, N.; Segal, N.H.; Jin, X.; Zheng, Y.; Narwal, R.; et al. Prognostic significance of liver metastasis in durvalumab-treated lung cancer patients. *Clin. Lung Cancer* **2019**, *20*, e601–e608. [CrossRef] [PubMed]
37. Juarez-Garcia, A.; Sharma, R.; Hunger, M.; Kayaniyil, S.; Penrod, J.R.; Chouaïd, C. Real-world effectiveness of immunotherapies in pre-treated, advanced non-small cell lung cancer Patients: A systematic literature review. *Lung Cancer* **2022**, *166*, 205–220. [CrossRef]
38. Huang, Y.; Zhu, L.; Guo, T.; Chen, W.; Zhang, Z.; Li, W.; Pan, X. Metastatic sites as predictors in advanced NSCLC treated with PD-1 inhibitors: A systematic review and meta-analysis. *Hum. Vaccines Immunother.* **2020**, *17*, 1278–1287.
39. Kitadai, R.; Okuma, Y.; Hakozaki, T.; Hosomi, Y. The efficacy of immune checkpoint inhibitors in advanced non-small-cell lung cancer with liver metastases. *J. Cancer Res. Clin. Oncol.* **2020**, *146*, 777–785. [CrossRef]
40. Komiya, T.; Takamori, S.; Shimokawa, M. Impact of Liver Metastasis on First-Line Immunotherapy in Stage IV Non-Small Cell Lung Cancer. *World J. Oncol.* **2023**, *14*, 234–245. [CrossRef]
41. Choi, M.G.; Choi, C.M.; Lee, D.H.; Kim, S.W.; Yoon, S.; Kim, W.S.; Ji, W.; Lee, J.C. Different prognostic implications of hepatic metastasis according to front-line treatment in non-small cell lung cancer: A real-world retrospective study. *Transl. Lung Cancer Res.* **2021**, *10*, 2551–2561. [CrossRef]
42. Shiroyama, T.; Suzuki, H.; Tamiya, M.; Tamiya, A.; Tanaka, A.; Okamoto, N.; Nakahama, K.; Taniguchi, Y.; Isa, S.I.; Inoue, T.; et al. Clinical characteristics of liver metastasis in nivolumab-treated patients with non-small cell lung cancer. *Anticancer Res.* **2018**, *38*, 4723–4729. [CrossRef] [PubMed]
43. Funazo, T.; Nomizo, T.; Kim, Y.H. Liver Metastasis Is Associated with Poor Progression-Free Survival in Patients with Non-Small Cell Lung Cancer Treated with Nivolumab. *J. Thorac. Oncol.* **2017**, *12*, e140–e141. [CrossRef] [PubMed]
44. Tournoy, K.G.; Thomeer, M.; Germonpré, P.; Derijcke, S.; De Pauw, R.; Galdermans, D.; Govaert, K.; Govaerts, E.; Schildermans, R.; Declercq, I.; et al. Does nivolumab for progressed metastatic lung cancer fulfill its promises? An efficacy and safety analysis in 20 general hospitals. *Lung Cancer.* **2018**, *115*, 49–55. [CrossRef]
45. Nishino, M.; Ramaiya, N.H.; Chambers, E.S.; Adeni, A.E.; Hatabu, H.; Jänne, P.A.; Hodi, F.S.; Awad, M.M. Immune-related response assessment during PD-1 inhibitor therapy in advanced non-small-cell lung cancer patients. *J. Immunother. Cancer.* **2016**, *4*, 84. [CrossRef] [PubMed]
46. Topalian, S.L.; Hodi, F.S.; Brahmer, J.R.; Gettinger, S.N.; Smith, D.C.; McDermott, D.F.; Powderly, J.D.; Sosman, J.A.; Atkins, M.B.; Leming, P.D.; et al. Five-Year Survival and Correlates Among Patients with Advanced Melanoma, Renal Cell Carcinoma, or Non-Small Cell Lung Cancer Treated with Nivolumab. *JAMA Oncol.* **2019**, *5*, 1411–1420. [CrossRef]
47. Maugeais, M.; Péron, J.; Dalle, S.; Boespflug, A.; Duruissaux, M.; Corbaux, P.; Reverdy, T.; Sahin, G.; Rabier, A.; Lopez, J.; et al. Impact of Liver Metastases and Number of Metastatic Sites on Immune-Checkpoint Inhibitors Efficacy in Patients with Different Solid Tumors: A Retrospective Study. *Biomedicines* **2022**, *11*, 83. [CrossRef]
48. Botticelli, A.; Cirillo, A.; Scagnoli, S.; Cerbelli, B.; Strigari, L.; Cortellini, A.; Pizzuti, L.; Vici, P.; De Galitiis, F.; Di Pietro, F.R.; et al. The Agnostic Role of Site of Metastasis in Predicting Outcomes in Cancer Patients Treated with Immunotherapy. *Vaccines* **2020**, *8*, 203. [CrossRef]

49. Bilen, M.A.; Shabto, J.M.; Martini, D.J.; Liu, Y.; Lewis, C.; Collins, H.; Akce, M.; Kissick, H.; Carthon, B.C.; Shaib, W.L.; et al. Sites of metastasis and association with clinical outcome in advanced stage cancer patients treated with immunotherapy. *BMC Cancer* **2019**, *19*, 857. [CrossRef]
50. Gadgeel, S.; Rodríguez-Abreu, D.; Speranza, G.; Esteban, E.; Felip, E.; Dómine, M.; Hui, R.; Hochmair, M.J.; Clingan, P.; Powell, S.F.; et al. Updated Analysis from KEYNOTE-189: Pembrolizumab or Placebo Plus Pemetrexed and Platinum for Previously Untreated Metastatic Nonsquamous Non-Small-Cell Lung Cancer. *J. Clin. Oncol.* **2020**, *38*, 1505–1517. [CrossRef]
51. Matsumoto, H.; Kobayashi, N.; Somekawa, K.; Fukuda, N.; Kaneko, A.; Kamimaki, C.; Kubo, S.; Tanaka, K.; Tagami, Y.; Teranishi, S.; et al. Pembrolizumab monotherapy versus pembrolizumab plus chemotherapy in patients with non-small-cell lung cancer: A multicenter retrospective trial. *Thorac. Cancer* **2022**, *13*, 228–235. [CrossRef]
52. Hong, L.; Aminu, M.; Li, S.; Lu, X.; Petranovic, M.; Saad, M.B.; Chen, P.; Qin, K.; Varghese, S.; Rinsurongkawong, W.; et al. Efficacy and clinicogenomic correlates of response to immune checkpoint inhibitors alone or with chemotherapy in non-small cell lung cancer. *Nat. Commun.* **2023**, *14*, 695. [CrossRef] [PubMed]
53. Chen, X.J.; Ren, A.; Zheng, L.; Zheng, E.D.; Jiang, T. Pan-Cancer Analysis Identifies Liver Metastases as Negative Predictive Factor for Immune Checkpoint Inhibitors Treatment Outcome. *Front. Immunol.* **2021**, *12*, 651086. [CrossRef] [PubMed]
54. Xia, H.D.; Zhang, W.G.; Zhang, Y.Q.; Shang, X.L.; Liu, Y.G.; Wang, X.W. Liver metastases and the efficacy of immune checkpoint inhibitors in advanced lung cancer: A systematic review and meta-analysis. *Front. Oncol.* **2022**, *12*, 978069. [CrossRef] [PubMed]
55. Yin, Q.; Dai, L.; Sun, R.; Ke, P.; Liu, L.; Jiang, B. Clinical Efficacy of Immune Checkpoint Inhibitors in Non-Small Cell Lung Cancer Patients with Liver Metastases: A Network Meta-Analysis of Nine Randomized Controlled Trials. *Cancer Res. Treat.* **2022**, *54*, 803–816. [CrossRef]
56. Shi, Y.; Chen, W.; Li, C.; Zhang, Y.; Bo, M.; Qi, S.; Li, G. Efficacy and safety of first-line treatments with immune checkpoint inhibitors plus chemotherapy for non-squamous non-small cell lung cancer: A meta-analysis and indirect comparison. *Ann. Palliat. Med.* **2021**, *10*, 2766–2775. [CrossRef]
57. Yang, K.; Li, J.; Bai, C.; Sun, Z.; Zhao, L. Efficacy of Immune Checkpoint Inhibitors in Non-small-cell Lung Cancer Patients with Different Metastatic Sites: A Systematic Review and Meta-Analysis. *Front. Oncol.* **2020**, *10*, 1098.
58. Menon, H.; Ramapriyan, R.; Cushman, T.R.; Verma, V.; Kim, H.H.; Schoenhals, J.E.; Atalar, C.; Selek, U.; Chun, S.G.; Chang, J.Y.; et al. Role of radiation therapy in modulation of the tumor stroma and microenvironment. *Front. Immunol.* **2019**, *10*, 193. [CrossRef]
59. Ko, E.C.; Raben, D.; Formenti, S.C. The integration of radiotherapy with immunotherapy for the treatment of non-small cell lung cancer. *Clin. Cancer Res.* **2018**, *24*, 5792–5806. [CrossRef]
60. Daly, M.E.; Monjazeb, A.M.; Kelly, K. Clinical trials integrating immunotherapy and radiation for non-small-cell lung cancer. *J. Thorac. Oncol.* **2015**, *10*, 1685–1693. [CrossRef]
61. Chen, Y.; Gao, M.; Huang, Z.; Yu, J.; Meng, X. SBRT combined with PD-1/PD-L1 inhibitors in NSCLC treatment: A focus on the mechanisms, advances, and future challenges. *J. Hematol. Oncol.* **2020**, *13*, 105. [CrossRef]
62. Corrao, G.; Marvaso, G.; Ferrara, R.; Lo Russo, G.; Gugliandolo, S.G.; Piperno, G.; Spaggiari, L.; De Marinis, F.; Orecchia, R.; Garassino, M.C.; et al. Stereotatic radiotherapy in metastatic non-small cell lung cancer: Combining immunotherapy and radiotherapy with a focus on liver metastases. *Lung Cancer* **2020**, *142*, 70–79. [CrossRef]
63. Welsh, J.; Menon, H.; Chen, D.; Verma, V.; Tang, C.; Altan, M.; Hess, K.; de Groot, P.; Nguyen, Q.N.; Varghese, R.; et al. Pembrolizumab with or without radiation therapy for metastatic non-small cell lung cancer: A randomized phase I/II trial. *J. Immunother. Cancer* **2020**, *8*, e001001. [CrossRef] [PubMed]
64. Bauml, J.M.; Mick, R.; Ciunci, C.; Aggarwal, C.; Davis, C.; Evans, T.; Deshpande, C.; Miller, L.; Patel, P.; Alley, E.; et al. Pembrolizumab After Completion of Locally Ablative Therapy for Oligometastatic Non-Small Cell Lung Cancer: A Phase 2 Trial. *JAMA Oncol.* **2019**, *5*, 1283–1290. [CrossRef] [PubMed]
65. Mahadevan, A.; Blanck, O.; Lanciano, R.; Peddada, A.; Sundararaman, S.; D'Ambrosio, D.; Sharma, S.; Perry, D.; Kolker, J.; Davis, J. Stereotactic Body Radiotherapy (SBRT) for liver metastasis-clinical outcomes from the international multi-institutional RSSearch® Patient Registry. *Radiat. Oncol.* **2018**, *13*, 26. [CrossRef]
66. Méndez Romero, A.; Schillemans, W.; van Os, R.; Koppe, F.; Haasbeek, C.J.; Hendriksen, E.M.; Muller, K.; Ceha, H.M.; Braam, P.M.; Reerink, O.; et al. The Dutch-Belgian Registry of Stereotactic Body Radiation Therapy for Liver Metastases: Clinical Outcomes of 515 Patients and 668 Metastases. *Int. J. Radiat. Oncol. Biol. Phys.* **2021**, *109*, 1377–1386. [CrossRef] [PubMed]
67. Al Farai, A.; Garnier, J.; Palen, A.; Ewald, J.; Delpero, J.R.; Turrini, O. Will Patients with Liver Metastasis from Aggressives Cancers Benefit from Surgical Resection? *World J. Oncol.* **2022**, *13*, 359–364. [CrossRef]
68. Ishige, F.; Takayama, W.; Chiba, S.; Hoshino, I.; Arimitsu, H.; Yanagibashi, H.; Iwatate, Y. Hepatectomy for oligo-recurrence of non-small cell lung cancer in the liver. *Int. J. Clin. Oncol.* **2018**, *23*, 647–651. [CrossRef]
69. Hakoda, H.; Sekine, Y.; Ichimura, H.; Ueda, K.; Aoki, S.; Mishima, H.; Sako, A.; Kobayashi, K.; Sakata, A.; Sato, Y. Hepatectomy for rapidly growing solitary liver metastasis from non-small cell lung cancer: A case report. *Surg. Case Rep.* **2019**, *5*, 71. [CrossRef]

70. Nagashima, A.; Abe, Y.; Yamada, S.; Nakagawa, M.; Yoshimatsu, T. Long-term survival after surgical resection of liver metastasis from lung cancer. *Jpn. J. Thorac. Cardiovasc. Surg.* **2004**, *52*, 311–313. [CrossRef]
71. Liao, Y.; Fan, X.; Wang, X. Effects of different metastasis patterns, surgery and other factors on the prognosis of patients with stage IV non-small cell lung cancer: A Surveillance, Epidemiology, and End Results (SEER) linked database analysis. *Oncol. Lett.* **2019**, *18*, 581–592. [CrossRef]

current issues in molecular biology

MDPI

Review

A Scoping Review on Hepatoprotective Mechanism of Herbal Preparations through Gut Microbiota Modulation

Chin Long Poo †, Mei Siu Lau †, Nur Liana Md Nasir, Nik Aina Syazana Nik Zainuddin, Mohd Rahimi Ashraf Abd Rahman, Siti Khadijah Mustapha Kamal, Norizah Awang and Hussin Muhammad *

Herbal Medicine Research Centre, Institute for Medical Research, National Institutes of Health, Ministry of Health Malaysia, Shah Alam 40170, Selangor, Malaysia; chinlong@moh.gov.my (C.L.P.); laums@moh.gov.my (M.S.L.); nurliana.md@moh.gov.my (N.L.M.N.); nikaina@moh.gov.my (N.A.S.N.Z.); rahimi.ashraf@moh.gov.my (M.R.A.A.R.); sitikhadijah.mk@moh.gov.my (S.K.M.K.); norizah@moh.gov.my (N.A.)

* Correspondence: hussin.m@moh.gov.my

† These authors contributed equally to this work.

Abstract: Liver diseases cause millions of deaths globally. Current treatments are often limited in effectiveness and availability, driving the search for alternatives. Herbal preparations offer potential hepatoprotective properties. Disrupted gut microbiota is linked to liver disorders. This scoping review aims to explore the effects of herbal preparations on hepatoprotective mechanisms, particularly in the context of non-alcoholic fatty liver disease (NAFLD), non-alcoholic steatohepatitis (NASH), and hepatic steatosis, with a focus on gut microbiota modulation. A systematic search was performed using predetermined keywords in four electronic databases (PubMed, Scopus, EMBASE, and Web of Science). A total of 55 studies were included for descriptive analysis, covering study characteristics such as disease model, dietary model, animal model, intervention details, comparators, and study outcomes. The findings of this review suggest that the hepatoprotective effects of herbal preparations are closely related to their interactions with the gut microbiota. The hepatoprotective mechanisms of herbal preparations are shown through their effects on the gut microbiota composition, intestinal barrier, and microbial metabolites, which resulted in decreased serum levels of liver enzymes and lipids, improved liver pathology, inhibition of hepatic fatty acid accumulation, suppression of inflammation and oxidative stress, reduced insulin resistance, and altered bile acid metabolism.

Keywords: hepatoprotection; herbal plants; gut microbiota; non-alcoholic fatty liver disease; steatohepatitis; scoping review

1. Introduction

The liver is the largest internal organ in the body and plays a vital role in physiological processes, making it susceptible to a wide range of toxic, microbial, metabolic, circulatory, and neoplastic insults [1]. Liver disease accounts for approximately 2 million deaths per year worldwide, which includes non-alcoholic fatty liver disease (NAFLD), hepatic steatosis, cirrhosis, viral hepatitis, and liver cancer [2], and is considered a serious health problem that significantly contributes to mortality and reduced quality of life. At present, available treatments such as colchicine, interferon, corticosteroids, and penicillamine are often limited in efficacy, are costly, and might cause adverse effects [3]. The search for alternative approaches to treat liver diseases using natural resources has become a global direction, particularly through investigating the hepatoprotective properties of herbal plants and their mechanism of actions [1].

In mammals, the gut microbiota consists of bacteria, viruses, and archaea and extends to unicellular eukaryotes like fungi and yeast [4]. The gut microbiome has a vital role in maintaining health by performing several functions, such as food fermentation, human energy metabolism, protection against pathogens, stimulating immune response, and vitamin production. It further serves to strengthen the biochemical barriers of the gut and intestine. Hence, microbiota density and composition changes can affect these functions [5]. The disruption of gut microbial homeostasis is associated with various diseases such as obesity, malnutrition, inflammatory bowel diseases, neurological disorders, cancer, liver diseases, and others [6]. Studies have shown that the disturbance of gut microbiota is related to the development and progression of liver disease. Hence, the gut microbiota could be the potential target in searching for the prevention and treatment of liver disease [7]. Interventions targeting gut microbiota for liver disease treatments include diet, prebiotics, probiotics, fecal microbiota transplantation (FMT), antibiotics, and symbiotics. However, there are no studies in the literature that have compared the direct outcomes and adverse events of prebiotics, probiotics, symbiotics, and FMT [8,9].

The use of herbal plants as an alternative source for disease treatment has been applied for many years. The presence of bioactive phytochemicals plays a major role in their therapeutic efficacy. The most commonly reported hepatoprotective mechanisms of herbal medicines include improving the cellular antioxidant defense system, exerting anti-inflammatory effects, and providing additional protection against hepatic cell injury [10]. Recent studies have demonstrated the interactions between the gut microbiota and herbal medicines in contributing to liver protection in animal models [11,12]. This review aims to systematically identify the effects of various herbal preparations on hepatoprotective mechanisms in vivo, focusing on NAFLD, non-alcoholic steatohepatitis (NASH), and hepatic steatosis involving gut microbiota modulation.

2. Materials and Methods

This review was conducted according to the York Framework of scoping studies by Arksey and O'Malley [13], advanced by Levac et al. [14]. This framework serves as a guide for a standardized and systematic approach to conducting scoping studies to address new or broad research questions of a complex or heterogeneous nature. As the hepatoprotective mechanisms of herbal preparations involving gut microbiota encompass heterogeneous evidence, this framework is suited to be applied. All five stages of the scoping review, namely (1) identification of research questions, (2) identification of relevant studies, (3) selection of studies, (4) data charting, and (5) collation, summarization, and reporting of findings, were undertaken. This scoping review has been registered with the Malaysian National Medical Research Register under the research ID NMRR ID-23-03336-JJR.

2.1. Research Questions

This review was conducted based on the primary research question, "What are the diverse effects of different herbal preparations on hepatoprotective mechanisms in vivo?" This primary question was further expanded to secondary research questions, including the following:

(i) How do these herbal preparations specifically impact liver conditions such as NAFLD, NASH, cirrhosis, and hepatic steatosis?
(ii) What role does gut microbiota modulation play in the hepatoprotective effects of these herbal preparations?
(iii) What are the challenges and limitations of the methodologies used in the in vivo animal studies?

A Population, Intervention, Comparison, and Outcome (PICO) framework was applied to address the study's research questions (Table 1). Three main population categories were targeted to answer the three secondary research questions.

Table 1. Population, Intervention, Comparison, and Outcome (PICO) framework.

PICO	Details
Population	Animal models
Intervention	Herbal preparations, either as a single herb or herbal mixture in any form of formulation
Comparator	Negative control
Outcome	Primary outcome: a. Effect on hepatobiliary system of herbal preparations b. Secondary outcome: c. Changes in liver enzyme levels d. Histological assessment of liver tissues e. Inflammation markers in the liver f. Lipid profile and metabolic parameters g. Gut microbiota composition analysis h. Oxidative stress markers i. Antioxidant enzyme activities j. Insulin resistance k. Bile acid l. Other relevant biochemical or molecular markers

2.2. Search Strategy

Two reviewers independently and systematically searched studies from electronic databases, namely PubMed, Scopus, EMBASE, and Web of Science, for inclusion in this review. The search terms used were a combination of keywords such as "herbal medicines", "medicinal plants", "medicinal herb", "herbal product", "herbal plant", "herbal extract", "plant extract", "phytotherapy", "botanical drug", "nutraceutical", "phytopharmaceutical", "natural product", "phytoconstituent", "phytochemical", "hepatoprotection", "liver protection", "liver disease", "cirrhosis", "liver damage", "liver injury", "hepatotoxicity", "gut microbiota", "gut microbiome", "gut bacteria", "gut flora", "gut axis", "gastrointestinal microbiota" and "gastrointestinal microbiome". The above keywords were used, catered, and adapted for each search engine. Examples of the keyword searches used for PubMed are presented in Supplementary Materials Table S1. The search strategy used a combination of free-text keywords and medical subject headings (MeSH) terms. Additional references for possible inclusion were obtained by manually searching the reference lists of eligible articles. We included all relevant studies from inception to 2023.

2.3. Article Inclusion

Studies were selected based on the inclusion and exclusion criteria shown in Table 2. The selection of articles was performed in two stages. In the first stage, two independent investigators screened the relevant titles and abstracts for inclusion. Titles or abstracts were screened based on inclusion and exclusion criteria. Those irrelevant articles that did not match the inclusion criteria were excluded. Any disagreement was referred to a third investigator for discussion. All relevant studies were exported into a spreadsheet, and duplicates were removed. Another two investigators retrieved full text from the titles and abstract list. In the second stage, full articles were screened to identify studies for inclusion. Another two investigators independently identified the full text for inclusion in this review. Any discrepancy was resolved through discussion with the third investigator. Data from both investigators were also compared to ensure the consistency of the review, and any discrepancies between the reviewers were discussed. Articles were excluded with reasons if they were irrelevant. Relevant articles were then assessed to answer the review questions. The results from the search were managed using Endnote X20 software, and data extracted from the full articles were documented in a Microsoft Excel spreadsheet 2016 MSO (Version 2407 Build 16.0.17830.20166).

Table 2. Inclusion and exclusion criteria.

Inclusion criteria	(a) Primary articles of in vivo (rodent) animal studies (b) Articles that investigated herbal preparations as an intervention in all types of formulations, including extracts, tablets, capsules, powders, liquids, decoctions, as a single herb, or a mixture of herbs (c) Articles that investigated the hepatoprotective mechanism through gut microbiota as the primary outcome
Exclusion criteria	(a) In vitro, in silico, and clinical studies (b) Review papers and book sections (c) Articles that investigated isolated compounds as the intervention (d) Compounds not isolated from herbal preparations or unmentioned sources (e) Any combination of herbs with other chemicals, such as sucrose, synthetic compounds, and prebiotics (f) Fermented herbal products (g) Partially or not accessible articles with abstract only and conference proceedings (h) Articles in languages other than English

2.4. Data Charting

Data extraction was carried out by two independent investigators (C.L.P. and M.S.L.), and any disparities were reviewed by a third investigator (H.M.). Two different data extraction tables were designed for in vivo articles to capture the required data for different studies and outcomes comprehensively. All investigators were briefed and trained on using the data extraction tables beforehand to ensure accurate and consistent data extraction. In general, the categories of main data extracted include the following:

(i) Article identifier: designated number, title, and author
(ii) Article characteristics: year; country; types of animal disease model; and objectives
(iii) Study population: number of animals (control groups, intervention groups); and details of study population (species and strain, age, sex, weight, and total numbers)
(iv) Intervention: plant species, plant part used, extract; formulation, route of administration, dose; duration and co-intervention
(v) Comparator: negative control, intervention description, formulation; dose and duration
(vi) Outcomes: hepatoprotective mechanisms (changes in liver enzyme levels, histological assessments of liver tissues, inflammation markers in the liver, lipid profile, and metabolic parameters; gut microbiota composition analysis, oxidative stress markers, antioxidant enzyme activities, insulin resistance, and other relevant biochemical or molecular markers)

2.5. Data Synthesis (Quality Assessment)

Data were synthesized qualitatively. All included studies are tabulated in Tables 3 and 4. This scoping review was conducted and reported according to the Preferred Reporting Items for Systematic Reviews and Meta-Analyses Extension for Scoping Reviews (PRISMA-SCRs) checklist.

Table 3. Summary of included studies using herbal preparations as hepatoprotective agents.

No.	Disease Models	Ref. Country	Dietary Models	Animal Model (Species; Gender)	Intervention Details (Herbs; Types of Extracts; Dose; Duration)	Comparator	Gut Microbiota Profiling Samples
1	NAFLD	[15] China	HFD	C57BL/6J mice; male	*Penthorum chinense* Pursh.; water extract; 2, 4, and 8 g/kg; 8 weeks	Untreated mice	Stool samples
2	NAFLD	[16] China	HFD	C57BL/6J mice; male	*Phyllanthus emblica*; aqueous extract; 0.9, 1.8, 3.6, and 6 g/kg; 6 weeks	Untreated mice	Stool samples
3	NAFLD	[17] China	HFD	Kunming mice; NS	*Polygoni Multiflori Radix Praeparata*; ethanol extract; 0.68, 1.35, 2.7, and 3 g/kg; 3 weeks	Untreated mice	Stool samples
4	NAFLD	[18] China	HFD	C57BL/6J mice; male	*Eriobotrya japonica*; NS; 25 and 50 mg/kg bw; 8 weeks	Oral gavage once daily, normal saline	Colonic mucosal samples
5	NAFLD	[19] China	HFD	SD rats; male	Walnut green husk; polyphenols; 100 and 200 mg/kg; 8 weeks	Untreated rats	Stool samples
6	NAFLD	[20] China	HFD	C57BL/6J mice; male	*Salviae miltiorrhizae*; polysaccharide; 10 and 20 mg/kg; 8 weeks	Untreated mice	Stool samples
7	NAFLD	[21] China	HFD	C57BL/6J mice; male	*Brassica juncea* var. tumida; water extract; 25 and 50 mg/kg; 8 weeks	Oral gavage once daily, normal saline	Colonic mucosal samples
8	NAFLD	[22] China	HFD	C57BL/6J mice; male	*Luffa cylindrica*; NS; 2 g/kg; 12 weeks	Oral gavage once daily, normal saline	Stool samples
9	NAFLD with low-grade colitis	[23] China	HFD	C57BL/6 mice; male	*Tetrastigma hemsleyanum* ; methanol; 100 and 500 mg/kg; 15 weeks	Untreated mice	Stool samples
10	NAFLD	[24] China	HFHC	C57BL/6 mice; male	Fufang Zhenzhu Tiaozhi (FTZ); ethanol–water extract; 12 g/kg BW/d; 7 weeks	Untreated mice	Stool samples
11	NAFLD	[25] China	HFD	C57BL/6J mice; male	Tianhuang Formula; ethanol extract; 100 mg/kg/day; 6 weeks	Untreated mice	Stool samples
12	Obesity, NAFLD	[26] China	HFD	SD rats; male	Walnut green husk; ethanol–water extract; 600 mg/kg BW; 50 days	Untreated rats	Stool samples
13	NAFLD	[27] China	HFFD	C3H mice; male	Yiqi-Bushen-Tiaozhi (YBT); NS; 26.76 g/kg/day; 12 weeks	Untreated mice	Stool samples

Table 3. *Cont.*

No.	Disease Models	Ref. Country	Dietary Models	Animal Model (Species; Gender)	Intervention Details (Herbs; Types of Extracts; Dose; Duration)	Comparator	Gut Microbiota Profiling Samples
14	NAFLD	[28] China	HFHS	C57BL/6J mice; male	White peony root, licorice, grape seed, and broccoli mixture; NS; 22.5 mg/kg of licorice extract, 37.5 mg/kg of white peony extract, 15 mg/d of broccoli extract, and 30 mg/d of grape seed extract (low dose), double low dose (high dose); 14 weeks	Untreated mice	Stool samples
15	NAFLD	[29] China	HFD	C57BL/6 mice; male	Si Miao Formula (SMF); decoction; 10 and 20 mg/kg; 16 weeks	Oral gavage once daily, distilled water	Stool samples
16	NAFLD	[30] Japan	Western diet	C57BL/6N mice; male	*Vaccinium myrtillus* L.; anthocyanins extract; 2%; 12 weeks	Untreated mice	Stool samples
17	NAFLD	[31] Israel	HFD	C57BL/6J mice; male	*Cannabis sativa* L.; ethanol extract; 5 mg/kg BW; 8 weeks	Untreated mice	Stool samples
18	NAFLD	[32] Taiwan	HFD	Wistar rats; male	*Camellia oleifera*; hot water extract; 4.5 and 9 g/L; 8 weeks	Untreated rats	Stool samples
19	NAFLD	[33] China	HFD	SD rats; male	Da-Chai-Hu (DCH) decoction; aqueous extract; 8 g/kg; 8 weeks	Oral gavage once daily, normal saline	Stool samples
20	NAFLD	[34] China	HFD	SD rats; male	Dahuang Zexie decoction (DZD) herbal mixture; water extract; 5.13 g/kg; 4 weeks	Oral gavage once daily, normal saline	Stool samples
21	NAFLD	[35] China	HFD	SD rats; male	919 Syrup herbal mixture; 10 mL/kg; 4 weeks	Untreated rats	Stool samples
22	NAFLD	[36] China	HFD	C57BL/6J mice; male	*Morinda citrifolia* L.; ethanol–water extract; 100 and 200 mg/kg BW/d; 10 weeks	Untreated mice	Stool samples
23	NAFLD	[37] China	HFD	C57BL/6 mice; male	Jiang Zhi Granule; aqueous extract; 479 and 944 mg/kg/day; 8 weeks	Untreated mice	Stool samples
24	NAFLD	[38] China	HFD	C57BL/6/J mice; male	*Myristica fragrans* Houtt; ethanol-petroleum ether; 125 mg/kg; 4 weeks	Untreated mice	Stool samples
25	NAFLD	[39] China	HFD	SD rats; male	Jian-Gan-Xiao-Zhi decoction; water extract; 8 and 16 g/kg/bw; 12 weeks	Oral gavage once daily, normal saline	Stool samples
26	NAFLD	[40] China	HFD	Wistar rats; male	*Mallotus furetiamus*; NS; 1.7, 2.5 and 3.3 g/kg; 8 weeks	Untreated rats	Stool samples

Table 3. *Cont.*

No.	Disease Models	Ref. Country	Dietary Models	Animal Model (Species; Gender)	Intervention Details (Herbs; Types of Extracts; Dose; Duration)	Comparator	Gut Microbiota Profiling Samples
27	NAFLD	[41] China	HFD	C57BL/6 mice; male	*Fagopyrum dibotrys*; ethanol extract; HFD mixed with FDE at a mass ratio of 9:1; 14 weeks	Untreated mice	Stool samples
28	NAFLD	[42] China	HFD	C57BL/6J mice; male	*Panax ginseng*; water extract; 100 and 200 mg/kg; 12 weeks	Untreated mice	Stool samples
29	NAFLD	[43] Japan	NS	ob/ob mice; male	Bofutsushosan (BTS) decoction; hot water extract; 5% BTS; 4 weeks	Untreated mice	Stool samples
30	NAFLD	[44] China	HFD	SD rats; male	Chinese Herbal Formula (CHF) (*Artemisia capillaries* Thunb., *Polygonum cuspidatum* Sieb. et Zucc, *Curcuma longa* L., *Hypericum japonicum*, *Gardenia jasminoides* Ellis) ethanol extract; 0.47 and 0.93 g/100 g body weight; 4 weeks	Untreated rats	Stool samples
31	NAFLD	[45] China	HFD	C57BL/6 mice; male	*Polygonatum cyrtonema* Hua; hot water extract; 200 and 400 mg/kg bw; 6 weeks	Untreated mice	Stool samples
32	NAFLD	[46] China	HFD with glucose and fructose	C57BL/6J mice; male	*Ficus hirta* Vahl. (FV); water extract; 5 and 10 g/kg BW; 17 weeks	Oral gavage once daily, normal saline	Stool samples
33	NAFLD	[47] China	HFD	C57BL/6J mice; male	*Pueraria lobata*; aqueous (PLS); 400 mg/kg bw; 8 weeks	Oral gavage once daily, distilled water	Stool samples
34	NAFLD	[48] China	HFD	Wistar rats; male	Qinghua Fang decoction (*Sedum sarmentosum* 15 g, *Radix salvia* 9 g, *Rhizoma atractylodis* macrocephalae 12 g, *Tangerine peel* 9 g, *Lotus leaf* 15 g, *Gynostemma pentaphyllum* 9 g, *Wolfiporia cocos* 9 g, *Eupatorium japonicum* 15 g, *Rhizoma alismatis* 15 g, and *Stigma maydis* 9 g); water extract; 0.80, 0.40, and 0.20 g/100 g body weight; 10 weeks	Oral gavage once daily, distilled water	Stool samples
35	NAFLD	[49] China	HFD	C57BL/6N mice; male	*Camellia sinensis*; ethanol extract; 50 and 100 mg/kg; 12 weeks	Oral gavage once daily, distilled water	Stool samples

Table 3. *Cont.*

No.	Disease Models	Ref. Country	Dietary Models	Animal Model (Species; Gender)	Intervention Details (Herbs; Types of Extracts; Dose; Duration)	Comparator	Gut Microbiota Profiling Samples
36	NAFLD	[50] Japan	HFD	STAM mice; male	Daisaikoto (DST) [Bupleurum root (26.09%), Pinellia tuber (17.39%), Scutellaria root (13.04%), peony root (13.04%), Jujube (13.04%), immature orange (8.7%), rhubarb (4.35%), and ginger (4.35%)]; hot water extract; 3%; 3 weeks	Untreated mice	Stool samples
37	NAFLD	[51] Republic of Korea	HFD	C57BL/6J mice; male	Yijin-Tang (YTJ) decoction [crow-dipper (*Pinellia ternata* (Thunb.) Makino), 4 g sun-dried tangerine peel (*Citrus unshiu* Markovich), 4 g china root (*Poria cocos* F.A. Wolf), 2 g licorice (*Glycyrrhiza uralensis* Fisch. ex DC.), and 2 g ginger (*Zingiber officinale* Roscoe)]; hot water; 50 mg/kg and 100 mg/kg; 6 weeks	Untreated mice	Stool samples
38	NAFLD	[52] China	HFD	C57BL/6J mice; male	*Zanthoxylum bungeanum* Maxim; dried; 120 mg/kg b.w; 12 weeks	Untreated mice	Stool samples
39	NAFLD	[53] Brazil	HFD	C57BL/6J mice; male	*Myrciaria jaboticaba*, NS; 5%, 10%, and 15%; 4 weeks	Untreated mice	Stool samples
40	NAFLD	[54] China	HFD	C57BL/6J mice; male	Zhishi Daozhi decoction (912.8 g of *Citrus aurantium*, 6.4 g of *Rheum palmatum*, 19.2 g of *Coptis chinensis*, 12.8 g of *Scutellaria baicalensis*, 19.2 g of *Massa Medicata Fermentata*, 19.2 g of *Atractylodes macrocephala*, 19.2 g of *Poria cocos*, and 12.8 g of *Alisma orientate*); water extract; 14.5 g/kg/bw; 4 weeks	Untreated mice	Stool samples
41	NAFLD	[55] Japan	Western diet	C57BL/6N mice; male	*Ampelopsis grossedentata*; polyphenol; 0.5, 1 and 2%; 12 weeks	Untreated mice	Stool samples
42	NAFLD	[56] China	HFD	C57BL/6 mice; male	*Juglans regia* L.; ethanol extract; 0.6 and 1.2 g/kg; 12 weeks	Oral gavage once daily, saline	Stool samples
43	NAFLD	[57] China	Western diet	C57BL/6N mice; male	*Areca catechu*: areca nut extract (ANE), areca nut polyphenols (ANPs), and arecoline (ARE); 0.005% ARE, 0.25% ANPs and 0.5% ANE; 12 weeks	Untreated mice	Stool samples

Table 3. *Cont.*

No.	Disease Models	Ref. Country	Dietary Models	Animal Model (Species; Gender)	Intervention Details (Herbs; Types of Extracts; Dose; Duration)	Comparator	Gut Microbiota Profiling Samples
44	NAFLD	[58] China	HFD	SD rats; male	Er-Chen decoction (15 g of *Pinellia ternata*, 15 g of *Citrus aurantium* L., 9 g of *Smilax glabra* Roxb., and 4.5 g of *Glycyrrhiza uralensis* Fisch.); water extract; 4.5 and 9 g/kg/d; 12 weeks	Oral gavage once daily, saline	Stool samples
45	NAFLD	[59] China	HFD	SD rats; male	*Gynostemma pentaphyllum* (GP) decoction; water extract; 1.5, 3, and 6 g/kg/d; 4 weeks	Oral gavage once daily, distilled water	Stool samples
46	NAFLD	[60] China	HFD	C57BL/6J mice; male	*Terminalia bellirica* (Gaertn.) Roxb; ethanol extract; 0.9, 1.8, and 3.6 g/kg; 14 weeks	Untreated mice	Stool samples
47	NASH	[61] China	HFD	SD rats; male	*Tylophora yunnanensis* Schltr; ethanol extracts, 80 and 160 mg/kg; 6 weeks	Untreated rats	Stool samples
48	NASH	[62] China	HFD	C57BL/6 J mice; male	*Citrus Varieties* (*Changshan Huyou*), flavonoids, 50 mg/kg; 12 weeks	Untreated mice	Stool samples
49	NASH	[63] China	MCD	C57BL/6 mice; male	Qian-Ghan decoction; water extract; 400 mg/kg; 4 weeks	Untreated mice	Stool samples
50	NASH	[64] China	MCD	C57BL/6 mice; male	Qingrequzhuo capsule (15 g of *Morus alba* L., 9 g of *Coptis chinensis* Franch., 12 g of *Anemarrhena asphodeloides* Bunge, 12 g of *Alisma plantago-aquatica* subsp. orientale (Sam.), 15 g of *Citrus aurantium* L., 9 g of *Carthamus tinctorius* L., 6 g of *Rheum palmatum* L., 15 g of *Smilax glabra* Roxb., 12 g of *Dioscorea oppositifolia* L., 12 g of *Cyathula officinalis*); water extract; 0.48, 0.96, and 1.92 g/kg; 6 weeks	Untreated mice	Stool samples
51	NASH	[65] China	MCD	C57BL/6 mice; male	*Salvia-Nelumbinis* naturalis; water–methanol extract; 750 mg/kg; 4 weeks	Untreated mice	Stool samples
52	Hepatic steatosis	[66] Mexico	HFD	C57BL/6J mice; male	*Agave salmiana;* saponin; 2.8 and 28 mg/kg; 8 weeks	Untreated mice	Stool samples
53	Hepatic steatosis	[67] China	HFD	C57BL/6J mice; male	*Malus domestica*; polyphenols; 125 and 500 mg/kg; 12 weeks	Untreated mice	Stool samples

Table 3. *Cont.*

No.	Disease Models	Ref. Country	Dietary Models	Animal Model (Species; Gender)	Intervention Details (Herbs; Types of Extracts; Dose; Duration)	Comparator	Gut Microbiota Profiling Samples
54	Hepatic steatosis	[68] Canada	HFHS	C57BL/6J mice; male	Arctic berries; ethanol extract; 200 mg/kg; 8 weeks	Untreated mice	Stool samples
55	Hepatic steatosis	[69] China	HFFD	Kunming mice; male	*Artemisia sphaerocephala* Krasch; polysaccharides; 800 mg/kg; 8 weeks	Untreated mice	Stool samples

HFD = high-fat diet; HFHC = high-fat and high-cholesterol diet; HFFD = high-fat and high-fructose diet; HFHS = high-fat and high-sucrose diet; NS = not specified; MCD = methionine- and choline-deficient; SD = Sprague–Dawley; STAM = STelic Animal Model.

Table 4. Effects and mechanisms of herbal preparations compared with control on various liver diseases through gut microbiota modulation.

No.	Ref.	Herbs	Hepatoprotective Mechanisms/Outcomes						
			Liver Enzymes	**Histopathological Parameters**	**Inflammation Markers**	**Lipid Profile**	**Antioxidant Enzyme Activity and Oxidative Stress Markers**	**Insulin Resistance/Bile Acid**	**Gut Microbiota Composition Analysis**
1	[15]	*Penthorum chinense* Pursh.	Decreased AST and ALT	Lowered NAFLD score, reduced fatty droplets, and degenerated hepatocyte ballooning	NR	Decreased TC, TGs, and LDL-c	NR	NR/NR	Decreased F/B ratio
2	[16]	*Phyllanthus emblica*	Decreased AST and ALT	Ameliorated hepatocyte swelling/necrosis, inflammatory cell infiltration, and improved numerous red lipid droplets	NR	Decreased LDL-c, TGs, and HDL-c	NR	NR/NR	Decreased F/B ratio
3	[17]	*Polygoni Multiflori Radix Praeparata*	NR	Suppressed hepatocellular steatosis with clear hepatic cord structure, dilated hepatic sinusoids, and decreased degree of redness in tissues	NR	Inhibited TC, TG, and LDL-c rise; restored HDL-c	NR	NR/Recovered bile acids genes	Decreased F/B ratio
4	[18]	*Eriobotrya japonica*	NR	Alleviated abnormal situation of fat deposition	Lowered hepatic SREBP-1c and ChREBP	Inhibited TC and TG rise	(1) Decreased MDA level (2) Reduced activities of GSH-Px and T-SOD	NR/NR	Decreased F/B ratio

Table 4. *Cont.*

No.	Ref.	Herbs	Hepatoprotective Mechanisms/Outcomes						
			Liver Enzymes	Histopathological Parameters	Inflammation Markers	Lipid Profile	Antioxidant Enzyme Activity and Oxidative Stress Markers	Insulin Resistance/Bile Acid	Gut Microbiota Composition Analysis
5	[19]	Walnut green husk	Inhibited an abnormal increase in AST and ALT	Prevented abnormal situation of liver fat deposition	(1) Reduced IL-1β, IL-6, TNF-α, and MCP-1 (2) Upregulated the expression of Nrf2 (3) Suppressed upregulation of the mRNA expression of colonic IL-1β, IL-6, and TNF-α	(1) Inhibited TC, TG, LDL-c, and NEFA rise (2) Prevented HDL-c decrease	Lowered GSH-Px and T-SOD in liver	NR/NR	(1) Decreased F/B ratio and reversed *Ruminococcaceae-UCG-014* and *Fusicatenibacter* (2) Increased expression levels of ZO-1, occludin, and MUC-2 mRNA in the colon
6	[20]	*Salviae miltiorrhizae*	Decreased ALT and AST	(1) Displayed much more normal hepatic histological morphology and steatosis grade (2) Decreased lipid droplets (3) Induced collagen deposition	(1) Suppressed hepatic expressions of TIMP1 and α-SMA (2) Reversed the decreased trend in the expressions of anti-inflammatory genes (IL-2, IL-10, TGF-B) (3) Reversed the increased trend in the expressions of pro-inflammatory genes (IL-6 and IL-23)	Decreased HDL-c, LDL-c, TGs, TC, NEFAs, and FBG	NR	NR/NR	Reversed increase in *Bifidobacterium* and *Lactobacillus*
7	[21]	*Brassica juncea var. tumida Tsen et Lee*	Decreased AST and ALT	Prevented dyslipidemia, oxidative stress, and hepatocyte damage by reducing hepatic vacuole and gap	Regulated the FOXO, MAPK, and the p53 signaling pathways	(1) Prevented TG and LDL-c increase; maintained HDL-c (2) Decreased hepatic fat	Increased GSH	NR/NR	Increased Bacteroidetes and Firmicutes
8	[22]	*Luffa cylindrica*	Decreased ALT and AST	Ameliorated lipid accumulation in hepatocytes and inhibited development of steatosis	Reduced MCP-1 expression in liver	Decreased TGs, TC, LDL, and FFAs	NR	Decreased fasting insulin in serum and enhanced insulin sensitivity/NR	NR

Table 4. *Cont.*

No.	Ref.	Herbs	Hepatoprotective Mechanisms/Outcomes						
			Liver Enzymes	Histopathological Parameters	Inflammation Markers	Lipid Profile	Antioxidant Enzyme Activity and Oxidative Stress Markers	Insulin Resistance/Bile Acid	Gut Microbiota Composition Analysis
9	[23]	*Tetrastigma hems-leyanum*	Decreased AST	Relieved symptoms of damaged liver structure	(1) Decreased IL-6, TNF-α (2) Lowered MCP-1	(1) Increased HDL-c; decreased LDL-c (2) Decreased PPAR-gamma relative expression (3) Decreased ACC-1 and SREBP-1c genes	(1) Increased SOD; decreased MDA (2) Increased Nrf2 expression (3) Upregulated NQO1 and HO-1 (4) Expressed NQO1, GCLC, and HO-1 genes	(1) Improved glucose clearance rate (2) Decreased serum TBA/NR	(1) Increased SCFAs (2) Increased gut microbiota abundance (3) Increased phylum level of Bacteroidetes, Firmicutes, and *Verrucomicrobia* (4) Enriched *Akkermansia* and Lanchnospiraceae (5) Upregulated occludin and ZO-1 expression
10	[24]	Fufang Zhenzhu Tiaozhi (FTZ)	Decreased ALT and AST	Ameliorated extent of steatosis and lipid droplet numbers in liver	NR	Decreased TC and TGs	NR	NR/NR	Increased the abundance of *Bacteroidetes* and reduced the ratio of *Firmicutes*/*Bacteroidales*
11	[25]	Tianhuang Formula	Decreased ALT and AST	Reduced hepatic steatosis and lipid accumulation	NR	Reduced TC, TGs, and LDL-c	(1) Increased GSH and SOD levels while decreasing MDA level (2) Increased mRNA expression of hepatic TRX1, GCLC, and NQO1 (3) Increased protein expressions of Nrf2, NQO1, and HO-1 (4) Decreased Keap1 protein expression	NR/NR	Increased *Lactobacillus* and its metabolites 5-MIAA
12	[26]	Walnut green husk	Decreased ALT and AST	(1) Reduced liver weight (2) Reduced adipocyte size (3) Improved abnormal liver lipid accumulation	Decreased IL-1β, IL-6, TNF-α, and MCP-1	Reduced TC, TGs, and LDL-c	(1) Improved liver MDA level (2) Decreased P-JNK/JNK ratio (3) Increased Nrf2 level (4) Enhanced activities of GSH-Px and T-SOD (5) Inhibited liver PPARα increase	NR/NR	(1) Increased bacterial generation of SCFAs (*Allobaculum*, *Prevotellaceae*, etc.) (2) Decreased F/B ratio (3) Reduced Lactobacillaceae and Lachnospiraceae

Table 4. *Cont.*

No.	Ref.	Herbs	Hepatoprotective Mechanisms/Outcomes						
			Liver Enzymes	**Histopathological Parameters**	**Inflammation Markers**	**Lipid Profile**	**Antioxidant Enzyme Activity and Oxidative Stress Markers**	**Insulin Resistance/Bile Acid**	**Gut Microbiota Composition Analysis**
13	[27]	Yiqi-Bushen-Tiaozhi (YBT)	Decreased ALT and AST	(1) Improved hepatic inflammation and steatosis (2) Reduced staining of collagen fibers to varying degrees	NR	Improved TGs, TC, HDL-c, and LDL-c	NR	NR/NR	Increased *Rikenellaceae*, *Erysipelotrichaceae_incertae_sedis*, and *Clostridium_XIVb*
14	[28]	White peony root, licorice, grape seed, broccoli	Decreased ALT and AST	Increased liver weight and hepatic fat deposition	Reduced LPS serum level	Reduced TC, TGs, and LDL-c	(1) Downregulated mRNA expression of TLR2, TLR4, and TLR9 (2) Reduced p-JNK/JNK ratio (3) Suppressed inflammatory pathways (MCP-1, ICAM-1, Adcy7, Dock2, Ccr2, Cxcl14, Cx3cr1, Mif, IL-6, and Ccl7)	NR/NR	(1) Increased *Ruminiclostridium_9* and *Blautia* (2) Reduced *Anaerotruncus*, *Bacteroides_Otu199*, *Desulfovibrio* genera (3) Reduced mRNA and protein expression of ZO-1 and occludin
15	[29]	Si Miao Formula (SMF)	Decreased ALT and AST	(1) Decreased adipocyte volume (2) Inhibited macrovascular steatosis	(1) Inhibited Nlrp-3 and IL-1α expression (2) Reduced Mcp-1 in eWAT expression	Decreased TC, LDL-c, and TGs	NR	(1) Reduced FBG and AUC values of IPGTT and IPITT (2) Upregulated glucose metabolism-related pathways/NR	(1) Increased *Akkermansia* genus within Verrumicrobia phylum and *Faecalibaculum* and *Caproiciproducens* genus (2) Decreased Firmicutes
16	[30]	*Vaccinium myrtillus* L.	Decreased AST and ALT	(1) Decreased final body weight, liver weight, liver weight/body weight ratio, and liver fat weight and content (2) Decreased liver lipid droplet accumulation	Reduced MCP-1	Decreased TC and LDL-c	(1) Decreased TBARS, α-SMA, and Keap-1 (2) Increased SOD2, Nrf2, and ubiquitinated Nrf2 (Ub-Nrf2)	Reduced insulin levels and insulin resistance/NR	(1) Decreased F/B ratio and *Deferribactere*, S24-7, *Prevotella*, *Lactobacillales*, and *Clostridiales* abundance (2) Increased *Verrucomicrobia*, *Bacteroides*, *acidifaciens*, *Parabacteroides*, *Akkermansia*, and *muciniphila*

Table 4. *Cont.*

No.	Ref.	Herbs	Hepatoprotective Mechanisms/Outcomes						
			Liver Enzymes	Histopathological Parameters	Inflammation Markers	Lipid Profile	Antioxidant Enzyme Activity and Oxidative Stress Markers	Insulin Resistance/Bile Acid	Gut Microbiota Composition Analysis
17	[31]	*Cannabis sativa* L.	NR	Decreased adipose tissue weight, severity of centrilobular hepatocytic vacuolation (fatty change)	NR	NR	NR	NR/NR	(1) Decreased Firmicutes Bacteroidetes (F/B) ratio, Bacteroidetes, and *Betaproteobacteria* abundance (2) Increased Firmicutes, *Clostridia*, *Flavobacteria*, Deferribacteraceae, *Bifidobacterium*, *Prevotella*, *Lactobacillus*, and *Akkermansia muciniphila*
18	[32]	*Camellia oleifera*	Decreased ALT and AST	(1) Lowered weight gain, abdominal fat, and liver weights (2) Suppressed hepatic fatty change and inflammation	Lowered TNF-α	Lowered TC and TAG concentrations	Lowered MDA	Lowered AUC of OGTT, FBG, and HOMA-IR levels/NR	Increased anaerobe/aerobe ratio
19	[33]	DCH decoction	Decreased liver index, serum AST, and ALT	(1) Decreased body weight, NAS score, lipid-loaded hepatocytes (2) Improved hepatic steatosis, hepatocyte ballooning, and lobular inflammation	NR	Decreased TGs and TC	(1) Increased SOD and GSH-Px (2) Decreased MDA	Lowered AUC of OGTT, FINS, and HOMA-IR levels/NR	(1) Increased *Romboutsia*, *Bacteroides*, *Lactobacillus*, *Akkermansia*, and *Turicibacter* (2) Decreased *Lachnoclostridium* and Enterobacteriaceae (3) Decreased F/B ratio
20	[34]	DZD herbal mixture	Decreased ALT and AST	(1) Decreased body weight and liver weight (2) Relieved macrosteatosis, hepatocyte ballooning	NR	Decreased TGs, TC, HDL-C, and LDL-C	NR	NR/NR	(1) Decreased Firmicutes, Desulfovibrionaceae, *Desulfovibrio*, and *Escherichia/Shigella* (2) Increased *Bacteroidetes*, *Ruminococcaceae*, *Bacteroides*, *Oscillibacter*, and *Butyricicoccus*

Table 4. *Cont.*

No.	Ref.	Herbs	Hepatoprotective Mechanisms/Outcomes						
			Liver Enzymes	Histopathological Parameters	Inflammation Markers	Lipid Profile	Antioxidant Enzyme Activity and Oxidative Stress Markers	Insulin Resistance/Bile Acid	Gut Microbiota Composition Analysis
21	[35]	919 Syrup herbal mixture	NR	NR	NR	NR	NR	NR/NR	(1) Increased Firmicutes, *Proteobacteria*, *Actinobacteria*, *Tenericutes*, *Lactobacillus*, and Ruminococcaceae_UCG-005 ratios (2) Reduced Bacteroidetes, *Spirochaetes*, *Cyanobacteria*, and *Epsilonbacteraeota* ratios
22	[36]	*Morinda citrifolia* L.	Suppressed ALT and AST	(1) Normalized body weight, liver index, and fat index (2) Reduced white adipose tissue (WAT) adipocyte sizes (3) Alleviated liver injury	(1) Decreased serum LPS, TNF-α, IL-β, IL-6, TLR4, MyD88, and NF-κB p65 (2) Improved liver inflammation by suppressing LPS/TLR4/NF-κB pathway	(1) Decreased TC, TGs, and LDL-c (2) Increased HDL-c	(1) Reduced MDA (2) Increased GSH and CAT activity	(1) Decreased FBG and FINS levels (2) Decreased blood glucose, AUC of OGTT, and HOMA-IR levels/NR	(1) Increased Bacteroidota and decreased Firmicutes, *Campilobacterota*, *Desulfobacterota*, *Proteobacteria*, *Parabacteroides*, *Prevotellaceae_UCG-001*, *Lactobacillus*, *Roseburia*, *Akkermansia* genera, *Akkermansia muciniphila*, and *Bacteroides acidifaciens* (2) Decreased F/B ratio
23	[37]	Jiang Zhi Granule	Decreased ALT	(1) Reduced liver lipid deposition (2) Ameliorated extensive micro/macrovascular steatosis and frequent incidence of hepatocyte ballooning	NR	Decreased TC, TGs, and FFAs	Decreased MDA	Decreased serum insulin and HOMA-IR levels/NR	(1) Decreased proteobacteria and F/B ratio (2) Decreased Ruminococcaceae, Desulfovibrionaceae, Rikenellaceae, Dehalobacteriaceae, Christensenellaceae, Peptococcaceae (3) Increased S24_7 and Lachnospiraceae

Table 4. *Cont.*

No.	Ref.	Herbs	Hepatoprotective Mechanisms/Outcomes						
			Liver Enzymes	Histopathological Parameters	Inflammation Markers	Lipid Profile	Antioxidant Enzyme Activity and Oxidative Stress Markers	Insulin Resistance/Bile Acid	Gut Microbiota Composition Analysis
24	[38]	*Myristica fragrans* Houtt (nutmeg)	Decreased AST and ALT	Reduced lipid accumulation	Reduced TNF-α, IL-6, and IL-1β	Decreased LDL-c, TC, and TGs	NR	NR/NR	(1) Regulated Bacteroidetes and Firmicutes (2) Increased probiotic species such as *Lactobacillus*, *Akkermansia*, and Bacteroides
25	[39]	Jian-Gan-Xiao-Zhi decoction	Decreased AST and ALT	Alleviated hepatocyte steatosis	Decreased IL-6, IL-1β, and TNF-α	Decreased TGs and TC	NR	NR/NR	(1) Decreased F/B ratio (2) Increased probiotics such as *Alloprevotella*, *Lactobacillus*, and *Turicibacter* (3) Increased ZO-1 and occludin
26	[40]	*Mallotus furetiamus*	Decreased AST, ALT, and GGT	Reduced liver steatosis	Reduced IL-1β, IL-6, TNFa, LEP, and ADP levels	Reduced TC, TGs, LDL-c, and HDL-c	NR	NR/NR	Increased *Bacteroides fragilis*, *Escherichia coli*, and *Staphylococcus xylosus*
27	[41]	*Fagopyrum dibotrys*	Decreased ALT, AST, and ALP	Reduced hepatic steatosis and lipid accumulation	NR	(1) Reduced TGs, TC, and LDL (2) Increased HDL	NR	Reduced AUCs of the IGT and IIT tests/NR	(1) Alleviated microbial community richness (2) Increased Bacteroidetes and *Verrucomicrobia* population (3) Decreased Firmicutes population and increased *Akkermansia*
28	[42]	Panax ginseng	Decreased AST and ALT	Decreased hepatic accumulation of lipids	Reduced TNF-α, IL-1β, and IL-6	Reduced TC, TGs, LDL-c, and LDL-C/HDL-c ratio	NR	NR/NR	(1) Increased richness and diversity of gut microbiota (2) Increased Bacteroidetes and decreased F/B ratio
29	[43]	Bofutsushosan (BTS) decoction	Decreased AST and ALT	Reduced cellular lipid accumulation, hepatocyte ballooning, and accumulation of inflammatory cells	NR	Decreased TC	NR	NR/NR	Increased *Bacteroides*, *Akkermansia*, and Enterobacteriaceae and decreased *Prevotella*

Table 4. *Cont.*

No.	Ref.	Herbs	Hepatoprotective Mechanisms/Outcomes						
			Liver Enzymes	**Histopathological Parameters**	**Inflammation Markers**	**Lipid Profile**	**Antioxidant Enzyme Activity and Oxidative Stress Markers**	**Insulin Resistance/Bile Acid**	**Gut Microbiota Composition Analysis**
30	[44]	CHF	NR	(1) Reduced fat deposition in hepatocytes (2) Reduced liver index	NR	Decreased TGs and FFAs	NR	NR/NR	(1) Reduced intensities of *Coprococcus catus GD/7*, *Ruminococcus gauvreauii* strain, and *Escherichia coli* strain CAIM 1647 (2) Increased *Actinobacteria*
31	[45]	*Polygonatum cyrtonema* Hua	Decreased AST and ALT	Reduced liver weight, vacuolation, fat accumulation, and liver damage	NR	Decreased TGs, TC, and LDL-c; increased HDL-c	Increased SOD and GSH; decreased MDA	Enhanced glucose regulation/NR	(1) Improved species richness and diversity of bacterial community (2) Reduced F/B ratio (3) Increased *Allobaculum*, *Ruminococcus*, and *Bifidobacterium* and decreased *Helicobacter* and *Acinebacter* (4) Ameliorated decrease in SCFAs, such as isobutyrate and isovalerate content, and promoted production of acetate, propionate, and valerate
32	[46]	*Ficus hirta* Vahl. (FV)	NR	(1) Reduced liver injury (2) Decreased accumulation of lipid droplets in hepatocytes	(1) Reduced the mRNA expression of Srebp-1, Acaca, Hmgcr, Fabp1, Pparγ, and Cd36 (2) Increased ACOX1 and CPT1α levels (3) Downregulated the levels of IL-6, IL-1β, and TNF-α in serum (4) Suppressed the upregulated proteins of TNF-α, IL-6, and IL-1β in the liver tissues (5) Suppressed the mRNA expression of Tnf-α, IL-1β, and Ccl5 in liver tissues	(1) Decreased TGs and TC in liver (2) Decreased TC and LDL-c in serum	NR	NR/NR	(1) Reduced F/B ratio (2) Produced SCFAs

Table 4. *Cont.*

No.	Ref.	Herbs	Hepatoprotective Mechanisms/Outcomes						
			Liver Enzymes	**Histopathological Parameters**	**Inflammation Markers**	**Lipid Profile**	**Antioxidant Enzyme Activity and Oxidative Stress Markers**	**Insulin Resistance/Bile Acid**	**Gut Microbiota Composition Analysis**
33	[47]	*Pueraria lobata*	Decreased AST and ALT	(1) Attenuated hepatic ballooning, inflammation, and fibrosis (2) Decreased histological scores	(1) Decreased IL-6 and TNF-alpha	Reduced TC, TGs, and LDL-c	NR	NR/NR	(1) Decreased Firmicutes (2) Increased Bacteroidota
34	[48]	Qinghua Fang decoction	NR	Improved liver injury	Reduced TNF-α, IL-6, IL-8, and IL-17 indexes	NR	NR	NR/NR	Increased richness and diversity of gut microbiota
35	[49]	*Camellia sinensis*	Decreased ALT, AST, and AKP	(1) Reduced, scattered, and made sparse the number of lipid droplets in liver cells (2) Alleviated fat lesions and the volume of lipid droplets	(1) Decreased IL-1β, IL-4, IL-6, IL-10, TNF-α, and IFN-γ	(1) Decreased D-LA, DAO, and LPS (2) Decreased TC, TGs, and LDL-c (3) Increased HDL-c	(1) Reduced ROS in liver tissue	NR/NR	(1) Decreased Firmicutes in feces (2) Increased *Bacteroides* and *Akkermansia* (3) Decreased F/B ratio
36	[50]	DST	NR	(1) Improved hepatic steatosis (2) Decreased hepatic ballooning	NR	(1) Decreased arachidonic acid, 13,14-dihydro-15-keto prostaglandin J2, and prostaglandin F2α (2) Decreased lipid mediators (derived from omega-3 fatty acids and omega-6 fatty acids)	NR	NR/NR	(1) Improved richness and evenness of microbiome (2) Improved percentage of microbiome composition at family level

Table 4. *Cont.*

No.	Ref.	Herbs	Hepatoprotective Mechanisms/Outcomes						
			Liver Enzymes	**Histopathological Parameters**	**Inflammation Markers**	**Lipid Profile**	**Antioxidant Enzyme Activity and Oxidative Stress Markers**	**Insulin Resistance/Bile Acid**	**Gut Microbiota Composition Analysis**
37	[51]	YTJ decoction	Reduced AST, ALT	Decreased hepatic steatosis and liver damage	NR	NR	NR	NR/NR	Increased *B. acidifaciens*
38	[52]	*Zanthoxylum bungeanum* Maxim	NR	(1) Improve lipid deposition (2) Relieved lipid accumulation	NR	Decreased TC, TGs, and LDL-c	NR	NR/NR	(1) Increased *Bacteroidota* (2) Decreased *Actinobacteria* (3) Decreased harmful microflora related to metabolic diseases (4) Increased beneficial bacteria
39	[53]	*Myrciaria jaboticaba*	Decreased AST and ALT	(1) Reduced lipid droplets (2) Completely reversed steatosis (3) Reduced hepatic nuclear binucleation	Decreased IL-6 and TNF-α	Reduced TC and LDL-c	NR	(1) Decreased glucose level and HOMA-IR index values (2) Lowered insulin level/NR	(1) Lowered bacteria number (2) Decreased F/B ratio (3) Decreased richness and abundance of bacteria community composition (4) Increased Lachnospiraceae and Enterobacteriaceae families and *Parabacteroides*, *Sutterella*, *Allobaculum*, and *Akkermansia* genera
40	[54]	Zhishi Daozhi decoction	Decreased AST and ALT	(1) Alleviated the arrangement of hepatocytes, fat vacuoles of varying sizes, and the marginalized nuclei (2) Decreased irregular cell shape with unclear border, marginalized nuclei, and lipid droplets of different sizes	NR	Reduced TGs and TC	NR	NR/Maintained high levels of acetic acid (AA), propionic acid (PA), butyric acid, and SCFAs	(1) Decreased trend in community richness (2) Altered intestinal flora (3) Decreased Firmicutes, increased Bacteroidetes (4) Reduced F/B ratio

Table 4. *Cont.*

No.	Ref.	Herbs	Hepatoprotective Mechanisms/Outcomes						
			Liver Enzymes	Histopathological Parameters	Inflammation Markers	Lipid Profile	Antioxidant Enzyme Activity and Oxidative Stress Markers	Insulin Resistance/Bile Acid	Gut Microbiota Composition Analysis
41	[55]	*Ampelopsis grossedentata*, vine tea polyphenol (VTP)	Decreased AST and ALT	(1) Decreased WD-induced body weight, liver weight, and epididymal fat weight (2) Reduced lipid droplets and prevented hepatocyte ballooning degeneration in a dose-dependent manner	NR	(1) Reduced TGs, TC, LDL, and VLDL (2) Reduced hepatic lipids content, hepatic TGs, and TC content	(1) Increased hepatic Nrf2 level and target enzymes of Nrf2, such as NQO1 and HO-1 (2) Decreased level of hepatic TBARS, a byproduct of lipid peroxidation	Decreased the serum insulin level/NR	Increased *Akkermansia* and reduced F/B ratio
42	[56]	*Juglans regia* L., defatted walnut powder extract (DWPE)	NR	(1) Lowered liver weights and Lee's index (2) Decreased lipid droplets and hepatocyte ballooning	Decreased expressions of proteins related to inflammation (p-p38, p-ERK, p-JNK, and p-NF-κB) in the liver	Decreased TGs, TC, and LDL; increased HDL	Reduced MDA level	NR/NR	(1) Reverted the profiles of gut microbiota (2) Decreased *Erysipelotrichia*, Firmicutes, and *Actinobacteria* while increased *Clostridiales*, *Bacteroidales*, Prevotellaceae, and *Bacteroides*
43	[57]	Areca catechu (ANE, ANPs, and ARE)	Decreased AST, ALT, and ALT/AST ratio	Decreased liver weight and epididymal fat	NR	(1) Reduced liver's total lipid rate (2) Reduced lipid droplets (3) Reduced T-Cho, HDL, and non-HDL (4) Activated p-AMPKα and decreased expression levels of SREBP2 and HMGCR	NR	NR/NR	Increased *Akkermansia* and decreased *Ruminococcus*

Table 4. *Cont.*

No.	Ref.	Herbs	Hepatoprotective Mechanisms/Outcomes						
			Liver Enzymes	**Histopathological Parameters**	**Inflammation Markers**	**Lipid Profile**	**Antioxidant Enzyme Activity and Oxidative Stress Markers**	**Insulin Resistance/Bile Acid**	**Gut Microbiota Composition Analysis**
44	[58]	Er-Chen decoction	Decreased TC, TGs, ALT, and AST	Alleviated steatosis and inflammatory cell infiltration and decreased lipid accumulation	Reduced IL-6 and IL-1β and mRNA expression of TNF-α	NR	Enhanced activities of SOD and GSH-Px and decreased MDA levels	NR/NR	Increased *Lactobacillus*, *Dubosiella*, Lachnospiraceae, *Akkermansia*, and *Intestinimonas* and decreased *Desulfovibrio* and *C._saccharimonas*
45	[59]	*Gynostemma pentaphyllum* (GP)	Decreased AST and ALT	Reduced the extent of steatosis	Decreased TNF-α, IL-1β, and IL-6	Decreased TC, TGs, and LDL-c; increased HDL	NR	Reduced endotoxemia and insulin resistance/NR	(1) Decreased F/B ratio (2) Increased beneficial bacteria, such as *Lactococcus* spp., and decreased pathogenic bacteria, including *Ruminococcus* spp.
46	[60]	*Terminalia bellirica* (Gaertn.) Roxb	Decreased AST and ALT	Attenuated lipid accumulation and hepatocellular necrosis	NR	Reduced TGs and LDL-c; increased HDL-c	NR	NR/NR	(1) Restored structural disturbances of gut microbiota (2) Reduced *Intestinimonas*, *Lachnoclostridium*, and *Lachnospiraceae_FCS020_*group and increased *Akkermansia* and *Bifidobacterium*
47	[61]	*Tylophora yunnanensis* Schltr	Decreased AST and ALT	Reduced infiltration of lipid droplets and hepatic steatosis	Decreased the protein expression levels of LPS, FFAs, VLDL, IL-6, IL-1β, TNF-α, TGF-β, caspase-1, ASC, and NLRP3 in the liver	(1) Decreased TC, TGs, and LDL-c; increased HDL-c	NR	NR/NR	(1) Increased *Lachnospiraceae*, *Muribaculaceae*, *Peptostreptococcaceae*, *Christenellaceae*, and *Coriobacteriaceae* (2) Inhibited NLRP3, thereby reversing composition of gut microbiota to improve NASH

Table 4. *Cont.*

No.	Ref.	Herbs	Hepatoprotective Mechanisms/Outcomes						
			Liver Enzymes	Histopathological Parameters	Inflammation Markers	Lipid Profile	Antioxidant Enzyme Activity and Oxidative Stress Markers	Insulin Resistance/Bile Acid	Gut Microbiota Composition Analysis
48	[62]	*Citrus Varieties* (*Chang-shan Huyou*)	Decreased ALT	Reduced degree of fatty, infiltration of inflammatory cells, and hepatocyte ballooning	NR	NR	NR	NR/NR	(1) Decreased Chao1, ACE (2) Increased *Christensenellaceace*, *Allobacullum*, and *Erysipelotrichaceae* (3) Decreased *Porphyromonadaceae*, *Streptococcaceae*, and *Eubacterium* and decreased *Porphyromonadaceae*
49	[63]	Qian-Ghan decoction	Decreased AST and ALT	Decreased infiltration of inflammatory cells and NAS value in liver sections	(1) Decreased the mRNA expression of TNF-α and IL-β and expression of TLR4, MYD88, and NF-κB (2) Decreased the protein expression of TLR4 and reduced the phosphorylation of IKBα and P65 in MCD diet-induced NASH mice	Decreased TGs and TC	NR	NR/(1) Decreased serum and liver BA concentration (2) Decreased serum TDCA and TLCA levels (3) Reversed liver CA increase but had no effect on liver βMCA and GCA, TCA, and TβMCA levels (4) Increased fecal levels of secondary BAs, such as α + ω MCA and LCA	Increased *Clostridia* and *Bacteroides*
50	[64]	Qingrequzhuo capsule	Decreased AST and ALT	Alleviated disorganized hepatic plate arrangement, inflammatory cell infiltration, and a large amount of vacuolar-like degeneration; reduced hepatic steatosis and reduced fat droplets	Decreased IL-1β, IL-6, and TNF-α	Decreased TC and TGs	Decreased MDA; increased SOD and GSH-Px activities	NR/NR	(1) Increased diversity of gut microbiota and affected relative abundances of *Dubosiella* and *Blautia* (2) Decreased Firmicutes, Bacteroidetes and *Lachnospiraceae_NK4A136_group*

Table 4. *Cont.*

No.	Ref.	Herbs	Hepatoprotective Mechanisms/Outcomes						
			Liver Enzymes	Histopathological Parameters	Inflammation Markers	Lipid Profile	Antioxidant Enzyme Activity and Oxidative Stress Markers	Insulin Resistance/Bile Acid	Gut Microbiota Composition Analysis
51	[65]	*Salvia-Nelumbinis* naturalis	Decreased ALT	Reduced liver steatosis, inflammation, ballooning, and total NAS score	(1) Decreased mRNA expression of SREBP1, ACC, FASN, and SCD1 (2) Decreased ratio of phosphorylated p65 to total p65 (3) Decreased mRNA expression of inflammatory TNF-α and IL-6	Decreased TGs	NR	NR/ Decreased TBA level, increased fecal norDCA and norCA, and decreased GHCA, wMCA, TwMCA, GCA, and GDCA levels	(1) Decreased F/B ratio (2) Decreased enrichment of *Coriobacteriaceae UCG-002*, *Corynebacterium 1*, *Clostridium sensu stricto 1*, and *Glutamicibacter*, whereas increased *Romboutsia*, *Ruminiclostridium 9*, *Alloprevotella*, *Candidatus Saccharimonas*, *Roseburia*, *Lachnospiraceae NK4A136 group*, and *Akkermansia*
52	[66]	*Agave salmiana*	Increased ALT	Decreased hepatic lipid accumulation	Increased TNF-α	Lowered LDL-c	Decreased MDA	Lowered insulin concentrations/NR	Increased *Akkermansia muciniphila*
53	[67]	Apple polyphenols	NR	NR	NR	NR	NR	NR/NR	(1) Increased *Akkermansia* and *Faecalibaculum* (2) Decreased *Enterococcus* and *Lactobacillus*
54	[68]	Arctic berry extracts	NR	NR	Lowered the mRNA expression of Tlr4 and liver inflammation	(1) Reduced hepatic triacylglycerol accumulation and triacylglycerolemia (2) Increased lipid oxidation	NR	NR/Increased tyrosine-phosphorylated CEACAM-1	(1) Decreased F/B ratio (2) Increased presence of *Peptostreptococcaceae*, *Akkermansia muciniphila*, and *Turicibacter* and lower presence of *Lactobacillus* and *Bifidobacterium* in fecal microbiota (3) Increased *Oscillibacter* and *Turicibacter* (4) Increased branched-chain SCFAs in feces

Table 4. *Cont.*

No.	Ref.	Herbs	Hepatoprotective Mechanisms/Outcomes						
			Liver Enzymes	Histopathological Parameters	Inflammation Markers	Lipid Profile	Antioxidant Enzyme Activity and Oxidative Stress Markers	Insulin Resistance/Bile Acid	Gut Microbiota Composition Analysis
55	[69]	*Artemisia sphaero-cephala* Krasch	NR	(1) Exhibited less microvesicular fatty changes and reduced adipocyte size (2) Reduced number and volume of hepatic fat granules	NR	(1) Decreased hepatic C16:1 and C18:1 levels, TC, TGs, and NEFAs (2) Lowered hepatic ACC, FAS, and SCD-1 and decreased expression of proteins involved in lipogenesis	NR	NR/NR	(1) Decreased Firmicutes (2) Increased Bacteroidetes and *Akkermansia*

NR: Not reported; AST: aspartate aminotransferase; ALT: alanine aminotransferase; TC: total cholesterol; TG: triglycerides; LDL-c: low-density lipoprotein cholesterol; F/B: Firmicutes/Bacteroidetes ratio; HDL-C: High-Density Lipoprotein Cholesterol; SREBP-1c: Sterol Regulatory Element-Binding Protein 1c; ChREBP: Carbohydrate-Responsive Element-Binding Protein; MDA: Malondialdehyde; GSH-Px: Glutathione Peroxidase; T-SOD: Total Superoxide Dismutase; IL-1β: Interleukin 1 Beta; IL-6: Interleukin 6; TNF-α: Tumor Necrosis Factor Alpha; MCP-1: monocyte chemoattractant protein 1; Nrf2: Nuclear Factor Erythroid 2-related Factor 2; NEFAs: Non-Esterified Fatty Acids; ZO-1: zonula occludens-1; MUC-2: Mucin 2; TIMP1: Tissue Inhibitor of Metalloproteinase 1; α-SMA: Alpha-Smooth Muscle Actin; IL-2: Interleukin 2; IL-10: Interleukin 10; TGF-β: Transforming Growth Factor Beta; FBG: fasting blood glucose; FOXO: Forkhead Box Protein O; MAPK: Mitogen-Activated Protein Kinase; p53: Tumor Protein 53; FFAs: free fatty acids; PPAR-gamma: Peroxisome Proliferator-Activated Receptor Gamma; ACC-1: Acetyl-CoA Carboxylase 1; NQO1: NAD(P)H Quinone Dehydrogenase 1; HO-1: Heme Oxygenase-1; GCLC: Glutamate-Cysteine Ligase Catalytic Subunit; TBA: total bile acid; SCFAs: short-chain fatty acids; TRX1: Thioredoxin 1; Keap1: Kelch-Like ECH-Associated Protein 1; 5-MIAA: 5-Methylthioadenosine; ICAM-1: Intercellular Adhesion Molecule 1; Adcy7: Adenylate Cyclase 7; Dock2: Dedicator of Cytokinesis 2; Ccr2: CC Chemokine Receptor 2; Cxcl14: C-X-C Motif Chemokine Ligand 14; Cx3cr1: C-X3-C Motif Chemokine Receptor 1; Mif: Macrophage Migration Inhibitory Factor; Ccl7: C-C Motif Chemokine Ligand 7; P-JNK/JNK: Phosphorylated c-Jun N-terminal Kinase (JNK); LPS: lipopolysaccharide; Nlrp-3: Nucleotide-binding oligomerization domain, Leucine rich Repeat and Pyrin domain containing 3; IPGTT: intraperitoneal glucose tolerance test; IPITT: intraperitoneal insulin tolerance test; TBARS: Thiobarbituric Acid Reactive Substances; TAG: Triacylglycerol; AUC: Area Under the Curve; OGTT: Oral Glucose Tolerance Test; HOMA-IR: Homeostatic Model Assessment for Insulin Resistance; FINS: Fasting Insulin; WAT: White Adipose Tissue; MyD88: Myeloid Differentiation Primary Response 88; NF-κB p65: nuclear factor kappa B p65 subunit; TLR4: Toll-like receptor 4; CAT: Catalase; GGT: gamma-glutamyl transferase; LEP: Leptin; ADP: Adenosine Diphosphate; Acaca: Acetyl-CoA Carboxylase Alpha; Hmgcr: 3-Hydroxy-3-Methylglutaryl-CoA Reductase; Fabp1: Fatty Acid-Binding Protein 1; Pparγ: Peroxisome Proliferator-Activated Receptor Gamma; Cd36: Cluster of Differentiation 36; ACOX1: Acyl-CoA Oxidase 1; Ccl5: Chemokine (C-C Motif) Ligand 5; AKP: alkaline phosphatase; IFN-γ: Interferon Gamma; D-LA: D-lactic Acid; DAO: Diamine Oxidase; AA: Arachidonic Acid; PA: Palmitic Acid; BA: Bile Acid; VLDL: very low-density lipoprotein; p-p38: Phosphorylated p38 (protein kinase); p-ERK: Phosphorylated Extracellular Signal-Regulated Kinase; p-JNK: Phosphorylated c-Jun N-terminal Kinase; p-NF-κB: Phosphorylated Nuclear Factor Kappa B.

3. Results

3.1. Study Inclusion

From a total of 1841 records identified from the initial search on four selected databases, 55 articles were included in this scoping review, as presented in the preferred reporting items for systematic review and meta-analysis (PRISMA) flow chart (Figure 1). In the first stage, 409 studies were removed due to duplication. The titles and abstracts of 1432 articles were then screened for relevance, resulting in the exclusion of 1151 articles. The full texts of the remaining 281 articles were retrieved and assessed for eligibility, and ultimately, 55 studies met the inclusion criteria and were included in the scoping review. Among the 226 excluded articles, 102 studies were not herbal preparations, 79 used isolated compounds, 20 were investigating different outcomes other than hepatoprotection, 9 used herbs in combination with other chemicals, full-text was not available for 7 articles, 7 were not studying gut microbiota, and 2 were review articles.

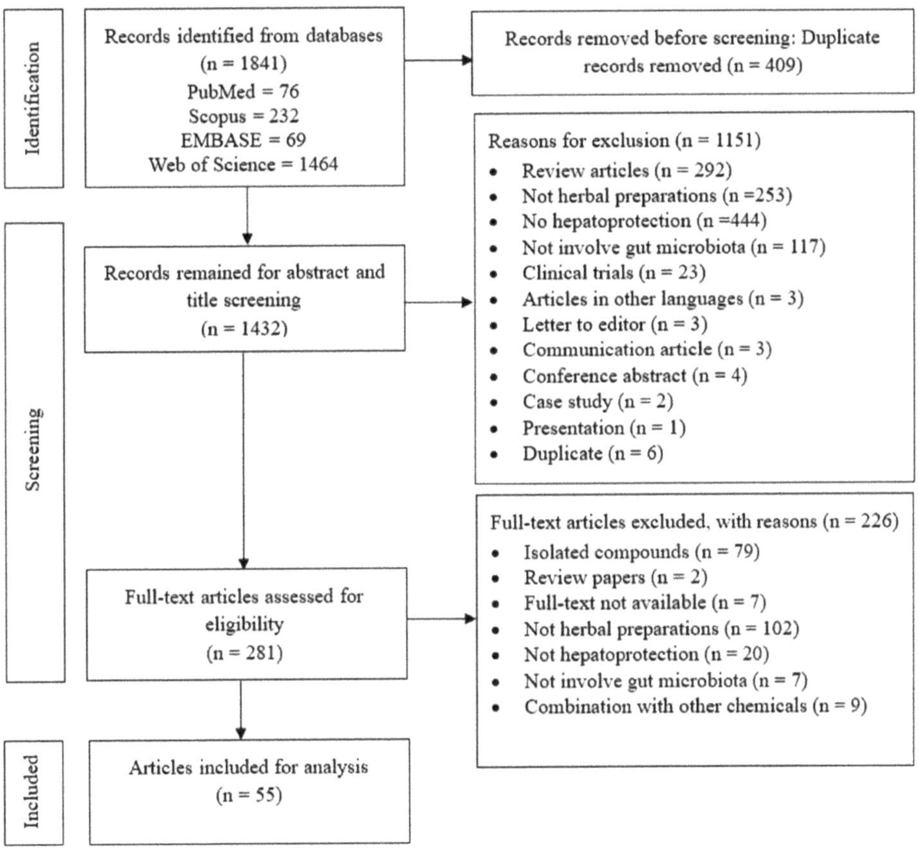

Figure 1. Preferred reporting items for systematic review and meta-analysis (PRISMA) flow chart of included studies.

3.2. Study Characteristics

Table 3 presents the characteristics of all included studies published between 2013 and 2023. All the studies were in vivo (rodent) studies. A total of 45 studies were conducted in China, with 4 studies in Japan and 1 study each in Korea, Brazil, Mexico, Canada, Israel, and Taiwan. Among the 55 articles included, most were in vivo studies reported on NAFLD (n = 46), while only 5 and 4 studies reported on NASH and hepatic steatosis, respectively. Among the 46 included studies, 38 used a high-fat diet (HFD) to induce NAFLD, 3 used a Western diet, and 4 used HFD combined with other dietary models such as cholesterol, fructose, sucrose, and glucose. There was only one study that used a mutant mouse model (ob/ob) without the use of any diet to induce NAFLD. In addition, two studies used HFD, and three studies used a methionine- and choline-deficient (MCD) diet to induce NASH, respectively. Two studies used HFD, and another two studies used a high-fat and high-sucrose diet (HFHS) and a high-fat and high-fructose diet (HFFD) to induce hepatic steatosis, respectively. In terms of the animal model, 42 studies used mice (76.4%), and 13 studies used rats (23.6%), making them the most common model organisms in the included studies. The most common mouse strains used in the 38 studies were the C57BL/6 strain (90.5%), followed by the Kunming strain (4.7%), C3H strain (2.4%), and ob/ob strain (2.4%). Among the 38 studies that used the C57BL/6 strains, one used the STAM mousee model, which was given streptozotocin. On the other hand, Sprague–Dawley rats (77%) were the most common rat strain used, followed by Wistar rats (23%). All animal models used in the studies were male.

Various herbs were used in the studies, with aqueous extracts being the most commonly used preparation, followed by enriched fractions containing specific classes of phytochemicals such as anthocyanins, polyphenols, flavonoids, and polysaccharides. Five articles reported using herbal preparations without specifying the type of extract. Generally, the dose used for herbal preparations ranged from 0.005 to 16 g/kg/day, while the dosage for isolated compounds was 10 to 800 mg/kg/day. In terms of duration, the herbs were administered for a short duration of 3 weeks and a longer duration of up to 16 weeks. All included studies were compared to the negative control groups, which were untreated HFD rats fed with mostly normal saline or distilled water. Furthermore, for assessing gut microbiota profiling, samples were collected from stool (96.4%) and colonic mucosal samples or tissue (3.6%).

3.3. *Hepatoprotective Mechanisms*

The effects and mechanisms of herbal preparations on various liver disease models by modulating gut microbiota are highlighted in Table 4. In general, herbal plants act as hepatoprotective agents by normalizing the liver biomarkers, reducing liver pathology, reducing inflammatory markers or lipid levels, modulating gut microbiota, enhancing antioxidant defense, reducing oxidative stress, mitigating insulin resistance, regulating bile acid (BA) metabolism, and other potential mechanisms. The most common mechanisms reported among all the included studies are modulation of gut microbiota (98.2%), followed by altering the liver pathology (94.5%), lipid profile (87.3%), liver enzymes (76.4%), inflammatory markers (54.5%), antioxidant and oxidative stress markers (34.6%), insulin resistance (23.6%), and BA metabolism (9.1%).

3.4. *Disease Models*

3.4.1. Non-Alcoholic Fatty Liver Disease (NAFLD)

NAFLD is one of the most common causes of chronic liver diseases worldwide, particularly in Western countries. It is characterized by an irregular deposition of excessive hepatic fat in the absence of chronic viral infection and alcohol abuse [70]. The mechanisms involved in the pathogenesis of NAFLD remain enigmatic, as it is a multifactorial disease that aligns with the 'multiple hit pathogenesis' model. Key factors or mechanisms within this model include insulin resistance (IR), lipotoxicity, endoplasmic reticulum (ER) stress, mitochondrial dysfunction, hepatocellular apoptosis, oxidative stress, perturbation of autophagy, gut microbiota alteration, dysregulation of microRNAs, and pro-inflammatory changes [71]. Existing evidence suggests the involvement of gut microbiota in the pathogenesis and progression of NAFLD, primarily through the imbalance of gut microbiota, known as dysbiosis. This could lead to disruption in the gut–liver axis that contributes to the development of NAFLD.

Studies have shown the protective effects of herbal preparations against NAFLD through the modulation of gut microbiota (Table 4). For instance, many of the studied herbal preparations effectively prevented NAFLD in mice by decreasing ALT and AST levels, improving liver pathology, attenuating inflammation, decreasing lipid markers, modulating the gut flora, enhancing the antioxidant defense system, suppressing oxidative stress, and mitigating insulin resistance. Regarding histopathological changes in the liver, most herbs reduced hepatic steatosis and lipid accumulation in mice. For instance, *Luffa cylindrica* (*L. cylindrica*) reduced hepatocyte lipid accumulation and ameliorated hepatic steatosis in HFD-fed mice. In addition, immunohistochemical staining showed reduced monocyte chemoattractant protein 1 (MCP-1) expression levels in the liver after *L. cylindrica* administration. The results showed that *L. cylindrica* could distinctly alleviate hepatic steatosis and hepatocyte inflammation [22]. In another study, treatment with Er-Chen decoction protected against HFD-induced NAFLD in rats by reducing the serum level of pro-inflammatory factors IL-6 and IL-1β and downregulating the mRNA expression of TNF-α in the liver tissue [58]. Additionally, *Ficus hirta* Vahl. (*F. hirta*) was found to decrease the serum levels of IL-6, IL-1β, and TNF-α and suppress the upregulated proteins of TNF-α, IL-6, and IL-1β in the liver tissues of mice fed with HFD [46]. These results indicated that attenuating inflammation is one of the hepatoprotective mechanisms of herbs.

Out of the 46 studies that reported on NAFLD, 41 suggested hepatoprotective mechanisms through decreasing lipid markers. For example, defatted walnut powder extract (DWPE) from *Juglans regia* L. (*J. regia*) protected the liver from lipid injury by decreasing the liver levels of triglycerides (TGs), total cholesterol (TC), and low-density lipoprotein (LDL) and also increasing the level of HDL in HFD-fed mice [56]. Apart from altering the lipid levels, *Tetrastigma hemsleyanum* (*T. hemsleyanum*) leaves decreased the relative expression of PPAR-gamma, ACC-1, and SREBP-1c in HFD-fed mice, which are involved in lipid metabolism. Furthermore, 17 studies have been reported to alleviate NAFLD by enhancing antioxidant defense systems and reducing oxidative stress. This mechanism was shown in a study investigating the hepatoprotective effect of *Polygonatum cyrtonema* Hua polysaccharides (PCPs). The results of the study showed that PCPs improved oxidative stress by increasing the GSH and SOD activities and decreasing the MDA level in the liver of HFD-fed mice [45]. A similar finding was found in *T. hemsleyanum*, which activated the Keap1/Nrf2 signaling pathways and upregulated the genes related to antioxidant enzymes, namely NQO1 and HO-1 [23].

There were 13 studies targeting insulin resistance to protect against NAFLD. Si Miao Formula (SMF) and *Morinda citrifolia* L. (*M. citrifolia*) fruit phenolic extract have shown protective effects on NAFLD by decreasing the fasting blood glucose (FBG) and the AUC values of the intraperitoneal glucose tolerance test (IPGTT) and the intraperitoneal insulin tolerance test (IPITT) in HFHS-fed mice. The results showed that SMF and *M. citrifolia* could improve glucose tolerance and insulin sensitivity [29,36]. Moreover, four studies reported that certain herbal preparations demonstrated beneficial effects on NAFLD by regulating BA metabolism. For example, *T. hemsleyanum* extracts decreased the serum total bile acid (TBA) level in HFD-treated mice [23].

All herbs tested in the 46 included studies demonstrated hepatoprotective effects by modulating the gut microbiota. The extract of *Penthorum chinense* Pursh. (*P. chinense*), for instance, was observed to modulate intestinal microbiota by decreasing the ratio of Firmicutes and Bacteriodetes and increasing the relative abundance of *Akkermansia*, *Parabacteroides*, and *Prevotella* [15]. Additionally, Jian-Gan-Xiao-Zhi decoction (JGXZ) has been reported to improve intestinal permeability, which was attributed to the increased expressions of zonula occludens-1 (ZO-1) and occludin in the colon [39]. The hepatoprotective effects against NAFLD were also observed in mice fed with PCPs and *F. hirta*, presenting restored intestinal barrier dysfunction as a result of increased levels of short-chain fatty acids (SCFAs) [45,46].

3.4.2. Non-Alcoholic Steatohepatitis (NASH)

NASH is a serious subtype of NAFLD, which is the second leading cause of liver transplant in the United States. It is characterized by hepatic inflammation and fibrosis that may progress to cirrhosis, hepatocellular carcinoma, and hepatic failure [72,73]. The pathogenesis of NASH results from a complex interplay between host and environmental factors, including the gut microbiome. Recent studies have shown that disruptions in intestinal microbiota may contribute to the pathogenesis of NASH. Altered gut permeability leads to bacterial translocation and upregulation of pro-inflammatory cytokines, ultimately leading to intestinal dysbiosis. The imbalance in the gut microbiome is believed to be the mechanism that causes the progression of NAFLD to NASH [74].

Out of 55 included studies, 5 reported the mechanisms of hepatoprotection on NASH. For example, the decreased levels of AST and ALT in the serum and liver of rats with NASH were shown by the treatment of *Tylophora yunnanensis* Schltr (*T. yunnanensis*) [61]. Furthermore, an in vivo study of pure total flavonoids from citrus (PTFC) was carried out in mice with NASH, and the results showed that PTFC reduced the degree of fat accumulation, infiltration of inflammatory cells, and ballooning of hepatocytes in the liver [62]. In another study, the Qingrequzhuo capsule (QRQZ) was administered to an MCD-induced NASH mouse model. The results showed reduced mRNA levels of pro-inflammatory cytokines (IL-1β, IL-6, and TNF-α) in the liver of the NASH mice, indicating the therapeutic potential of QRQZ on NASH [64]. Additionally, Qiang-Gan extract (QGE) was reported to decrease liver TG content and liver cholesterol in MCD-induced NASH mice [63]. Moreover, PTFC could attenuate NASH by increasing the relative abundances of Bacteroidaceae and Christensenellaceae and decreasing the relative abundance of Porphyromonadaceae and Streptococcaceae [62]. In addition, QRQZ reduced colonic permeability by increasing the expression of ZO-1 and occludin in the colon. In terms of oxidative stress indicators, the SOD and GSH-Px activities increased, and the MDA levels decreased after QRQZ treatment [64]. Furthermore, the liver and serum BA concentrations were decreased in the QGE-treated mice [63]. The same finding was observed in another study using the Salvia-Nelumbinis naturalis (SNN) formula. The results of that study reported that SNN decreased the total fecal BA level and increased the colonic nuclear BA receptor and FXR expression, suggesting that modulating BA concentration might be a novel strategy to protect against NASH [65].

3.4.3. Hepatic Steatosis

Hepatic steatosis is defined as intrahepatic fat of at least 5% of liver weight, and it can be characterized by the increased accumulation of lipids in the liver. The prolonged accumulation of triacylglycerol may lead to hepatic inflammation, liver metabolic dysfunction, and advanced forms of NAFLD. Several mechanisms are involved in hepatic steatosis, including increased lipolysis from the fat cells, increased flux of fatty acids to the liver, increased de novo lipogenesis, and/or reduced clearance through β-oxidation or very low-density lipoprotein secretion [75]. The development of hepatic steatosis has also been associated with dysbiosis [76].

Various herbs have shown beneficial effects on hepatic steatosis via gut flora alteration (Table 4). For instance, aguamiel concentrate (AC) from *Agave salmiana* (*A. salmiana*) was shown to reduce hepatic steatosis by increasing plasma ALT, decreasing hepatic lipid accumulation, reducing hepatic TNF-α, and decreasing MDA [66]. Arctic berry extracts (ABEs) were found to ameliorate steatosis in mice by alleviating hepatic inflammation, as evidenced by decreased mRNA expression of Tlr4. Additionally, ABEs were reported to reduce HFHS-induced hepatic triacylglycerol accumulation and triacylglycerolemia. In terms of gut microbiota composition, ABEs decreased the Firmicutes/Bacteroidetes (F/B) ratio, increased the abundance of Peptostreptococcaceae, *Akkermansia muciniphila*, and *Turicibacter*, and lowered the presence of *Lactobacillus* and *Bifidobacterium* in the fecal microbiota. In addition, an increased tyrosine-phosphorylated CEACAM-1 level was observed in ABE treatment. This result suggests that ABEs protect the liver from the detrimental effects of HFHS feeding, thus preventing hyperinsulinemia from diet-induced obesity (DIO) [68]. Furthermore, apple polyphenol extract (APE) improved hepatic steatosis induced by a high-fat diet in C57BL/6 mice. This effect was achieved by reducing the total fecal BA content, particularly primary BA levels, such as cholic acid, chenodeoxycholic acid, and muricholic acid [67].

4. Discussion

4.1. Study Characteristics

This is the first review on herbal preparations, consisting of 55 included studies published from 2013 to 2023, summarizing existing in vivo research on their hepatoprotective effects via modulation of gut microbiota. Notably, NAFLD, followed by NASH and hepatic steatosis, emerged as the primary disease models studied, likely due to NAFLD's high prevalence [77]. NAFLD progresses to NASH in 20–30% of cases and may further advance to cirrhosis and hepatocellular carcinoma (HCC). Previous studies have shown that intestinal dysbiosis has been linked to NAFLD progression to NASH [78], with hepatic steatosis serving as an early stage of NAFLD [79].

NAFLD's multifactorial etiologies, often stemming from unhealthy lifestyles like hypercaloric diets, are frequently associated with high fat and fructose intakes [80]. There are many dietary models available to induce liver diseases, such as high-fat diets, high-fructose or -sucrose diets, combined diets, and nutrient-depleted diets (e.g., MCD diet) [81]. The current review revealed that HFD with fat content ranging from 30% to 60% is prevalent in inducing NAFLD, NASH, and hepatic steatosis [82]. A study by Xu et al. [83] has shown that an HFD consisting of 30% fat successfully induced NASH, as indicated by the increased levels of liver weight and liver index. Similarly, our results indicated a remarkable increase in liver index and lipid accumulation in HFD-induced NAFLD mice and rat model groups [41,58].

Apart from HFD, high-fructose or -sucrose and MCD diets are used to induce liver diseases in this review. Excessive fructose or sucrose consumption drives de novo lipogenesis and is linked to NAFLD, obesity, and insulin resistance [84]. Yustisia et al. [85] demonstrated that a high-fat and high-fructose diet induces hepatic steatosis, dyslipidemia, renal lesions, and hyperuricemia in non-obese rats, akin to our study demonstrating hepatic steatosis in Kunming mice fed with a high-fructose diet [69]. MCD diets comprise high sucrose (40%) and fat (10%) but lack methionine and choline, which are essential for hepatic mitochondrial β-oxidation and very low-density lipoprotein (VLDL) synthesis [86]. When animals are fed an MCD diet, the rapid accumulation of TGs in the liver occurs, leading to NASH followed by fibrosis [87]. The study by De Lima et al. [88] used an MCD diet and reported the development of histological NASH with cirrhosis and CK-19-positive hepatocellular carcinoma. Moreover, Kirsch et al. [89] demonstrated that C57/BL6 male mice exhibited histological characteristics closely resembling those observed in humans with NASH after feeding with an MCD diet. Our review similarly found NASH modeled in C57BL/6 mice via MCD diets [63].

This scoping review revealed that the majority of the studies used aqueous extract against liver diseases, especially NAFLD, as it is environmentally friendly and easily accessible [90]. Water is typically used to extract polar phytochemicals such as phenolics, alkaloids, and flavonoids, generally known for hepatoprotective effects [91]. Nevertheless, the extraction methods did not interfere with the hepatoprotective effect of the herbal preparations. For instance, both of the ethanolic and aqueous extracts of *Terminalia belerica* significantly prevented the physical and biochemical changes in ethanol-induced hepatotoxicity in rats [23,60,92]. Furthermore, in order to study the therapeutic effects of herbal preparations, the duration of the treatment plays an important role in hepatoprotective potential. The selected studies utilized different treatment durations, and a short treatment duration of 3 weeks was able to demonstrate hepatoprotective effects [17]. The different treatment durations are believed to be related to the concentration of the herbal preparations, with varying doses observed across the selected studies.

Recent studies emphasize the close relationship between the gut microbiota and health. Microbiota distribution varies based on health, diet, and lifestyle, with changes reflecting disease states [11]. For 16s RNA sequencing analysis, various types of samples, including stool, mucosal swabs, and colonic swabs, are used to determine gut microbiota profiling. The current review revealed that stool samples were predominantly used for gut microbiota profiling due to their non-invasive collection and convenience for repeated sampling [93]. In contrast, only two studies opted for colonic mucosal samples, primarily due to the costliness and invasiveness of this sampling method, which might cause unexpected bleeding and infection [94].

4.2. Regulation of Gut Microbiota

Based on current studies, herbal preparations can regulate the composition of gut microbiota and the levels of gut microbial metabolites to improve diseases. In this review, some herbal preparations have been shown to reduce the F/B ratio. Firmicutes and Bacteriodetes are the most abundant bacteria at the phylum level. An elevated ratio of F/B in the gut microbiota has been associated with metabolic diseases such as obesity, hyperlipidemia, and NAFLD [95]. Therefore, decreasing the F/B ratio would improve the disease conditions. Moreover, *Bacteroidetes*, *Lactobacillus*, *Bacillus*, *Akkermansia*, *Parabacteroides*, and *Prevotella* are shown to be beneficial to the host and referred to as beneficial gut microbiota (BGM) [11]. *Akkermansia*, for instance, protects the intestinal mucosal barrier by degrading mucin and producing propionic acid [96]. *Lactobacillus* bacteria possess probiotic properties that help regulate host health and are used to treat dyslipidemia [97]. *Parabacteroides* and *Prevotella* are anti-inflammatory symbionts that affect immune system metabolic pathways [98]. Our results demonstrated that *Cannabis sativa* L. (*C. sativa*), DCH decoction, *M. citrifolia*, *Myristica fragrans* Houtt (*M. fragrans*), *P. chinense*, JGXZ decoction, and other herbs significantly increased BGM abundance in HFD-induced NAFLD mice [15,31,33,36,38,39].

In addition to increasing the relative abundance of BGM, herbal preparations decreased harmful gut microbiota (HGM) such as *Lachnoclostridium*, Enterobacteriaceae, *Bacillus*, *Shigella*, *Helicobacter*, and other microbiomes in our findings [11]. *Lachnoclostridium* is a mucin-degrading bacteria that can impact both glycan composition and mucus thickness, and its increased abundance is associated with NAFLD in rat models [99]. Enterobacteriaceae has been recently demonstrated to be associated with the progression of NAFLD [100]. A previous study demonstrated that the abundance of *Bacillus*/*Shigella* was significantly increased in HFD rats but decreased in the Qushi Huayu Fang-treated rats [44]. Moreover, sulforaphane, a major chemical constituent of *Raphanus sativus*, was reported to inhibit the growth of HGM such as *Bacillus coli*, *Shigella*, and *Helicobacter pylori*, without affecting the abundance of *Lactobacillus* levels [101]. All these findings indicate that decreasing the abundance of HGM could be the ameliorative mechanism of herbs on NAFLD.

A disrupted intestinal barrier can lead to the pathogenesis of NAFLD. When the intestinal barrier is damaged, endotoxins such as lipopolysaccharide (LPS) from gut microbes could escape from the intestinal lumen into the bloodstream via the hepatic portal system [34]. LPS, a crucial component of the outer membrane of Gram-negative bacteria, increases gut permeability, binds to CD14 and Toll-like receptor 4 on the surface of immune cells, activates the nuclear factor kappa B (NF-κB) signaling pathway, and then triggers a series of inflammatory processes that ultimately lead to chronic low-grade inflammation. Thus, improving the function of the intestinal barrier and suppressing inflammation are the mechanisms to ameliorate NAFLD [102]. Certain herbal preparations have been shown to restore the intestinal barrier function by inhibiting LPS production, primarily by targeting LPS-producing bacteria such as *Desulfovibrio*, *Escherichia–Shigella*, *Enterococcus*, *Klebsiella*, and *Enterobacter* [95]. In our findings, *M. citrifolia* fruit phenolic extract caused a significant reduction in the abundance of *Helicobacter*, Desulfovibrionaceae, *Desulfovibrio*, and *Bacteroides vulgatus* species in HFD-fed mice [36]. Desulfovibrionaceae, a family of sulfate-reducing bacteria (SRB), and Desulfovibrio, a genus of SRB, are reported to be a type of LPS-producing bacteria that are capable of inducing an inflammatory response and damaging intestinal epithelial cells. It has been reported that *Desulfovibrio* is associated with NAFLD, obesity, and inflammation [34]. Our results were in corroboration with previous research conducted by others who discovered that *Dendrobium officinale* improved liver recovery in NAFLD mice by reducing the relative abundance of *Desulfovibrio* by 76.1% and the level of microbial LPS [103]. Additionally, the number of Desulfovibrionaceae bacteria was significantly reduced by *Momordica charantia* L. (*M. charantia*) in obese rats [104]. In the present research, the abundance of *Escherichia/Shigella* was reduced in HFD-induced NAFLD rats after DZD treatment [34]. *Escherichia/Shigella* is another group of Gram-negative bacteria that contain LPS in their cell walls and contribute to the high serum LPS and TNF-α expression usually found in NAFLD patients. An increase in *Escherichia/Shigella* can trigger intestinal inflammation, causing intestinal barrier dysfunction [44].

The interaction between gut microbiota and intestinal epithelial cells plays a crucial role in gastrointestinal tract homeostasis, modulated by tight junctions [105]. Tight junctions are the dynamic permeability barrier of intestinal epithelial cells that prevent the transmission of harmful substances from causing liver damage and inflammation. Decreased expression of tight junction proteins, such as ZO-1, occludin, and claudin-1, can impair intestinal barrier function and increase gut permeability, contributing to the inflammatory response [95]. ZO-1 is a cytoplasmic plaque protein that can connect the transmembrane proteins to the cytoskeleton, while occludin can influence the tight junction of intestinal epithelial cells by regulating macromolecule flux [106,107]. In this study, JGXZ protected the intestinal mucosal barrier by upregulating the expressions of ZO-1 and occludin [39]. These results were in accordance with a study that used polyphenol-rich loquat fruit extract (LFP), which improved HF-induced breakage of the intestinal barrier by increasing the protein expression of ZO-1 and occludin [18]. In addition, increased levels of ZO-1 and occludin were observed in HFD rats after WGHP, Simiao decoction, and Quzhuo Tongbi decoction treatments, indicating that herbs could prevent the accumulation of endotoxin in blood by protecting the intestinal physical barrier [19,29,108].

Gut microbiota produces bioactive metabolites such as SCFAs and BAs. SCFAs are gut microbiota-derived metabolites produced from the fermentation of polysaccharides. SCFAs, including acetic acid, propionic acid, and butyric acid, are beneficial in improving metabolic diseases [95]. SCFAs alleviate gut inflammation and improve intestinal permeability. Studies have shown that SCFAs, specifically butyrate, can reduce the production of IL-6, MCP-1, and TNF-α and inhibit the NF-κB signaling pathway [109] Furthermore, butyric acid and propionic acid can improve glucose homeostasis by activating intestinal gluconeogenesis through the cAMP-dependent pathway and thus reducing hepatic glucose production [110]. Acetic acid improves lipid metabolism by inhibiting liver lipid accumulation through the upregulation of the peroxisome proliferator-activated receptor α (PPARα) gene and several fatty acid oxidation-related proteins [111]. In addition, SCFAs promote insulin secretion and sensitivity while also reducing energy intake and insulin resistance [112,113].

In a recent study, PCPs and *F. hirta* and *T. hemsleyanum* leaf extracts improved intestinal barrier function by targeting the gut microbiota and thus increasing SCFA levels [23,45,46]. All these herbs increased the relative abundance of *Ruminococcaceae, Bifidobacteriales, Allobaculum, Akkermansia, Faecalibaculum, and Butyricicoccus*, which have been identified as SCFA-producing bacteria [114,115]. This is consistent with our findings that showed DZD treatment decreased the abundance of SCFA-producing bacteria (*Ruminococcaceae, Oscillibacter*, and *Butyricicoccus*) in NAFLD rats [34]. These SCFA-producing bacteria prevent metabolic endotoxemia by strengthening the intestinal barrier. Changes in these gut bacteria induce an increment of SCFAs, decreasing intestinal permeability and contributing to intestinal homeostasis [116]. In addition, Han et al. [29] demonstrated that the Simiao decoction could increase the abundance of SCFA-producing bacteria (*Bifidobacterium* and *Faecalibaculum*) and improve lipid metabolism and inflammation in NAFLD rats. Moreover,

Roseburia, which is another SCFA-producing bacterium, has been proven to be associated with type 2 diabetes mellitus (T2DM), obesity, and inflammation [117].

4.3. Liver Biomarkers

Hepatoprotection is often reflected in the normalization of liver biomarkers and a reduction in liver pathology. Biomarkers are measurable indicators that provide information about the physiological state of an organ. Many herbal interventions aim to restore the levels of liver biomarkers to their normal range. These biomarkers include enzymes such as alanine aminotransferase (ALT) and aspartate aminotransferase (AST), which, when elevated, suggest liver cell damage. Alkaline phosphatase (ALP) and gamma-glutamyl transferase (GGT) are other enzymes indicative of liver function and biliary health. Chronic liver diseases often involve inflammation, fibrosis, and, in advanced stages, cirrhosis. All these biomarkers, especially ALT and AST, were decreased in our findings. A reduction in ALT and AST is usually observed after herbal intervention, and it is the hepatoprotective mechanism of herbs to improve liver function. In addition, the gut microbiota plays an important regulatory role in hepatic function and affects the severity of liver diseases. A previous study showed that the abundance of Firmicutes was negatively correlated with ALT and AST levels. A reduction in the abundance of Firmicutes might be associated with worsened hepatic function [118]. This contradicts our findings, which show that the abundance of Firmicutes was reduced, resulting in low ALT and AST levels.

4.4. Liver Histopathology

Herbal interventions that lead to a reduction in liver pathology demonstrate their ability to mitigate these adverse structural changes caused by liver diseases. This can be assessed through histological examinations of liver tissue, where a decrease in inflammatory infiltrates, fibrotic tissue, and other pathological features indicates a positive response to the herbal treatment. This was observed in our findings, where the herbal preparations improved the liver histopathological changes of mice with liver diseases. This hepatoprotective mechanism is closely related to the regulation of the gut microbiota, as gastrointestinal dysbiosis is one of the factors connected to the liver pathological changes, especially fat deposition in the liver [119]. In NAFLD, the decreased abundance of Bacteroidetes and increased abundance of Firmicutes and *Proteobacteria* were reversed by the herbal preparations in our findings, which results in improved liver histopathological changes such as reduced liver lipid droplet accumulation and improved steatosis [120].

4.5. Lipid Metabolism

Disruptions in lipid metabolism, common among NASH patients, lead to the accumulation of various lipids like TC and TGs, alongside increased levels of ALT and AST in the liver, exacerbating liver damage and elevating the risk of cardiovascular diseases, particularly in individuals with T2DM and NAFLD [121,122]. Research indicates that Christensenellaceae levels are reduced in individuals with metabolic syndrome (MetS) compared to healthy counterparts, implying its potential significance in metabolic health and suggesting potential avenues for therapeutic exploration [123]. TGs, the primary form of fat accumulated in NAFLD, play a crucial role in the pathogenesis of the condition. Dysregulation in hepatic de novo lipogenesis, free fatty acid (FFA) availability, and lipid export contribute to hepatic fat deposition, with TG levels associated with insulin resistance, a key factor in NAFLD development. While TG accumulation alone may not be hepatotoxic, the enzyme DGAT2 exacerbates steatohepatitis by reducing TGs and enhancing FFA oxidation. The dysregulation of lipolysis and de novo lipogenesis are particularly critical in the association between insulin resistance and NAFLD. Lycium barbarum oligosaccharides (LBOs) demonstrated efficacy against hepatic steatosis by significantly reducing TGs and improving liver lipid accumulation compared to the HFD group in mouse models [124].

The gut microbiota has been associated with cardiovascular diseases (CVDs) due to their involvement in cholesterol metabolism. It can change the blood lipid composition through its role in BA metabolism and the generation of microbial products such as primary and secondary BAs, trimethylamine N-oxide (TMAO), and SCFAs [125]. A higher abundance of *Lactobacillus* and *Enterobacteriaceae*, a lower abundance of Bacteroidetes, and an increased F/B ratio correlate with CVD. Most of these gut microbiomes may increase gut permeability, leading to increased systemic levels of gut microbial metabolites, which, in turn, cause low-grade chronic inflammation. This inflammation may then alter plasma lipid and lipoprotein levels [126]. Therefore, the herbal preparations in our findings restored the microbial population by decreasing the abundance of *Lactobacillus* and *Enterobacteriaceae*, as well as the F/B ratio, causing a decrease in the serum levels of TC, TGs, and LDL-c. LBOs regulated the modulation of HFD-induced shifts in the intestinal flora, decreasing the abundance of *Barnesiellaceae*, *Barnesiella*, and CHKCI001 while increasing the proportion of *Dubosiella*, *Eubacterium*, and *Lactobacillus* [124].

4.6. Insulin Resistance

Insulin resistance, a central pathogenic feature of metabolic syndrome and a common risk factor for NAFLD, exists in systemic and hepatic forms. Systemic insulin resistance impedes glucose uptake, and hepatic insulin resistance disrupts insulin's suppression of hepatic glucose production while stimulating lipogenesis. Genetic variations in insulin signaling, such as the ENPP1 Lys121Gln and IRS1 Gly972Arg polymorphisms, are associated with NAFLD pathogenesis, adding complexity to the condition. Herbal interventions targeting insulin resistance and regulating BA metabolism show promise in ameliorating NAFLD progression [127]. Moreover, the gut microbiota has been reported to contribute to insulin resistance. The changes in the gut microbiota increased the absorption of LPS, which activates the Toll-like receptor (TLR), leading to increased activation of the innate immune system and inflammatory pathways [128]. As a result, insulin signaling is impaired with decreased phosphorylation of the insulin receptor (IR), insulin receptor substrate (IRS), and Akt, as well as decreased tyrosine phosphorylation of IRS-1 [129]. Previous studies showed that the increased presence of Bacteroidetes in the gut microbiota was negatively correlated with insulin resistance. Bacteroidetes was found to be enriched in the gut of insulin-sensitive individuals, and it reduced fecal monosaccharide accumulation, resulting in lowered lipid accumulation and inflammation, which contributes to reduced insulin resistance [130]. This is in agreement with our findings that the F/B ratio was reduced and Bacteroidetes was increased, which resulted in decreased serum insulin and Homeostatic Model Assessment of Insulin Resistance (HOMA-IR) levels and enhanced insulin sensitivity after herbal intervention.

4.7. Bile Acid Metabolism

BAs are essential molecules in modulating lipid absorption and cholesterol homeostasis. BAs also play a key role in regulating not only their synthesis but also hepatic lipid and energy homeostasis. They are produced by the liver to form the primary BAs and converted to secondary BAs by the gut microbiota [131]. The interactions of BAs and the gut microbiota are bidirectional. Changes in gut microbiota abundance can alter BA composition, influencing the microbiota composition and leading to dysbiosis, with impaired BA signaling evident in patients with NAFLD and dysbiosis. Since BAs and gut microbiota interact reciprocally, modulating them might be a promising herbal strategy for treating liver diseases [120]. In the present study, MCD diet-induced NASH mice who received herbal treatment showed a significant decrease in serum and TBA levels, indicating that herbal treatment modulated the BA metabolism. Furthermore, bile salt hydrolase (BSH) catalyzes the conversion of primary BAs to secondary BAs. BSH-producing microbiomes such as *Bacteroidetes*, *Clostridium*, *Bifidobacterium*, *Listeria*, *Lactobacillus*, *Eubacterium*, *Escherichia*, and *Ruminococcus* have been reported to be involved in BA conversion [132]. Our findings show that the abundance of Bacteroidetes and Clostridium increased in herbal plant-treated mice, contributing to the increased lithocholic acid (LCA) production, suggesting that the underlying possible mechanisms of herbal treatment in preventing NASH may be mediated by the interaction of BAs and the gut microbiota [63].

4.8. Antioxidant Activity and Oxidative Stress

Beyond biomarkers and tissue-level changes, herbal interventions may exert protective effects at the cellular level by preventing oxidative damage, promoting cell survival, and enhancing the regenerative capacity of liver cells. Liver diseases often involve increased cell death through apoptosis or necrosis. Herbal treatments may influence these cell death pathways to promote a balance that supports cell survival. Herbal interventions with anti-inflammatory properties can attenuate hepatic inflammation, which is particularly relevant in conditions like NASH and viral hepatitis, where inflammation is a key driver of disease progression [127]. Studies have shown that altered gut microbiota or intestinal dysbiosis can lead to inflammation and increase oxidative stress. Dysbiosis causes increased intestinal permeability, and elevated permeability triggers the liver's inflammatory response [131, 133]. An alteration in the gut microbiome can cause a shift in the redox status of an organism, as gut bacteria are capable of inducing ROS production that leads to oxidative stress [134]. Thus, the decrease in inflammatory and oxidative stress markers shown in our findings might be the outcome of the hepatoprotective effects of herbal preparations through the regulation of gut microbiota. The reduced level of liver MDA found in rats fed an HF diet after herbal treatment suggests that herbal treatment inhibited the formation of ROS. Several studies have demonstrated that Nrf2 acts as a regulator of cellular antioxidant response. In response to oxidative stress, Nrf2 translocates to the nucleus and activates the transcription of several genes, such as GSH-Px and SOD. In our findings, herbal preparations increased the protein expression of Nrf2 and GSH-Px and SOD activities in the liver of HF-fed rats, indicating that herbal preparations prevent liver oxidative stress by modulating the Nrf2 signaling pathway [25,135]. As discussed earlier, SCFAs have been shown to inhibit oxidative stress and inflammation. Propionic acid and butyric acid can decrease the production of pro-inflammatory factors such as MCP-1 and pro-inflammatory cytokines such as TNF-α and IL-6 [109]. Our findings show that herbal preparations increased the production of SCFAs and increased the relative abundance of SCFA-producing bacteria such as *Ruminococcaceae*, *Bifidobacteriales*, *Allobaculum*, *Akkermansia*, and *Prevotellaceae* [19, 23]. These data suggested that herbal preparations modulated gut microbiota and SCFA composition, resulting in reduced inflammation and oxidative stress.

4.9. Inflammation

A large number of cytokines have been demonstrated to play important roles in regulating liver injury, inflammation, fibrosis, and regeneration [136]. Culture broth extract of *Penicillium polonicum* (endophytic fungi of *P. nigrum* fruit) was found to inhibit pro-inflammatory cytokines by controlling the TNF-α, IL-1β, iNOS, MPO, and NO levels [135,137]. The primary cause of liver-related mortality in NAFLD patients is the accumulation of extracellular matrix in the liver, leading to progressive fibrosis, cirrhosis, portal hypertension, and liver failure. The fibrogenesis process is initiated by signaling from stressed or injured hepatocytes and activated macrophages, specifically Kupffer cells in the liver, prompting the transformation of resident hepatic stellate cells into myofibroblasts. These myofibroblasts then produce matrix proteins at a rate faster than degradation, contributing to fibrosis [138]. There is a growing understanding of NASH-specific fibrogenic pathways [139]. In mouse models of NASH, heightened signaling by the transcriptional activator TAZ in NASH hepatocytes is observed, promoting the secretion of Indian hedgehog ligand. This secretion leads to paracrine fibrogenic signaling to stellate cells [140]. Another NASH-specific pathway involves the PNPLA3-I148M variant directly affecting stellate cell fibrogenesis [141]. This suggests that therapies targeting PNPLA3 function could have antifibrotic effects [142]. Herbal preparations downregulate pro-inflammatory cytokines (IL-1β, IL-6, and TNF-α) in the liver. They also reduce colonic permeability, enhance the expression of tight junction proteins (ZO-1 and occludin), and modulate oxidative stress indicators, as evidenced by increased SOD and GSH-Px activities and decreased MDA levels.

4.10. Future Perspectives on Gut Microbiota-Related Strategies

Gut microbiota-related strategies are emerging as potential approaches to treating liver diseases, given the strong connection between the gut and liver, known as the gut–liver axis. These strategies include diet, probiotics, prebiotics, synbiotics, and fecal microbiota transplantation (FMT) [143]. FMT is a procedure that restores a balanced gut microbiota by transferring feces from a healthy donor to a patient's gastrointestinal tract (GI) [144]. FMT has been widely utilized in the treatment of several liver diseases, such as NAFLD, autoimmune hepatitis (AIH), and hepatocellular carcinoma (HCC). Zhou et al. showed that an 8-week FMT intervention could restore gut microbiota balance and reverse steatohepatitis in mice fed with an HFD, as indicated by a significant decrease in intrahepatic lipid accumulation, triglycerides, and pro-inflammatory cytokines. Additionally, body weight, liver index, fasting blood glucose (FBG), and HOMA-IR were improved, suggesting a beneficial effect of FMT on HFD-induced metabolic disturbances [145]. Subsequently, a clinical study indicated that the clinical symptoms of NAFLD in patients undergoing FMT were significantly improved compared to patients in the no-FMT group [146]. However, standards related to microbiota quality and safety control have not been widely accepted. FMT is still in the preclinical stage for liver disease patients, thereby deserving further clinical investigation. The safety of FMT, including the short-term and long-term adverse effects on liver diseases and the parameters required for post-transplant follow-up, remains unknown [144]. Increasing the availability of clinical data on FMT for liver diseases will enhance our mechanistic understanding in a field that has previously lacked substantial information.

Studies have shown the interaction of herbs and the gut microbiota in treating liver diseases. Herbs can regulate the composition of the gut microbiota and the levels of gut microbial metabolites to improve disease outcomes [95]. Hence, they can be considered part of gut microbiota-related strategies to manage and treat liver disease. Many herbs have been shown to not only regulate gut microbiota but also to possess antioxidant and anti-inflammatory properties. For instance, *Salviae miltiorrhizae* (*S. miltorrhizae*), *Brassica juncea* (*B. juncea*), *T. hemsleyanum*, and other herbs in our review were reported to modulate the gut microbiota, leading to decreased hepatic lipid accumulation, triglycerides, and pro-inflammatory cytokines and thus improving NAFLD conditions. Improvement in liver index and insulin resistance were also observed in HFD-induced NAFLD rats following herbal treatment [20,21,23]. These findings are similar to the results shown in the HFD-induced steatohepatitis mouse model after FMT intervention. This suggests that herbal preparations are as effective as FMT in modulating gut microbiota for treating liver diseases, with additional medical properties that could benefit liver health. Herbs can complement other gut microbiota-related treatments, especially FMT, by providing anti-inflammatory and antioxidant benefits for enhanced efficacy. Nevertheless, more preclinical and clinical investigations are required to confirm the safety and efficacy of herbal preparations and FMT in treating liver diseases through gut microbiota modulation.

5. Study Challenges and Limitations

This review has several limitations. Firstly, only English articles with full-text accessibility were included, potentially excluding valuable findings published in other languages and leading to an incomplete understanding of the subject matter. Secondly, due to insufficient in vitro studies and clinical trials meeting the inclusion criteria, in vitro and clinical studies were excluded. Although the limited number of such studies identified might overlook useful in vitro and human data concerning gut microbiota, the substantial inclusion of animal studies is believed to adequately represent the bulk of available evidence on hepatoprotective mechanisms of herbs specific to gut microbiota regulation. Thirdly, the scope of the review was limited to herbal extracts and mixtures, omitting compound-based interventions, as the therapeutic effect of herbal preparations often stems from the combined effects of multiple phytochemicals present in the whole plant. Therefore, studies on single or isolated compounds were excluded. Fourthly, articles investigating herbal preparations in combination with other chemicals, such as synthetic compounds and prebiotics, were also excluded. Lastly, the quantitative analysis of adverse reactions was not feasible in this review due to the lack of reported adverse effects data in the included studies.

Despite advances in the research of herbal medicines and their effects on the gut microbiota, the standardization of herbal preparations and the identification of the active constituents responsible for hepatoprotective effects are intrinsically complex. The interactions between individual phytochemicals and diverse gut microbiota constituents remain significant hurdles. Further analysis of more chemical composition is necessary to better understand the overall impact of herbal medicines on the gut microbiota. Humans, mice, and rats share some similarities in their gut microbiota composition, particularly at the phylum level, with Firmicutes and Bacteroidetes being dominant in all three. However, there are significant differences in the specific genera and species present, as well as in the microbiota's overall diversity and functional capacity. These differences are influenced by diet, environment, and host genetics. As a result, the effects observed in rodent models may not fully translate to humans. Moreover, translating preclinical results into clinical efficacy poses a challenge. Humans, mice, and rats share similarities in their gut microbiota composition at the phylum level but not at the species level. These differences are influenced by diet, environment, and host genetics. As a result, the effects observed in rodent models may not fully translate to humans [147]. Clinical trials are essential to provide practical and safe strategies to treat liver disease using herbal medicines by modulating the gut microbiota. In addition, the interindividual variability in the gut microbiota and host responsiveness makes it difficult to predict how herbal interventions may influence gut microbiota composition and host outcomes. This may, in turn, present a challenge in creating generalized treatments and underscores the need for personalized approaches. The consistency of results among studies and the efficacy of a herbal intervention in eliciting beneficial changes to the gut microbiota might also be affected. Furthermore, advancing high-throughput sequencing technologies and metabolomics will provide a deeper understanding of the gut microbiota and elucidate the complex mechanisms underlying the action of herbal medicine on the gut microbiota.

6. Conclusions

In conclusion, the gut microbiota closely correlates with liver diseases, especially NAFLD, NASH, and hepatic steatosis. Changes in gut microbiota influence the pathogenesis and progression of liver diseases, making targeting the gut microbiota a promising treatment strategy. Herbal preparations have shown hepatoprotective effects and are important in treating liver disease. This review demonstrated that the hepatoprotective effects of herbal preparations are closely related to their interaction with the gut microbiota. Herbal preparations protect the liver from hepatic diseases by ameliorating gut dysbiosis, restoring gut barrier integrity, promoting gut-derived SCFA production, attenuating increased serum liver enzymes and lipids, improving liver pathology, inhibiting hepatic fatty acid accumulation, suppressing inflammation and insulin resistance, restoring antioxidant status, reducing oxidative stress, and altering BA metabolism. Overall, the roles of herbal preparations in the hepatoprotection of liver diseases, specifically NAFLD through the modulation of the gut microbiota, is encouraging. A comprehensive understanding of the hepatoprotective mechanisms of herbal preparations targeting intestinal microbiota will help develop therapeutics for preventing and treating liver diseases. Further studies are warranted to study the mechanisms of action of effective components in herbal preparations with hepatoprotective effects. This exploration can help identify new targets for managing liver diseases. Additionally, clinical trials are necessary to provide practical and economical approaches for managing liver diseases using herbal preparations that modulate the gut microbiota.

Supplementary Materials: The following supporting information can be downloaded at: https://www.mdpi.com/article/10.3390/cimb46100682/s1, Table S1: Search strategy and keywords.

Author Contributions: Conceptualization, C.L.P. and M.S.L.; methodology, C.L.P. and M.S.L.; validation, H.M.; formal analysis, C.L.P. and M.S.L.; data curation, C.L.P., M.S.L., N.L.M.N., N.A.S.N.Z., M.R.A.A.R., S.K.M.K. and N.A.; writing—original draft preparation, C.L.P. and M.S.L.; writing—review and editing, C.L.P., M.S.L. and H.M.; visualization, C.L.P. and M.S.L.; supervision, H.M. All authors have read and agreed to the published version of the manuscript.

Funding: This research was funded by National Institutes of Health, Ministry of Health, Malaysia.

Acknowledgments: The authors would like to thank the Director General of Health Malaysia, the Deputy Director of Health Malaysia, and the Director of the Institute for Medical Research, Head Centre of Herbal Medicine Research Centre for their support and permission to publish this article.

Conflicts of Interest: The authors declare no conflicts of interest.

References

1. Shamsi-Baghbanan, H.; Sharifian, A.; Esmaeili, S.; Minaei, B. Hepatoprotective herbs, avicenna viewpoint. *Iran. Red. Crescent Med. J.* **2014**, *16*, e12313. [CrossRef]
2. Devarbhavi, H.; Asrani, S.K.; Arab, J.P.; Nartey, Y.A.; Pose, E.; Kamath, P.S. Global burden of liver disease: 2023 update. *J. Hepatol.* **2023**, *79*, 516–537. [CrossRef]
3. Ghosh, N.; Ghosh, R.; Mandal, V.; Mandal, S.C. Recent advances in herbal medicine for treatment of liver diseases. *Pharm. Biol.* **2011**, *49*, 970–988. [CrossRef] [PubMed]
4. D'Argenio, V.; Salvatore, F. The role of the gut microbiome in the healthy adult status. *Clin. Chim. Acta* **2015**, *451*, 97–102. [CrossRef] [PubMed]
5. Ma, T.; Shen, X.; Shi, X.; Sakandar, H.A.; Quan, K.; Li, Y.; Jin, H.; Kwok, L.-Y.; Zhang, H.; Sun, Z. Targeting gut microbiota and metabolism as the major probiotic mechanism—An evidence-based review. *Trends Food Sci. Technol.* **2023**, *138*, 178–198. [CrossRef]
6. Afzaal, M.; Saeed, F.; Shah, Y.A.; Hussain, M.; Rabail, R.; Socol, C.T.; Hassoun, A.; Pateiro, M.; Lorenzo, J.M.; Rusu, A.V.; et al. Human gut microbiota in health and disease: Unveiling the relationship. *Front. Microbiol.* **2022**, *13*, 999001. [CrossRef]
7. Porras, D.; Nistal, E.; Martínez-Flórez, S.; Pisonero-Vaquero, S.; Olcoz, J.L.; Jover, R.; González-Gallego, J.; García-Mediavilla, M.V.; Sánchez-Campos, S. Protective effect of quercetin on high-fat diet-induced non-alcoholic fatty liver disease in mice is mediated by modulating intestinal microbiota imbalance and related gut-liver axis activation. *Free Radic. Biol. Med.* **2017**, *102*, 188–202. [CrossRef]
8. Liu, J.; Yang, D.; Wang, X.; Asare, P.T.; Zhang, Q.; Na, L.; Shao, L. Gut microbiota targeted approach in the management of chronic liver diseases. *Front. Cell. Infect. Microbiol.* **2022**, *12*, 774335. [CrossRef]
9. Shah, Y.R.; Ali, H.; Tiwari, A.; Guevara-Lazo, D.; Nombera-Aznaran, N.; Pinnam, B.S.M.; Gangwani, M.K.; Gopakumar, H.; Sohail, A.H.; Kanumilli, S. Role of fecal microbiota transplant in management of hepatic encephalopathy: Current trends and future directions. *World J. Hepatol.* **2024**, *16*, 17. [CrossRef]
10. Pandey, B.; Baral, R.; Kaundinnyayana, A.; Panta, S. Promising hepatoprotective agents from the natural sources: A study of scientific evidence. *Egypt. Liver J.* **2023**, *13*, 14. [CrossRef]
11. An, X.; Bao, Q.; Di, S.; Zhao, Y.; Zhao, S.; Zhang, H.; Lian, F.; Tong, X. The interaction between the gut Microbiota and herbal medicines. *Biomed. Pharmacother.* **2019**, *118*, 109252. [CrossRef]
12. Meng, X.; Li, S.; Li, Y.; Gan, R.Y.; Li, H.B. Gut microbiota's relationship with liver disease and role in hepatoprotection by dietary natural products and probiotics. *Nutrients* **2018**, *10*, 1457. [CrossRef]
13. Arksey, H.; O'Malley, L. Scoping studies: Towards a methodological framework. *Int. J. Soc. Res. Methodol.* **2005**, *8*, 19–32. [CrossRef]
14. Levac, D.; Colquhoun, H.; O'Brien, K.K. Scoping studies: Advancing the methodology. *Implement. Sci.* **2010**, *5*, 69. [CrossRef]
15. Li, X.; Zhao, W.; Xiao, M.; Yu, L.; Chen, Q.; Hu, X.; Zhao, Y.; Xiong, L.; Chen, X.; Wang, X.; et al. *Penthorum chinense* Pursh. extract attenuates non-alcholic fatty liver disease by regulating gut microbiota and bile acid metabolism in mice. *J. Ethnopharmacol.* **2022**, *294*, 115333. [CrossRef]
16. Luo, X.; Zhang, B.; Pan, Y.; Gu, J.; Tan, R.; Gong, P. *Phyllanthus emblica* aqueous extract retards hepatic steatosis and fibrosis in NAFLD mice in association with the reshaping of intestinal microecology. *Front. Pharmacol.* **2022**, *13*, 893561. [CrossRef]
17. Dai, X.; He, L.; Hu, N.; Guo, C.; Zhou, M.; Zhao, X.; Wang, C.; Gong, L.; Ma, C.; Xue, X.; et al. Polygoni Multiflori Radix Praeparata Ethanol Extract Exerts a Protective Effect Against High-Fat Diet Induced Non-Alcoholic Fatty Liver Disease in Mice by Remodeling Intestinal Microbial Structure and Maintaining Metabolic Homeostasis of Bile Acids. *Front. Pharmacol.* **2021**, *12*, 734670. [CrossRef]
18. Li, W.; Yang, H.; Zhao, Q.; Wang, X.; Zhang, J.; Zhao, X. Polyphenol-rich loquat fruit extract prevents fructose-induced nonalcoholic fatty liver disease by modulating glycometabolism, lipometabolism, oxidative stress, inflammation, intestinal barrier, and gut microbiota in mice. *J. Agric. Food Chem.* **2019**, *67*, 7726–7737. [CrossRef]

19. Wang, G.; Han, Q.; Yan, X.; Feng, L.; Zhang, Y.; Zhang, R.; Zhang, Y. Polyphenols-rich extracts from walnut green husk prevent non-alcoholic fatty liver disease, vascular endothelial dysfunction and colon tissue damage in rats induced by high-fat diet. *J. Funct. Foods* **2021**, *87*, 104853. [CrossRef]
20. Li, L.; Lan, X.; Peng, X.; Shi, S.; Zhao, Y.; Liu, W.; Luo, Q.; Jia, L.; Feng, B.; Chen, Z. Polysaccharide from *Salviae miltiorrhizae* Radix et Rhizoma Attenuates the Progress of Obesity-Induced Non-Alcoholic Fatty Liver Disease through Modulating Intestinal Microbiota-Related Gut–Liver Axis. *Int. J. Mol. Sci.* **2022**, *23*, 10620. [CrossRef]
21. Li, C.; Li, W.; Yang, H.; Mi, Z.; Tan, S.; Lei, X. Polysaccharides from Tumorous stem mustard prevented high fructose diet-induced non-alcoholic fatty liver disease by regulating gut microbiota, hepatic lipid metabolism, and the AKT/FOXO1/MAPK signaling pathway. *J. Funct. Foods* **2023**, *102*, 105448. [CrossRef]
22. Zhang, L.; Wang, P.; Shi, M.; Fang, Z.; Ji, J.; Liao, X.; Hu, X.; Chen, F. The modulation of *Luffa cylindrica* (L.) Roem supplementation on gene expression and amino acid profiles in liver for alleviating hepatic steatosis via gut microbiota in high-fat diet-fed mice: Insight from hepatic transcriptome analysis. *J. Nutr. Biochem.* **2020**, *80*, 108365. [CrossRef]
23. Xiao, L.; Xiong, H.; Deng, Z.; Peng, X.; Cheng, K.; Zhang, H.; Jiang, L.; Sun, Y. *Tetrastigma hemsleyanum* leaf extracts ameliorate NAFLD in mice with low-grade colitis via the gut-liver axis. *Food Funct.* **2023**, *14*, 500–515. [CrossRef]
24. Ren, X.; Ye, D.; He, W.; Wu, X.; Guo, J. Therapeutic roles of a natural herbal product on ameliorating hepatic steatosis via gut dysbacteriosis modulating in treatment of nonalcoholic fatty liver disease in mice. *Phytomed. Plus* **2021**, *1*, 100030. [CrossRef]
25. Luo, D.; Yang, L.; Pang, H.; Zhao, Y.; Li, K.; Rong, X.; Guo, J. Tianhuang formula reduces the oxidative stress response of NAFLD by regulating the gut microbiome in mice. *Front. Microbiol.* **2022**, *13*, 984019. [CrossRef]
26. Wang, G.; Yang, X.; Wang, J.; Zhong, D.; Zhang, R.; Zhang, Y.; Feng, L.; Zhang, Y. Walnut green husk polysaccharides prevent obesity, chronic inflammatory responses, nonalcoholic fatty liver disease and colonic tissue damage in high-fat diet fed rats. *Int. J. Biol. Macromol.* **2021**, *182*, 879–898. [CrossRef]
27. Yan, J.; Nie, Y.; Liu, Y.; Li, J.; Wu, L.; Chen, Z.; He, B. Yiqi-Bushen-Tiaozhi recipe attenuated high-fat and high-fructose diet induced nonalcoholic steatohepatitis in mice via gut microbiota. *Front. Cell. Infect. Microbiol.* **2022**, *12*, 824597. [CrossRef]
28. Chen, L.; Kan, J.; Zheng, N.; Li, B.; Hong, Y.; Yan, J.; Tao, X.; Wu, G.; Ma, J.; Zhu, W. A botanical dietary supplement from white peony and licorice attenuates nonalcoholic fatty liver disease by modulating gut microbiota and reducing inflammation. *Phytomedicine* **2021**, *91*, 153693. [CrossRef]
29. Han, R.; Qiu, H.; Zhong, J.; Zheng, N.; Li, B.; Hong, Y.; Ma, J.; Wu, G.; Chen, L.; Sheng, L.; et al. Si Miao Formula attenuates non-alcoholic fatty liver disease by modulating hepatic lipid metabolism and gut microbiota. *Phytomedicine* **2021**, *85*, 153544. [CrossRef]
30. Nakano, H.; Wu, S.; Sakao, K.; Hara, T.; He, J.; Garcia, S.; Shetty, K.; Hou, D.-X. Bilberry anthocyanins ameliorate NAFLD by improving dyslipidemia and gut microbiome dysbiosis. *Nutrients* **2020**, *12*, 3252. [CrossRef]
31. Zandani, G.; Anavi-Cohen, S.; Assa-Glazer, T.; Gorelick, J.; Nyska, A.; Sela, N.; Bernstein, N.; Madar, Z. Cannabis Extract Effects on Metabolic Parameters and Gut Microbiota Composition in a Mice Model of NAFLD and Obesity. *Evidence-Based Complement. Altern. Med.* **2022**, *2022*, 7964018. [CrossRef]
32. Yeh, W.J.; Ko, J.; Huang, W.C.; Cheng, W.Y.; Yang, H.Y. Crude extract of Camellia oleifera pomace ameliorates the progression of non-alcoholic fatty liver disease via decreasing fat accumulation, insulin resistance and inflammation. *Br. J. Nutr.* **2020**, *123*, 508–515. [CrossRef] [PubMed]
33. Cui, H.; Li, Y.; Wang, Y.; Jin, L.; Yang, L.; Wang, L.; Liao, J.; Wang, H.; Peng, Y.; Zhang, Z.; et al. Da-Chai-Hu Decoction Ameliorates High Fat Diet-Induced Nonalcoholic Fatty Liver Disease Through Remodeling the Gut Microbiota and Modulating the Serum Metabolism. *Front. Pharmacol.* **2020**, *11*, 584090. [CrossRef] [PubMed]
34. Fang, J.; Sun, X.; Xue, B.; Fang, N.; Zhou, M. Dahuang Zexie Decoction Protects against High-Fat Diet-Induced NAFLD by Modulating Gut Microbiota-Mediated Toll-Like Receptor 4 Signaling Activation and Loss of Intestinal Barrier. *Evidence-Based Complement. Altern. Med.* **2017**, *2017*, 2945803. [CrossRef]
35. Chen, M.; Xing, J.; Pan, D.; Gao, P. Effect of Chinese herbal medicine mixture 919 syrup on regulation of the ghrelin pathway and intestinal microbiota in rats with non-alcoholic fatty liver disease. *Front. Microbiol.* **2021**, *12*, 793854. [CrossRef]
36. Wang, R.; Wang, L.; Wu, H.; Zhang, L.; Hu, X.; Li, C.; Liu, S. Noni (*Morinda citrifolia* L.) fruit phenolic extract supplementation ameliorates NAFLD by modulating insulin resistance, oxidative stress, inflammation, liver metabolism and gut microbiota. *Food Res. Int.* **2022**, *160*, 111732. [CrossRef]
37. Wang, R.-R.; Zhang, L.-F.; Chen, L.-P.; Wang, J.-Y.; Zhang, L.; Xu, Y.-S.; Yang, P.-L.; Ji, G.; Liu, B.-C. Structural and functional modulation of gut microbiota by Jiangzhi granules during the amelioration of nonalcoholic fatty liver disease. *Oxidative Med. Cell. Longev.* **2021**, *2021*, 2234695. [CrossRef]
38. Zhao, W.; Guo, M.; Feng, J.; Gu, Z.; Zhao, J.; Zhang, H.; Wang, G.; Chen, W. Myristica fragrans Extract Regulates Gut Microbes and Metabolites to Attenuate Hepatic Inflammation and Lipid Metabolism Disorders via the AhR–FAS and NF-κB Signaling Pathways in Mice with Non-Alcoholic Fatty Liver Disease. *Nutrients* **2022**, *14*, 1699. [CrossRef]
39. Liao, J.; Xie, X.; Gao, J.; Zhang, Z.; Qu, F.; Cui, H.; Cao, Y.; Han, X.; Zhao, J.; Wen, W.; et al. Jian-Gan-Xiao-Zhi Decoction Alleviates Inflammatory Response in Nonalcoholic Fatty Liver Disease Model Rats through Modulating Gut Microbiota. *Evidence-Based Complement. Altern. Med.* **2021**, *2021*, 5522755. [CrossRef] [PubMed]
40. Lin, D.; Ding, Y.; Cheng, Y.; Chen, Y.; Tang, Y.; Wu, X.; Cheng, Y. Efficacy and mechanism of *Mallotus furetianus* müll. arg. extract on nonalcoholic fatty liver disease. *Evid.-Based Complement. Altern. Med.* **2022**, *2022*, 4897463. [CrossRef]

41. Zhang, D.; Xu, Y.; Chen, H.; Wang, D.; Geng, Z.; Chen, Y.; Chen, Y.; Xiong, D.; Yang, R.; Liu, X.; et al. *Fagopyrum dibotrys* extract alleviates hepatic steatosis and insulin resistance, and alters autophagy and gut microbiota diversity in mouse models of high-fat diet-induced non-alcoholic fatty liver disease. *Front. Nutr.* **2022**, *9*, 993501. [CrossRef]
42. Liang, W.; Zhou, K.; Jian, P.; Chang, Z.; Zhang, Q.; Liu, Y.; Xiao, S.; Zhang, L. Ginsenosides improve nonalcoholic fatty liver disease via integrated regulation of gut microbiota, inflammation and energy homeostasis. *Front. Pharmacol.* **2021**, *12*, 622841. [CrossRef] [PubMed]
43. Nishiyama, M.; Ohtake, N.; Kaneko, A.; Tsuchiya, N.; Imamura, S.; Iizuka, S.; Ishizawa, S.; Nishi, A.; Yamamoto, M.; Taketomi, A. Increase of *Akkermansia muciniphila* by a diet containing Japanese traditional medicine Bofutsushosan in a mouse model of non-alcoholic fatty liver disease. *Nutrients* **2020**, *12*, 839. [CrossRef] [PubMed]
44. Yin, X.; Peng, J.; Zhao, L.; Yu, Y.; Zhang, X.; Liu, P.; Feng, Q.; Hu, Y.; Pang, X. Structural changes of gut microbiota in a rat non-alcoholic fatty liver disease model treated with a Chinese herbal formula. *Syst. Appl. Microbiol.* **2013**, *36*, 188–196. [CrossRef]
45. Liu, W.; Shao, T.; Tian, L.; Ren, Z.; Gao, L.; Tang, Z.; Fang, Z.; Yuan, P.; Liu, C.; Li, J.; et al. Structural elucidation and anti-nonalcoholic fatty liver disease activity of *Polygonatum cyrtonema* Hua polysaccharide. *Food Funct.* **2022**, *13*, 12883–12895. [CrossRef]
46. Quan, T.; Zhou, F.; Chen, H.; Jian, L.; Yang, Y.; Xia, F.; Xiang, S.; Zhou, B.; Li, S. Ficus hirta Vahl. Ameliorates Nonalcoholic Fatty Liver Disease through Regulating Lipid Metabolism and Gut Microbiota. *Oxidative Med. Cell. Longev.* **2022**, *2022*, 3474723. [CrossRef] [PubMed]
47. Yang, Y.; Li, M.; Wang, Q.; Huang, H.; Zhao, Y.; Du, F.; Chen, Y.; Shen, J.; Luo, H.; Zhao, Q. Pueraria lobata starch regulates gut microbiota and alleviates high-fat high-cholesterol diet induced non-alcoholic fatty liver disease in mice. *Food Res. Int.* **2022**, *157*, 111401. [CrossRef]
48. Wang, Y.; Lv, S.; Shen, T.; Ye, M.; Wang, D.; Zhou, P.; Li, Y.; Zhang, W. Qinghua Fang inhibits high-fat diet-induced non-alcoholic fatty liver disease by modulating gut microbiota. *Ann. Palliat. Med.* **2021**, *10*, 3219–3234. [CrossRef]
49. Liu, B.; Zhang, J.; Sun, P.; Yi, R.; Han, X.; Zhao, X. Raw bowl tea (Tuocha) polyphenol prevention of nonalcoholic fatty liver disease by regulating intestinal function in mice. *Biomolecules* **2019**, *9*, 435. [CrossRef]
50. Ishizawa, S.; Nishi, A.; Kaifuchi, N.; Shimobori, C.; Nahata, M.; Yamada, C.; Iizuka, S.; Ohbuchi, K.; Nishiyama, M.; Fujitsuka, N. Integrated analysis of effect of daisaikoto, a traditional Japanese medicine, on the metabolome and gut microbiome in a mouse model of nonalcoholic fatty liver disease. *Gene* **2022**, *846*, 146856. [CrossRef]
51. Lee, J.-E.; Lee, S.M.; Jung, J. Integrated omics analysis unraveled the microbiome-mediated effects of Yijin-Tang on hepatosteatosis and insulin resistance in obese mouse. *Phytomedicine* **2020**, *79*, 153354. [CrossRef] [PubMed]
52. Huang, X.; Yuan, Z.; Liu, X.; Wang, Z.; Lu, J.; Wu, L.; Lin, X.; Zhang, Y.; Pi, W.; Cai, D. Integrative multi-omics unravels the amelioration effects of Zanthoxylum bungeanum Maxim. on non-alcoholic fatty liver disease. *Phytomedicine* **2023**, *109*, 154576. [CrossRef] [PubMed]
53. Soares, E.; Soares, A.C.; Trindade, P.L.; Monteiro, E.B.; Martins, F.F.; Forgie, A.J.; Inada, K.O.; de Bem, G.F.; Resende, A.; Perrone, D. Jaboticaba (*Myrciaria jaboticaba*) powder consumption improves the metabolic profile and regulates gut microbiome composition in high-fat diet-fed mice. *Biomed. Pharmacother.* **2021**, *144*, 112314. [CrossRef]
54. Bi, C.-R.; Sun, J.-T.; Du, J.; Chu, L.-Y.; Li, Y.-J.; Jia, X.-Y.; Liu, Y.; Zhang, W.-P.; Li, Y.-C.; Liu, Y.-J. Effects of Zhishi Daozhi Decoction on the intestinal flora of nonalcoholic fatty liver disease mice induced by a high-fat diet. *Front. Cell. Infect. Microbiol.* **2023**, *12*, 1005318. [CrossRef]
55. Xie, K.; He, X.; Chen, K.; Sakao, K.; Hou, D.-X. Ameliorative effects and molecular mechanisms of vine tea on western diet-induced NAFLD. *Food Funct.* **2020**, *11*, 5976–5991. [CrossRef] [PubMed]
56. Ren, S.-M.; Zhang, Q.-Z.; Chen, M.-L.; Jiang, M.; Zhou, Y.; Xu, X.-J.; Wang, D.-M.; Pan, Y.-N.; Liu, X.-Q. Anti-NAFLD effect of defatted walnut powder extract in high fat diet-induced C57BL/6 mice by modulating the gut microbiota. *J. Ethnopharmacol.* **2021**, *270*, 113814. [CrossRef]
57. Yi, S.; Chen, K.; Sakao, K.; Ikenaga, M.; Wang, Y.; Hou, D.-X. Assessment of Areca Nut Bioactivities in Western Diet-Induced Mice NAFLD Model. *Nutrients* **2023**, *15*, 2403. [CrossRef] [PubMed]
58. Miao, J.; Guo, L.; Cui, H.; Wang, L.; Zhu, B.; Lei, J.; Li, P.; Jia, J.; Zhang, Z. Er-Chen Decoction Alleviates High-Fat Diet-Induced Nonalcoholic Fatty Liver Disease in Rats through Remodeling Gut Microbiota and Regulating the Serum Metabolism. *Evidence-Based Complement. Altern. Med.* **2022**, *2022*, 6221340. [CrossRef]
59. Shen, S.-H.; Zhong, T.-Y.; Peng, C.; Fang, J.; Lv, B. Structural modulation of gut microbiota during alleviation of non-alcoholic fatty liver disease with *Gynostemma pentaphyllum* in rats. *BMC Complement. Med. Ther.* **2020**, *20*, 34. [CrossRef]
60. Zhang, B.; Luo, X.; Han, C.; Liu, J.; Zhang, L.; Qi, J.; Gu, J.; Tan, R.; Gong, P. *Terminalia bellirica* ethanol extract ameliorates nonalcoholic fatty liver disease in mice by amending the intestinal microbiota and faecal metabolites. *J. Ethnopharmacol.* **2023**, *305*, 116082. [CrossRef]
61. Lin, Y.P.; Fang, Q.L.; Xue, Y.M.; Fu, S.N.; Hu, C.Y.; Huang, F.; Wang, M.M.; Qiao, X.; Yin, X.Q.; Zeng, Y.C.; et al. Effects of Tylophora yunnanensis Schltr on regulating the gut microbiota and its metabolites in non-alcoholic steatohepatitis rats by inhibiting the activation of NOD-like receptor protein 3. *J. Ethnopharmacol.* **2023**, *305*, 116145. [CrossRef] [PubMed]
62. He, B.; Jiang, J.; Shi, Z.; Wu, L.; Yan, J.; Chen, Z.; Luo, M.; Cui, D.; Xu, S.; Yan, M.; et al. Pure total flavonoids from citrus attenuate non-alcoholic steatohepatitis via regulating the gut microbiota and bile acid metabolism in mice. *Biomed. Pharmacother.* **2021**, *135*, 111183. [CrossRef] [PubMed]

63. Li, Q.; Li, M.; Li, F.; Zhou, W.; Dang, Y.; Zhang, L.; Ji, G. Qiang-Gan formula extract improves non-alcoholic steatohepatitis via regulating bile acid metabolism and gut microbiota in mice. *J. Ethnopharmacol.* **2020**, *258*, 112896. [CrossRef]
64. Lv, S.; Zhang, Z.; Su, X.; Li, W.; Wang, X.; Pan, B.; Li, H.; Zhang, H.; Wang, Y. Qingrequzhuo capsule alleviated methionine and choline deficient diet-induced nonalcoholic steatohepatitis in mice through regulating gut microbiota, enhancing gut tight junction and inhibiting the activation of TLR4/NF-kappa B signaling pathway. *Front. Endocrinol.* **2023**, *13*, 1106875. [CrossRef] [PubMed]
65. Li, C.; Zhou, W.; Li, M.; Shu, X.; Zhang, L.; Ji, G. Salvia-Nelumbinis naturalis extract protects mice against MCD diet-induced steatohepatitis via activation of colonic FXR-FGF15 pathway. *Biomed. Pharmacother.* **2021**, *139*, 111587. [CrossRef]
66. Leal-Díaz, A.M.; Noriega, L.G.; Torre-Villalvazo, I.; Torres, N.; Alemán-Escondrillas, G.; López-Romero, P.; Sánchez-Tapia, M.; Aguilar-López, M.; Furuzawa-Carballeda, J.; Velázquez-Villegas, L.A.; et al. Aguamiel concentrate from *Agave salmiana* and its extracted saponins attenuated obesity and hepatic steatosis and increased *Akkermansia muciniphila* in C57BL6 mice. *Sci. Rep.* **2016**, *6*, 34242. [CrossRef]
67. Li, D.; Cui, Y.; Wang, X.; Liu, F.; Li, X. Apple Polyphenol Extract Improves High-Fat Diet-Induced Hepatic Steatosis by Regulating Bile Acid Synthesis and Gut Microbiota in C57BL/6 Male Mice. *J. Agric. Food Chem.* **2021**, *69*, 6829–6841. [CrossRef]
68. Anhe, F.F.; Varin, T.V.; Le Barz, M.; Pilon, G.; Dudonne, S.; Trottier, J.; St-Pierre, P.; Harris, C.S.; Lucas, M.; Lemire, M.; et al. Arctic berry extracts target the gut-liver axis to alleviate metabolic endotoxaemia, insulin resistance and hepatic steatosis in diet-induced obese mice. *Diabetologia* **2018**, *61*, 919–931. [CrossRef] [PubMed]
69. Zhang, B.; Ren, D.; Zhao, Y.; Liu, Y.; Zhai, X.; Yang, X. Artemisia sphaerocephala Krasch polysaccharide prevents hepatic steatosis in high fructose-fed mice associated with changes in the gut microbiota. *Food Funct.* **2019**, *10*, 8137–8148. [CrossRef]
70. Nassir, F. NAFLD: Mechanisms, Treatments, and Biomarkers. *Biomolecules* **2022**, *12*, 824. [CrossRef]
71. Lian, C.Y.; Zhai, Z.Z.; Li, Z.F.; Wang, L. High fat diet-triggered non-alcoholic fatty liver disease: A review of proposed mechanisms. *Chem. Biol. Interact.* **2020**, *330*, 109199. [CrossRef] [PubMed]
72. Peng, C.; Stewart, A.G.; Woodman, O.L.; Ritchie, R.H.; Qin, C.X. Non-Alcoholic Steatohepatitis: A Review of Its Mechanism, Models and Medical Treatments. *Front. Pharmacol.* **2020**, *11*, 603926. [CrossRef] [PubMed]
73. Nagaya, T.; Tanaka, N.; Kimura, T.; Kitabatake, H.; Fujimori, N.; Komatsu, M.; Horiuchi, A.; Yamaura, T.; Umemura, T.; Sano, K.; et al. Mechanism of the development of nonalcoholic steatohepatitis after pancreaticoduodenectomy. *BBA Clin.* **2015**, *3*, 168–174. [CrossRef] [PubMed]
74. Tsay, C.J.; Lim, J.K. NASH and the Gut Microbiome: Implications for New Therapies. *Clin. Liver Dis.* **2022**, *19*, 97–100. [CrossRef]
75. Nassir, F.; Rector, R.S.; Hammoud, G.M.; Ibdah, J.A. Pathogenesis and Prevention of Hepatic Steatosis. *Gastroenterol. Hepatol.* **2015**, *11*, 167–175.
76. Campo, L.; Eiseler, S.; Apfel, T.; Pyrsopoulos, N. Fatty Liver Disease and Gut Microbiota: A Comprehensive Update. *J. Clin. Transl. Hepatol.* **2019**, *7*, 56–60. [CrossRef]
77. Teng, M.L.; Ng, C.H.; Huang, D.Q.; Chan, K.E.; Tan, D.J.; Lim, W.H.; Yang, J.D.; Tan, E.; Muthiah, M.D. Global incidence and prevalence of nonalcoholic fatty liver disease. *Clin. Mol. Hepatol.* **2023**, *29*, S32–S42. [CrossRef]
78. Xiang, H.; Sun, D.; Liu, X.; She, Z.G.; Chen, Y. The Role of the Intestinal Microbiota in Nonalcoholic Steatohepatitis. *Front. Endocrinol.* **2022**, *13*, 812610. [CrossRef]
79. AlShaalan, R.; Aljiffry, M.; Al-Busafi, S.; Metrakos, P.; Hassanain, M. Nonalcoholic fatty liver disease: Noninvasive methods of diagnosing hepatic steatosis. *Saudi J. Gastroenterol.* **2015**, *21*, 64–70. [CrossRef]
80. Neuschwander-Tetri, B.A. Non-alcoholic fatty liver disease. *BMC Med.* **2017**, *15*, 45. [CrossRef]
81. Rodrigues, A.A.; Andrade, R.S.B.; Vasconcelos, D.F.P. Relationship between Experimental Diet in Rats and Nonalcoholic Hepatic Disease: Review of Literature. *Int. J. Hepatol.* **2018**, *2018*, 9023027. [CrossRef] [PubMed]
82. Recena Aydos, L.; Aparecida do Amaral, L.; Serafim de Souza, R.; Jacobowski, A.C.; Freitas Dos Santos, E.; Rodrigues Macedo, M.L. Nonalcoholic Fatty Liver Disease Induced by High-Fat Diet in C57bl/6 Models. *Nutrients* **2019**, *11*, 3067. [CrossRef] [PubMed]
83. Xu, Z.J.; Fan, J.G.; Ding, X.D.; Qiao, L.; Wang, G.L. Characterization of high-fat, diet-induced, non-alcoholic steatohepatitis with fibrosis in rats. *Dig. Dis. Sci.* **2010**, *55*, 931–940. [CrossRef] [PubMed]
84. Lim, J.S.; Mietus-Snyder, M.; Valente, A.; Schwarz, J.M.; Lustig, R.H. The role of fructose in the pathogenesis of NAFLD and the metabolic syndrome. *Nat. Rev. Gastroenterol. Hepatol.* **2010**, *7*, 251–264. [CrossRef]
85. Yustisia, I.; Tandiari, D.; Cangara, M.H.; Hamid, F.; Daud, N.A. A high-fat, high-fructose diet induced hepatic steatosis, renal lesions, dyslipidemia, and hyperuricemia in non-obese rats. *Heliyon* **2022**, *8*, e10896. [CrossRef]
86. Anstee, Q.M.; Goldin, R.D. Mouse models in non-alcoholic fatty liver disease and steatohepatitis research. *Int. J. Exp. Pathol.* **2006**, *87*, 1–16. [CrossRef]
87. Kučera, O.; Garnol, T.; Lotková, H.; Staňková, P.; Mazurová, Y.; Hroch, M.; Bolehovská, R.; Roušar, T.; Červinková, Z. The effect of rat strain, diet composition and feeding period on the development of a nutritional model of non-alcoholic fatty liver disease in rats. *Physiol. Res.* **2011**, *60*, 317–328. [CrossRef]
88. de Lima, V.M.; Oliveira, C.P.; Alves, V.A.; Chammas, M.C.; Oliveira, E.P.; Stefano, J.T.; de Mello, E.S.; Cerri, G.G.; Carrilho, F.J.; Caldwell, S.H. A rodent model of NASH with cirrhosis, oval cell proliferation and hepatocellular carcinoma. *J. Hepatol.* **2008**, *49*, 1055–1061. [CrossRef]

89. Kirsch, R.; Clarkson, V.; Shephard, E.G.; Marais, D.A.; Jaffer, M.A.; Woodburne, V.E.; Kirsch, R.E.; Hall Pde, L. Rodent nutritional model of non-alcoholic steatohepatitis: Species, strain and sex difference studies. *J. Gastroenterol. Hepatol.* **2003**, *18*, 1272–1282. [CrossRef]
90. Tavares, W.R.; Barreto, M.d.C.; Seca, A.M. Aqueous and ethanolic plant extracts as bio-insecticides—Establishing a bridge between raw scientific data and practical reality. *Plants* **2021**, *10*, 920. [CrossRef]
91. Truong, D.-H.; Nguyen, D.H.; Ta, N.T.A.; Bui, A.V.; Do, T.H.; Nguyen, H.C. Evaluation of the Use of Different Solvents for Phytochemical Constituents, Antioxidants, In VitroAnti-Inflammatory Activities of *Severinia buxifolia*. *J. Food Qual.* **2019**, *2019*, 8178294. [CrossRef]
92. Jain, R.; Nandakumar, K.; Srivastava, V.; Vaidya, S.K.; Patet, S.; Kumar, P. Hepatoprotective activity of ethanolic and aqueous extract of *Terminalia belerica* in rats. *Pharmacologyonline* **2008**, *2*, 411–427.
93. Tang, Q.; Jin, G.; Wang, G.; Liu, T.; Liu, X.; Wang, B.; Cao, H. Current Sampling Methods for Gut Microbiota: A Call for More Precise Devices. *Front. Cell Infect. Microbiol.* **2020**, *10*, 151. [CrossRef] [PubMed]
94. Miyauchi, E.; Taida, T.; Kawasumi, M.; Ohkusa, T.; Sato, N.; Ohno, H. Analysis of colonic mucosa-associated microbiota using endoscopically collected lavage. *Sci. Rep.* **2022**, *12*, 1758. [CrossRef] [PubMed]
95. Wang, L.; Gou, X.; Ding, Y.; Liu, J.; Wang, Y.; Wang, Y.; Zhang, J.; Du, L.; Peng, W.; Fan, G. The interplay between herbal medicines and gut microbiota in metabolic diseases. *Front. Pharmacol.* **2023**, *14*, 1105405. [CrossRef] [PubMed]
96. Cui, H.X.; Zhang, L.S.; Luo, Y.; Yuan, K.; Huang, Z.Y.; Guo, Y. A Purified Anthraquinone-Glycoside Preparation From Rhubarb Ameliorates Type 2 Diabetes Mellitus by Modulating the Gut Microbiota and Reducing Inflammation. *Front. Microbiol.* **2019**, *10*, 1423. [CrossRef]
97. Kadooka, Y.; Sato, M.; Imaizumi, K.; Ogawa, A.; Ikuyama, K.; Akai, Y.; Okano, M.; Kagoshima, M.; Tsuchida, T. Regulation of abdominal adiposity by probiotics (*Lactobacillus gasseri* SBT2055) in adults with obese tendencies in a randomized controlled trial. *Eur. J. Clin. Nutr.* **2010**, *64*, 636–643. [CrossRef]
98. Giri, S.; Mangalam, A.K. The Gut Microbiome and Metabolome in Multiple Sclerosis. In *Microbiome and Metabolome in Diagnosis, Therapy, and Other Strategic Applications*; Elsevier: London, UK, 2019; pp. 333–340.
99. Quesada-Vázquez, S.; Bone, C.; Saha, S.; Triguero, I.; Colom-Pellicer, M.; Aragonès, G.; Hildebrand, F.; Del Bas, J.M.; Caimari, A.; Beraza, N. Microbiota dysbiosis and gut barrier dysfunction associated with non-alcoholic fatty liver disease are modulated by a specific metabolic cofactors' combination. *Int. J. Mol. Sci.* **2022**, *23*, 13675. [CrossRef]
100. Liang, Y.; Zhang, Y.; Deng, Y.; Liang, S.; He, Y.; Chen, Y.; Liu, C.; Lin, C.; Han, L.; Tu, G.; et al. Chaihu-Shugan-San Decoction Modulates Intestinal Microbe Dysbiosis and Alleviates Chronic Metabolic Inflammation in NAFLD Rats via the NLRP3 Inflammasome Pathway. *Evid. Based Complement Altern. Med.* **2018**, *2018*, 9390786. [CrossRef]
101. Li, M.; Gao, J.; Tang, Y.; Liu, M.; Wu, S.; Qu, K.; Long, X.; Li, H.; Liu, M.; Liu, Y.; et al. Traditional Herbal Medicine-Derived Sulforaphene LFS-01 Reverses Colitis in Mice by Selectively Altering the Gut Microbiota and Promoting Intestinal Gamma-Delta T Cells. *Front. Pharmacol.* **2017**, *8*, 959. [CrossRef]
102. He, L.H.; Yao, D.H.; Wang, L.Y.; Zhang, L.; Bai, X.L. Gut Microbiome-Mediated Alteration of Immunity, Inflammation, and Metabolism Involved in the Regulation of Non-alcoholic Fatty Liver Disease. *Front. Microbiol.* **2021**, *12*, 761836. [CrossRef] [PubMed]
103. Lei, S.-S.; Li, B.; Chen, Y.-H.; He, X.; Wang, Y.-Z.; Yu, H.-H.; Zhou, F.-C.; Zheng, X.; Chen, X.; Zhang, N.-Y.; et al. Dendrobii Officinalis, a traditional Chinese edible and officinal plant, accelerates liver recovery by regulating the gut-liver axis in NAFLD mice. *J. Funct. Foods* **2019**, *61*, 103458. [CrossRef]
104. Bai, J.; Zhu, Y.; Dong, Y. Response of gut microbiota and inflammatory status to bitter melon (*Momordica charantia* L.) in high fat diet induced obese rats. *J. Ethnopharmacol.* **2016**, *194*, 717–726. [CrossRef]
105. Allam-Ndoul, B.; Castonguay-Paradis, S.; Veilleux, A. Gut Microbiota and Intestinal Trans-Epithelial Permeability. *Int. J. Mol. Sci.* **2020**, *21*, 6402. [CrossRef]
106. Ranganathan, R.; Ross, E.M. PDZ domain proteins: Scaffolds for signaling complexes. *Curr. Biol.* **1997**, *7*, R770–R773. [CrossRef]
107. Al-Sadi, R.; Khatib, K.; Guo, S.; Ye, D.; Youssef, M.; Ma, T. Occludin regulates macromolecule flux across the intestinal epithelial tight junction barrier. *Am. J. Physiol. Gastrointest. Liver Physiol.* **2011**, *300*, G1054–G1064. [CrossRef]
108. Wen, X.; Lou, Y.; Song, S.; He, Z.; Chen, J.; Xie, Z.; Shi, X.; Wen, C.; Shao, T. Qu-Zhuo-Tong-Bi Decoction Alleviates Gouty Arthritis by Regulating Butyrate-Producing Bacteria in Mice. *Front. Pharmacol.* **2020**, *11*, 610556. [CrossRef]
109. Ohira, H.; Fujioka, Y.; Katagiri, C.; Mamoto, R.; Aoyama-Ishikawa, M.; Amako, K.; Izumi, Y.; Nishiumi, S.; Yoshida, M.; Usami, M.; et al. Butyrate attenuates inflammation and lipolysis generated by the interaction of adipocytes and macrophages. *J. Atheroscler. Thromb.* **2013**, *20*, 425–442. [CrossRef] [PubMed]
110. De Vadder, F.; Kovatcheva-Datchary, P.; Goncalves, D.; Vinera, J.; Zitoun, C.; Duchampt, A.; Bäckhed, F.; Mithieux, G. Microbiota-generated metabolites promote metabolic benefits via gut-brain neural circuits. *Cell* **2014**, *156*, 84–96. [CrossRef]
111. Kondo, T.; Kishi, M.; Fushimi, T.; Kaga, T. Acetic acid upregulates the expression of genes for fatty acid oxidation enzymes in liver to suppress body fat accumulation. *J. Agric. Food Chem.* **2009**, *57*, 5982–5986. [CrossRef]
112. Canfora, E.E.; Jocken, J.W.; Blaak, E.E. Short-chain fatty acids in control of body weight and insulin sensitivity. *Nat. Rev. Endocrinol.* **2015**, *11*, 577–591. [CrossRef] [PubMed]
113. Mandaliya, D.K.; Seshadri, S. Short Chain Fatty Acids, pancreatic dysfunction and type 2 diabetes. *Pancreatology* **2019**, *19*, 280–284. [CrossRef] [PubMed]

114. Morrison, D.J.; Preston, T. Formation of short chain fatty acids by the gut microbiota and their impact on human metabolism. *Gut Microbes* **2016**, *7*, 189–200. [CrossRef] [PubMed]
115. Gomes, A.C.; Hoffmann, C.; Mota, J.F. The human gut microbiota: Metabolism and perspective in obesity. *Gut Microbes* **2018**, *9*, 308–325. [CrossRef]
116. Zhou, D.; Pan, Q.; Xin, F.Z.; Zhang, R.N.; He, C.X.; Chen, G.Y.; Liu, C.; Chen, Y.W.; Fan, J.G. Sodium butyrate attenuates high-fat diet-induced steatohepatitis in mice by improving gut microbiota and gastrointestinal barrier. *World J. Gastroenterol.* **2017**, *23*, 60–75. [CrossRef]
117. Li, J.; Chen, C.; Yang, H.; Yang, X. Tea polyphenols regulate gut microbiota dysbiosis induced by antibiotic in mice. *Food Res. Int.* **2021**, *141*, 110153. [CrossRef]
118. Wu, Z.; Zhou, H.; Liu, D.; Deng, F. Alterations in the gut microbiota and the efficacy of adjuvant probiotic therapy in liver cirrhosis. *Front. Cell Infect. Microbiol.* **2023**, *13*, 1218552. [CrossRef]
119. Yao, Y.; Shen, Y. Cross-talk between gut microbiota and liver steatosis: Complications and therapeutic target. *Open Life Sci.* **2023**, *18*, 20220699. [CrossRef]
120. Chen, J.; Vitetta, L. Gut Microbiota Metabolites in NAFLD Pathogenesis and Therapeutic Implications. *Int. J. Mol. Sci.* **2020**, *21*, 5214. [CrossRef]
121. Deprince, A.; Haas, J.T.; Staels, B. Dysregulated lipid metabolism links NAFLD to cardiovascular disease. *Mol. Metab.* **2020**, *42*, 101092. [CrossRef]
122. Lu, D.; Xia, Q.; Yang, Z.; Gao, S.; Sun, S.; Luo, X.; Li, Z.; Zhang, X.; Han, S.; Li, X.; et al. ENO3 promoted the progression of NASH by negatively regulating ferroptosis via elevation of GPX4 expression and lipid accumulation. *Ann. Transl. Med.* **2021**, *9*, 661. [CrossRef] [PubMed]
123. Waters, J.L.; Ley, R.E. The human gut bacteria Christensenellaceae are widespread, heritable, and associated with health. *BMC Biol.* **2019**, *17*, 83. [CrossRef]
124. Li, M.; Zhang, Z.; Yu, B.; Jia, S.; Cui, B. *Lycium barbarum* Oligosaccharides Alleviate Hepatic Steatosis by Modulating Gut Microbiota in C57BL/6J Mice Fed a High-Fat Diet. *Foods* **2023**, *12*, 1617. [CrossRef] [PubMed]
125. Vourakis, M.; Mayer, G.; Rousseau, G. The Role of Gut Microbiota on Cholesterol Metabolism in Atherosclerosis. *Int. J. Mol. Sci.* **2021**, *22*, 8074. [CrossRef]
126. Yoshida, N.; Yamashita, T.; Hirata, K.I. Gut Microbiome and Cardiovascular Diseases. *Diseases* **2018**, *6*, 56. [CrossRef]
127. Chao, H.W.; Chao, S.W.; Lin, H.; Ku, H.C.; Cheng, C.F. Homeostasis of Glucose and Lipid in Non-Alcoholic Fatty Liver Disease. *Int. J. Mol. Sci.* **2019**, *20*, 298. [CrossRef]
128. Caricilli, A.M.; Saad, M.J. The role of gut microbiota on insulin resistance. *Nutrients* **2013**, *5*, 829–851. [CrossRef]
129. Schinner, S.; Scherbaum, W.A.; Bornstein, S.R.; Barthel, A. Molecular mechanisms of insulin resistance. *Diabet. Med.* **2005**, *22*, 674–682. [CrossRef] [PubMed]
130. Semo, D.; Reinecke, H.; Godfrey, R. Gut microbiome regulates inflammation and insulin resistance: A novel therapeutic target to improve insulin sensitivity. *Signal Transduct. Target. Ther.* **2024**, *9*, 35. [CrossRef]
131. Wang, C.; Zhu, C.; Shao, L.; Ye, J.; Shen, Y.; Ren, Y. Role of Bile Acids in Dysbiosis and Treatment of Nonalcoholic Fatty Liver Disease. *Mediat. Inflamm.* **2019**, *2019*, 7659509. [CrossRef]
132. Zheng, X.; Huang, F.; Zhao, A.; Lei, S.; Zhang, Y.; Xie, G.; Chen, T.; Qu, C.; Rajani, C.; Dong, B.; et al. Bile acid is a significant host factor shaping the gut microbiome of diet-induced obese mice. *BMC Biol.* **2017**, *15*, 120. [CrossRef]
133. Arslan, N. Obesity, fatty liver disease and intestinal microbiota. *World J. Gastroenterol.* **2014**, *20*, 16452–16463. [CrossRef]
134. Kunst, C.; Schmid, S.; Michalski, M.; Tümen, D.; Buttenschön, J.; Müller, M.; Gülow, K. The Influence of Gut Microbiota on Oxidative Stress and the Immune System. *Biomedicines* **2023**, *11*, 1388. [CrossRef] [PubMed]
135. Arifah, S.N.; Atho'illah, M.F.; Lukiati, B.; Lestari, S.R. Herbal Medicine from Single Clove Garlic Oil Extract Ameliorates Hepatic Steatosis and Oxidative Status in High Fat Diet Mice. *Malays. J. Med. Sci.* **2020**, *27*, 46–56. [CrossRef] [PubMed]
136. He, Y.; Hwang, S.; Ahmed, Y.A.; Feng, D.; Li, N.; Ribeiro, M.; Lafdil, F.; Kisseleva, T.; Szabo, G.; Gao, B. Immunopathobiology and therapeutic targets related to cytokines in liver diseases. *Cell Mol. Immunol.* **2021**, *18*, 18–37. [CrossRef] [PubMed]
137. Paramita Pal, P.; Sajeli Begum, A.; Ameer Basha, S.; Araya, H.; Fujimoto, Y. New natural pro-inflammatory cytokines (TNF-α, IL-6 and IL-1β) and iNOS inhibitors identified from *Penicillium polonicum* through in vitro and in vivo studies. *Int. Immunopharmacol.* **2023**, *117*, 109940. [CrossRef] [PubMed]
138. Tsuchida, T.; Friedman, S.L. Mechanisms of hepatic stellate cell activation. *Nat. Rev. Gastroenterol. Hepatol.* **2017**, *14*, 397–411. [CrossRef]
139. Lade, A.; Noon, L.A.; Friedman, S.L. Contributions of metabolic dysregulation and inflammation to nonalcoholic steatohepatitis, hepatic fibrosis, and cancer. *Curr. Opin. Oncol.* **2014**, *26*, 100–107. [CrossRef]
140. Wang, X.; Zheng, Z.; Caviglia, J.M.; Corey, K.E.; Herfel, T.M.; Cai, B.; Masia, R.; Chung, R.T.; Lefkowitch, J.H.; Schwabe, R.F.; et al. Hepatocyte TAZ/WWTR1 Promotes Inflammation and Fibrosis in Nonalcoholic Steatohepatitis. *Cell Metab.* **2016**, *24*, 848–862. [CrossRef]
141. Bruschi, F.V.; Claudel, T.; Tardelli, M.; Caligiuri, A.; Stulnig, T.M.; Marra, F.; Trauner, M. The PNPLA3 I148M variant modulates the fibrogenic phenotype of human hepatic stellate cells. *Hepatology* **2017**, *65*, 1875–1890. [CrossRef]
142. Valenti, L.; Dongiovanni, P. Mutant PNPLA3 I148M protein as pharmacological target for liver disease. *Hepatology* **2017**, *66*, 1026–1028. [CrossRef] [PubMed]

143. Zhao, Y.; Gong, C.; Xu, J.; Chen, D.; Yang, B.; Chen, Z.; Wei, L. Research progress of fecal microbiota transplantation in liver diseases. *J. Clin. Med.* **2023**, *12*, 1683. [CrossRef] [PubMed]
144. Boicean, A.; Birlutiu, V.; Ichim, C.; Anderco, P.; Birsan, S. Fecal Microbiota Transplantation in Inflammatory Bowel Disease. *Biomedicines* **2023**, *11*, 1016. [CrossRef] [PubMed]
145. Zhou, D.; Pan, Q.; Shen, F.; Cao, H.-X.; Ding, W.-J.; Chen, Y.-W.; Fan, J.-G. Total fecal microbiota transplantation alleviates high-fat diet-induced steatohepatitis in mice via beneficial regulation of gut microbiota. *Sci. Rep.* **2017**, *7*, 1529. [CrossRef]
146. Xue, L.; Deng, Z.; Luo, W.; He, X.; Chen, Y. Effect of fecal microbiota transplantation on non-alcoholic fatty liver disease: A randomized clinical trial. *Front. Cell. Infect. Microbiol.* **2022**, *12*, 759306. [CrossRef]
147. Nagpal, R.; Wang, S.; Solberg Woods, L.C.; Seshie, O.; Chung, S.T.; Shively, C.A.; Register, T.C.; Craft, S.; McClain, D.A.; Yadav, H. Comparative microbiome signatures and short-chain fatty acids in mouse, rat, non-human primate, and human feces. *Front. Microbiol.* **2018**, *9*, 2897. [CrossRef]

MDPI AG
Grosspeteranlage 5
4052 Basel
Switzerland
Tel.: +41 61 683 77 34

Current Issues in Molecular Biology Editorial Office
E-mail: cimb@mdpi.com
www.mdpi.com/journal/cimb

www.ingramcontent.com/pod-product-compliance
Lightning Source LLC
Chambersburg PA
CBHW061443210926
51850CB00008B/1597
* 9 7 8 3 7 2 5 8 5 2 1 7 8 *